In-Space Manufacturing and Resources

In-Space Manufacturing and Resources

Earth and Planetary Exploration Applications

Edited by
Volker Hessel
Jana Stoudemire
Hideaki Miyamoto
Ian D. Fisk

WILEY-VCH

Editors

Prof. Dr. Volker Hessel
The University of Adelaide
School of Chemical Engineering and Advance Materials
Engineering North Bld. N2
5005 Adelaide
Australia

Ms. Jana Stoudemire
Axiom Space
Corporate Headquarters
1290 Hercules Avenue
Houston, Texas 77058

Prof. Hideaki Miyamoto
University of Tokyo
Dept. of Systems Innovation
7-3-1 Honog, Bunkyo-ku
113-8656 Tokyo
Japan

Prof. Ian D. Fisk
University of Nottingham
School of Biosciences
Sutton Bonington Campus
Food Sciences
LE12 5RD Sutton Bonington
United Kingdom

Cover Image: © Pike-28/Shutterstock

Library of Congress Card No.: applied for

British Library Cataloguing-in-Publication Data
A catalogue record for this book is available from the British Library.

Bibliographic information published by the Deutsche Nationalbibliothek
The Deutsche Nationalbibliothek lists this publication in the Deutsche Nationalbibliografie; detailed bibliographic data are available on the Internet at <http://dnb.d-nb.de>.

Print ISBN: 978-3-527-34853-4
ePDF ISBN: 978-3-527-83089-3
ePub ISBN: 978-3-527-83091-6
oBook ISBN: 978-3-527-83090-9

Typesetting Straive, Chennai, India
Printing and Binding CPI Group (UK) Ltd, Croydon, CR0 4YY

C9783527348534_270924

Contents

Introductory Statement

Tara Ruttley

The National Aeronautics and Space Administration (NASA) has spent decades implementing research programs in Low-Earth Orbit (LEO), starting with Skylab, NASA's first sustained space station. A rudimentary research lab outfitted inside of a repurposed Saturn V propellant stage, Skylab paved the way for experiments ranging from solar and Earth observations to physical and biological sciences and human physiology over its brief 24-week lifetime. Then, in the 1980s, space shuttle flights housed more capable Spacelab research facilities, and the microgravity science yielded breakthroughs in all disciplines with each short-duration flight. I joined NASA in 2001, a few months after the first International Space Station (ISS) crew was launched to the partially built, newest, most capable long-duration laboratory in LEO. Since then, the ISS has evolved into the premier microgravity research facility it is today, enabling over 3000 scientists from over 100 countries to perform over 4000 investigations to date across every imaginable discipline. If there were ever an ideal time to be a researcher designing microgravity experiments, this has been it.

In those twenty years of ISS access, NASA has learned how to better prepare astronauts for long-duration space exploration. We've developed the optimal countermeasures for preventing bone and muscle loss and keeping the heart healthy, and we've learned that long-duration stays can significantly affect astronaut vision. We've refined emergency medicine and ultrasound procedures, evaluated astronaut behavioral health, and improved nutrition. We better learned how to ignite and extinguish flames, and leveraged surface tension and capillary action to better manage fluid flow. We've tested the newest radiation dosimeters and we have harvested plants that have gone from the on-orbit Veggie "farm to table" facility for astronaut taste tests.

While tackling those challenges for human exploration of space, NASA was given an additional mission: use the ISS to commercialize LEO. This required us to ask of American industry, "what can we do for you?" Companies responded with ideas to use the unique microgravity environment that only ISS can offer for in-space production experiments such as cutting-edge tissue engineering, 3D bioprinting of cells and tissues, retinal implant development, and exotic glass optical fiber manufacturing.

This is only the beginning. With unprecedented access to LEO, businesses are now racing to design, develop, and test in-space manufacturing and production facilities that will one day deliver state-of-the-art materials and technology that will improve the lives of those of us on Earth. They're also making plans for their own commercial space stations independent of the ISS that will create a busy, flourishing new LEO economy in ways that humanity has never before been a part of. With so much to look forward to in the future of human space exploration and development of the LEO economy, what a time to be part of the space revolution!

Dr. Tara Ruttley
Associate Chief Scientist for Exploration and Applied Research
National Aeronautics and Space Administration Headquarters

Introductory Statement

Nicol Caplin

Since 1975, ESA has become established as a key enabler for European access to space. With 22 member states, three associate states, and a growing number of agreements with other countries, ESA is a truly international organization that possesses "strength in numbers" and can therefore carry out activities in space on a scale beyond that of any single European country's capability.

ESA's core values embedded via exclusively peaceful means promote European space research, technology, and applications. Alongside implementing mandatory activities, ESA also conducts a selection of optional programs which is notionally funded by each member state that wishes to contribute. Within this, the Human and Robotic Exploration division has rapidly expanded to provide cutting-edge scientific and technological research applications for the over-arching goal of exploring space for the benefit of humankind.

The present-day focus for the HRE research efforts centers on two themes: Research that is enabled by space and research that enables the achievement of further exploration goals. The Science in the Space Environment (SciSpacE) Team coordinates research in the areas of life (including human research) and physical sciences across multiple platforms and facilities. These include platforms for microgravity research such as sounding rockets, drop towers, parabolic flights, and of course the International Space Station.

In November 2000 the first astronaut crew entered the very early two-module configuration of the ISS, and less than a year later the first European experiment facility[1] was up and running. ESA's single largest contribution to the ISS is the Columbus laboratory. Launched in 2008, it provides European researchers with a permanent laboratory in Low-Earth orbit, for conducting research across a range of disciplines.

1 https://www.esa.int/Science_Exploration/Human_and_Robotic_Exploration/Research/First_European_experiment_facility_up_and_running_on_the_ISS.

Fast forward to today and insights from European research and the international community are gearing up to explore further into space with human and robotic missions. The ISS will continue to serve as an orbiting laboratory, delivering crucial information to researchers that enable deeper exploration while simultaneously improving life on our home planet.

ESA are a key player in the planned Gateway[2] facility; a spacecraft in Lunar orbit that will host scientific and technological projects that will provide a test bed in a much harsher space environment, away from the Earth's protective layer. This environment is more representative of deep space, with far more challenging conditions to ensure spacecraft, experiments, and indeed crews can exist within.

If humanity's goals of once again reaching the Moon and someday landing human crews on Mars are to be achieved, carrying out research in such an intrepid environment will be vital in fully understanding the implications and risks to embarking on such a mission. In addition to this, we should be aiming to explore space in a sustainable manner.

In-space manufacturing is a growing area of research and a welcome one. Our planet is already suffering the effects of being depleted of materials, so it makes not only sense from an environmental perspective but also has huge potential to be an economically effective way of building what humans need to go further. This can take on many forms, from 3D printing of plastic and metals right the way through to biomining processes and chemical extraction of precious resources from regolith.

We are making the preparations to go further into space, sustainably, for the benefit of everyone – Who knows what wonders we will discover? The book *In Space Manufacturing and Resources* is a timely addition to international literature in an important niche field that is evolving at a remarkable pace. It effectively compiles the current relevant outputs for space utilization and exploration.

Dr. Nicol Caplin
Deep Space Exploration Scientist
European Space Agency

2 https://www.esa.int/Science_Exploration/Human_and_Robotic_Exploration/Exploration/Gateway.

Introductory Statement

Masaki Fujimoto

Dr. Masaki Fujimoto (Deputy Director General of ISAS, JAXA) greeting Dr. Lori Glaze (Director of Science Mission Directorate's Planetary Science Division, NASA) for JAXA's participation to the Solar System Exploration Research Virtual Institute.

We are in an exciting era of space exploration. International space agencies are working together to advance space science and technologies, leading to new scientific discoveries announced almost every day. Discoveries always inspire further fundamental questions, where new technologies are developed to promote new explorations for answering. Such chain activities gain more and more momentum in various fields, accelerating international collaborations as a natural consequence. I feel very honored to be a part of the current global and international endeavor for space developments.

Japan launched its first satellite, Ohsumi, in 1970. Numerous experimental and scientific challenges followed, including the 1985 Halley's comet observation satellites Sakigake and Suisei and the first Japanese interplanetary mission, Mars Orbiter Nozomi, launched in 1998 and passed Mars in 2003. In that year, Japan Aerospace Exploration Agency (JAXA) was established as an independent administrative institution, integrating the Institute of Space and Astronautical Science (ISAS), the National Space Development Agency of Japan (NASDA), and the National Aerospace Laboratory of Japan (NAL). Since then, JAXA has been the core performance agency to support the Japanese government's aerospace development and utilization.

JAXA has completed many science missions and significantly contributed to advance astronomical and planetary sciences. AKEBONO (aurora observation satellite), HITOME and SUZAKU (X-ray astronomy satellite), AKARI (infrared imaging satellite), HAYABUSA (asteroid explorer), and KAGUYA (lunar explorer) are good examples of previous missions. Hayabusa 2 (asteroid explorer), IKAROS (small solar power sail demonstrator), AKATSUKI (Venus climate orbiter) have successfully performed their primary objectives and are now in their extended phases. Currently ongoing missions include BepiColombo (Mercury exploration with ESA), HISAKI (planetary atmosphere), HINODE (solar physics), REIMEI (technology demonstration), GEOTAIL (magnetospheric observation), and ARASE (energization and radiation in geospace). XRISM (X-ray imaging and spectroscopy mission), MMX (Martian Moons exploration), and SLIM (smart lander for investigating Moon) are under development and scheduled to be launched in a few years.

We will continue exploring scientific questions regarding the origin and the evolution of the solar system, where a broader range of international collaborations and private associations will be required. JAXA provides a wide range of opportunities including to industries to help private companies enter the space business. We have programs to share our intellectual property acquired through aerospace research and experience, allow industries to use our test facilities, offer a system for cooperative research, and provide launch opportunities for small secondary satellites.

We expect so many new research fields to be involved in future explorations, including human space activities and the development of the lunar economy. We are looking forward to new activities related to space manufacturing and space resources as crucial technologies in this decade and the next with significant pioneer splits and international collaborations.

Dr. Masaki Fujimoto
Deputy Director General
Institute of Space and Astronautical Science
Japan Aerospace Exploration Agency

Introductory Statement

Enrico Palermo

The Australian Government established the Australian Space Agency in 2018 to leverage the rapid growth of the global space economy and its shift to a commercial stage. Space is an emerging industry for both government and the private sector that presents a wealth of economic opportunities–and space manufacturing has a key role to play in harnessing those opportunities, as well as supporting our national strategic interests. Space manufacturing is an important element of the Australian Government's goal to triple the size of the national civil space sector to $12 billion by 2030 and continue making progress in the years beyond[1].

Soon after it was established, the Agency began implementing programs to build capacity and capability in the sector. One of our first steps was the Moon to Mars initiative–the result of a partnership between NASA and the Agency to support Australian space manufacturing efforts that align with NASA's exciting future plans for space exploration. This initiative has had great success with manufacturing businesses across the value chain that already work in the space sector and want to increase space heritage, as well as businesses in adjacent industries looking to diversify their manufacturing outputs to include products with space applications. Under the Moon to Mars initiative we have also reached an agreement for an Australian designed, built and operated semi-autonomous rover to be included in a future lunar mission with NASA. In addition, we are delivering a National Space Mission for Earth Observation which will include Australian industry designing, constructing, launching, and operating four new Earth observation satellites. It will take Australian from a consumer to a contributor of critical Earth observation data.

1 https://www.industry.gov.au/data-and-publications/australian-civil-space-strategy-2019–2028.

Space is also one of six manufacturing priority areas the Australian Government is investing heavily in to drive the national economy. This is coupled with a Space Manufacturing Road Map[2] that sets the future vision for manufacturing investment in the space sector and targets key growth opportunities–including products that go into space, space components, and associated products and infrastructure. This ambition is backed by the Agency's own series of technical roadmaps, which identify diverse space manufacturing opportunities corresponding to our seven National Civil Space Priority Areas[1].

Australia's competitive advantages give us great enthusiasm for the future. For example, we have world-leading expertise in remote asset management in the mining and resources sector, and a range of remote and extreme environments ideal for testing space-focused technology. Australian companies are combining robotics technology with satellite communications to remotely service pipelines underwater, and trialling Satellite-Based Augmentation Systems to aid precision control of mining vehicles. Mine sites in the Pilbara region in Western Australia can be operated remotely from 1600 kilometers away - so imagine the opportunities to operate activities in space, which starts just 100 kilometers above us. By leveraging existing capabilities like these, Australia can become a major player in a global space economy tipped to be worth US$1 trillion by 2040[3].

Dr. Enrico Palermo
Head of the Australian Space Agency

2 https://www.industry.gov.au/data-and-publications/space-national-manufacturing-priority-road-map.
3 https://www.morganstanley.com/ideas/investing-in-space/.

Preface

I made my last major move, scientifically and privately, in 2018 from the Netherlands to Australia. At that time, Australia had started to engage with satellite and diverse computing-related space activities. This turned within just three years into a broad development including the "new space economy" (on resources and human space exploitation) and having the Australian Space Agency settled in Adelaide. When I started with space research in early 2019, it was no more than a curiosity for me. I had such moment of change already in 1994, when starting with microreactors and microfluidics. This "curiosity" has turned into commercial business and success after 20 years of research. Being a child of the 1960s, I am fascinated by the moon landings and Apollo program. My curiosity had now led me into a laboratory at the International Space Station. I am so passionate, seeing more far-fetching goals, such as a moon village, and for any resource and sustainability engineer, the journey to Mars must be utmost exciting and challenging. It is a privilege reporting about in the progress that is being made in this book at this pivotal turning point, with some of the most agile global players in space science and space industry; sincere appreciation to all those. This book highlights current developments of which we are unable to imagine the full range of future implications. Einstein once said "Think truly to the future and make those dreams come true."
Volker Hessel, University of Adelaide

When I joined the space industry, just over seven years ago, the idea of a robust commercial space economy in low-Earth orbit (LEO) – that included anything other than launch vehicles and satellites – was just a vision. Today we are witnessing emerging markets for in-space manufacturing take shape at a pace that seems to grow exponentially each year – with no signs of slowing down. Public–private partnerships driving the transition from government-funded space agencies to commercial companies providing launch services offering expanded access to LEO, platforms that maintain a permanent presence in space beyond the International Space Station, and services to sustain future cities in space, are creating an expanded definition of global that includes not just the circumference of Earth but 250 miles overhead. And the expanded exploration opportunities – to the moon, Mars, and perhaps beyond – that this transition provides are a legacy and a gift we are leaving for future generations. The chapters in this book will hopefully not only provide a new

perspective but inspire you to dream as well, perhaps envisioning possibilities that have yet to be imagined. Sincere thanks to all of those who have shared their work and their vision in these pages, and to their courage as trailblazers to lead the way.
Jana Stoudemire, Axiom Space

I worked on Venusian geology in my early career in mid-90's, when the accessibility to spacecraft data was very limited. I had to spend a significant amount of time to obtain the Magellan mission data generously distributed by NASA as CD-ROMs, which made my work more theoretical than observational. However, thanks to the evolution of communication technologies, the situation has been significantly improved. Also, the science community now prefers to disclose more and more data to the public with the strong belief that it is the best to enhance science. Now we can access growing (and already huge) amounts of raw data of numerous space missions of about 100 extraterrestrial bodies. Also, online international meetings regarding space missions have become important routines of my daily schedule; we can work together with world-leading experts in various fields for operating ongoing space missions and preparing future missions. There is no doubt that our scientific interests will be further accelerated through new findings from the current and the near-future missions, which stimulate curiosities in outer space and incubate new businesses including tourism and settlements. As innovative technologies, including AI, have helped democratize many different markets on Earth, the same is happening in space exploration and development. This book is full of such aspects, and I'm very proud to be a part.
Hideaki Miyamoto, University of Tokyo

The rapidly emerging opportunities offered through the new economy of extended space travel have opened up numerous exciting areas of research. My area of interest, the development of healthy nutritious foods that can be produced and processed safely and easily, raises many key questions that can only be addressed by taking a cross-disciplinary view on these grand challenges. How to develop foods that respond to the nutritional needs of astronauts and take account of metabolic challenges of extended space missions; how to ensure a balanced energy expenditure and intake; how to ensure the immune system is not compromised and the gut microbiota continue to thrive in a healthy state; how to select, develop, and grow foods in microgravity; how to engineer plants in readiness for spaceflight application; how to create foods in situ through space farming; how to develop food processing technologies to ensure processability, minimize loss of nutrients, and maximize storability. Finally, it is critically important that the produced foods are well designed in taste, aroma, and flavor, such that they are palatable and would be enjoyed even after repeat consumption. If these challenges can be solved, then we are well placed to ensure future space travelers have the nutrients and foods that they will require to achieve extended space travel. Furthermore, these in-space approaches to manufacturing essential resources for food will have many terrestrial applications and as the space industry develops further, so too will the terrestrial equivalents.
Ian D. Fisk, University of Nottingham

Part I

Space Medicine and Human Health

1

Human Health in Space

Jennifer Fogarty

Chief Scientist, Translational Research Institute for Space Health, Baylor College of Medicine Director, Human Health and Performance, Sophic Synergistics, LLC, Houston, TX, USA

1.1 Introduction

The human is capable of surviving in ordinary and extreme environments on every continent on Earth. This robustness indicates that humans adapt to various diverse environments and use resources to endure conditions that challenge our survival. The most extreme and austere environments that humans have inhabited are low-Earth orbit, lunar orbit, and the lunar surface otherwise known as spaceflight. Being prepared to enable survival and performance during spaceflight requires an understanding and assessment of the natural course of human health as we age and are exposed to environmental and biological stressors, specifically spaceflight stressors such as microgravity, isolation and confinement, closed loop environmental systems, and space radiation.

Human health and performance form a continuum and data used to quantify and describe each requires context to be adequately interpreted for risk assessment to provide a rational foundation for space exploration medical and countermeasure systems [1–4]. This concept is especially important in the extreme environment of spaceflight where small or unanticipated changes may have critical impacts to the individual and the mission. Loss of crew (LOC) and loss of mission (LOM) are common endpoints that are used to assess the severity of acute and off-nominal events. All mission planning and space vehicle development from the time of conception are designed to avoid or mitigate LOC or LOM events and parallels the occupational health model (engineer out the exposure, provide personal protective equipment for unavoidable exposures, and perform surveillance for untoward effects for early intervention) used to manage the human system risk. This is challenging given the independent and intersecting components of the spaceflight environment, vehicle design, and missions requirements. This challenge is compounded during human spaceflight due to the human changing over time in response to the natural course of their biology (aging?) and exposure to the spaceflight environment that includes novel stressors such as microgravity, extreme isolation and confinement, and space radiation (chronic galactic cosmic rays and acute solar particle events).

In-Space Manufacturing and Resources: Earth and Planetary Exploration Applications, First Edition.
Edited by Volker Hessel, Jana Stoudemire, Hideaki Miyamoto, and Ian D. Fisk.

Establishing a comprehensive understanding and record of an individual's baseline health is extremely important to risk assessment and risk acceptance processes of human spaceflight. Each human spaceflight participant represents a unique combination of genetics and environmental exposures including lifestyle factors (diet, smoking, alcohol consumption, exercise training, etc.) that contribute to the risk of disease and impaired function that might affect the mission, LOC, or LOM. However, requirements that drive decisions about vehicle design, inclusion of large countermeasure equipment (treadmill, cycle ergometer, resistive exercise, lower body negative pressure (LBNP), etc.), and mission objectives are determined long before specific participants are identified. This temporal chasm between knowing an individual's risk and needs versus locking-in a vehicle design and capability has been addressed by using an epidemiological approach to human spaceflight requirements. This approach has been refined over time and adjusted for healthier populations based on updates to terrestrial medical and public health standards and practices. It allows the development of requirements based on a generic representation of humans as an average and standard deviation with occasional allowance for accommodations of ranges such as that associated with anthropometrics, and a risk management approach that uses current evidence for both the likelihood and consequence of a medical event. In addition, historical human spaceflight and dedicated research data are used to assess the effect of exposure to spaceflight specific stressors on the risk of a medical event or a performance decrement. The implication of this process is that participants that do not fall into an acceptable risk posture based on the requirements that were used to generate design and capability of the vehicle may not be permitted to fly on a mission. Since we are entering the era of commercial spaceflight and opening access to spaceflight to a broader population, this undesirable and restrictive potential outcome can be mitigated by actions on the human, the vehicle, and the mission. The human risk can be affected by aggressive risk factor and disease management that may shift the risk benefit assessment in favor of prophylactic surgery or use of additional pharmaceuticals, but each action must have a thorough assessment to understand the inadvertent introduction of risk. The vehicle and mission can mitigate risk by accommodating specific medical and countermeasure hardware, custom protocols for existing hardware, providing personalized pharmaceuticals in custom kits, and adjusting mission duration so that pre-flight screening and treatment can effectively mitigate the medical risk during the mission. This approach is human system integration (HSI) and requires significant communication, evaluation of data, and integration by many specialists from the medical, engineering, and risk management communities.

1.2 Preflight Determination of Health and Performance Risk

The approach to assess health and medical risk is rooted in preventive medicine's clinical standard of care and focuses on gathering multi-system quality evidence, characterization of health risk factors and presence of disease, comparison to relevant populations (terrestrial and previous spaceflight) to assess probability of medical event, and quantification of the health and medical risk for the mission. Assessing performance begins with understanding the health of the individual and collecting fitness metrics, aerobic capacity, power output, endurance, etc. that can be used to characterize and benchmark functional performance and provide a comparison for in-mission assessment to detect decrements and develop an action plan if an unacceptable threshold is reached. Actual evaluation of mission-relevant performance that includes physical capabilities and executive function is done by the flight operations community during mission training and includes intra-vehicular activity (IVA; telerobotic, vehicle maintenance and repair, emergency procedures, medical officer functions, etc.) and extra-vehicular activity (EVA; microgravity and planetary) objectives to establish fitness for duty and a baseline for comparison during the mission. Establishing the pre-flight health and performance baseline provides evidence to calculate risk and also the ability to estimate the impact of changes over the course of the mission. The duration of a mission plays crucial role in estimating risk since many medical conditions have a limited window of prediction based on current screening techniques. This principle is illustrated by the terrestrial medical guidance for annual exams and regular screening tests such as blood work to assess high-density lipids (HDL) and low-density lipids (LDL), prostate specific antigen, fasting blood glucose, mammography, colonoscopy, and pap smears. Screening in-line with these recommendations allows for early and pre-symptomatic disease identification and treatment often remediating the pathological process and reducing the risk of a medical outcome to an acceptable baseline level. Most human spaceflight missions have not exceeded the recommended terrestrial screening durations and coupled with intensive preventive medicine the spaceflight community has been confident that most critical medical events of a natural origin are very unlikely to occur during the mission. [1, 2].

Exploration class missions require mission durations, three years or more, that are outside the prediction window of current medical screening and will require additional screening tools such as precision medicine 31 or more robust in-mission diagnostic and interventional capability to manage medical risk [32]. Missions that do not require extreme durations but have high tempo EVAs, such as lunar exploration, represent additional medical risk due to injuries associated with repetitive activities in the suit, use of heavy equipment, and lunar dust exposure. Therefore, pre-flight the overall health and performance risk for a mission is a complex matrix of medical event probability for each crewmember based on pre-existing risk factors and health risks induced by the operational environment. The element of human spaceflight that has been very challenging to include in the quantitative health and performance risk assessment is how the hazards of spaceflight (altered gravity, closed environment, isolation and confinement, and space radiation) influence how the human changes overtime in response to those unique exposures and how those changes alter the probability of a health and performance risk ultimately influencing LOC and LOM. [3, 4].

1.3 Physiologic Responses to Spaceflight

Human response to these hazards ranges from adaptive to maladaptive and has been documented through evaluation of data from medical operations and dedicated research performed during spaceflight and terrestrial analog missions. The interpretation and application of the data varies depending on the mission duration, time to effect of the adaptive or maladaptive process, and the ability of the individual and mission to tolerate the effect.

The following sections describe the known human responses to the spaceflight hazards and will provide context for the relevance to risk assessment [4]. The most challenging aspect to this section is that the human experiences spaceflight in its totality and begins adapting immediately. The hazard exposures are overlapping and influence each other leading to additive or synergistic effects that continue to be elucidated. Acute responses manifest over minutes and hours and depending on the sustained presence of the exposure may lead to a more chronic response process that includes compensatory mechanisms that may delay or mask overt symptoms and performance impacts [5] (Figure 1.1). The chronic responses to spaceflight hazards have more significant implications for late in-mission and long-term, post-mission health and performance consequences.

It can be very challenging to identify the first response to spaceflight because of intrinsic dependencies and continuous interaction between molecular and physiologic systems. Given this challenge, it seems logical to describe the human response to spaceflight based on the order in which they appear at a phenotypic level, i.e. cause symptoms or behaviors making the manifestation obvious. The issues are described in summary with in-mission and post-flight implications provided for context and relevancy.

1.4 Vestibular and Sensorimotor Disruption and Adaptation

One of the first issues that humans may experience when exposed to novel and new motion is motion sickness. Motion sickness is commonly associated with air travel and sailing, and spaceflight, a combination of aviation and sailing, is no exception. During the first 2–3 days of spaceflight 60–80% [6] of crewmembers experience symptoms that vary from mild nausea to severe vomiting requiring medication to reduce the risk of dehydration and allow crewmembers to adapt. Motion sickness may pose a significant risk within the first few days of the mission when severe but even mild cases of motion sickness may be accompanied by spatial disorientation, difficulty acquiring and tracking visual targets, and alterations of proprioception [7]. While overt motion sickness was not deemed a serious issue during the Apollo missions due to mitigating circumstances such as the small capsule and quick transit times, the spatial disorientation and tracking of visual targets was recognized as a potential issue by Apollo lunar module (LM) pilots compounded by visual illusions created by the angle of view from the LM windows and sunlight.

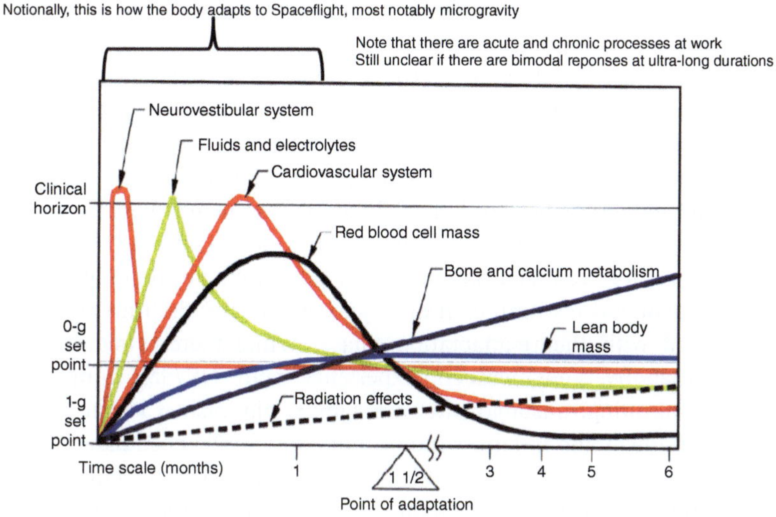

Figure 1.1 Time Course of Physiological Changes During Weightlessness. Source: Adapted from Ref. [5].

[8] To mitigate operational risk, EVAs are not scheduled within the first three days of a mission given the seriousness of the disruption to the performance capability of most crewmembers from the motion sickness and the medications used to treat it. This operational approach began with the Space Shuttle era and continues today for arriving crew on the International Space Station (ISS).

Once crewmembers adapt to the constant acceleration and lack of gravity during spaceflight, their symptoms abate, they function and navigate the environment well. While films from Apollo surface operations often captured astronauts falling, it is challenging to identify the role of contributing factors such as sensorimotor deficit, EVA suit configuration (high center of gravity), and EVA tool function to determine the most responsible contributor. Experience during microgravity operations on the Space Shuttle and ISS indicates that crewmembers adapt well within the first few weeks to the new environment and successfully accomplish tasks that require fine motor (e.g. cell culture and dissections experiments, telerobotics) and gross motor (e.g. EVA, unloading cargo) activities. The current countermeasures consist of medications to manage symptoms early in-mission and physical exercise such as walking and running on a treadmill and resistive training to engage sensorimotor and proprioceptive organs that lack stimulus in microgravity. While hard to quantify, the role of physical exercise is imperative to maintain physical, physiological, and psychological balance throughout the mission including landing and immediately postlanding which can be some of the most physically demanding aspects of the mission. This will be especially true during exploration class missions with landings on a distant planet such as Mars after a long transit in microgravity.

The process of re-entering Earth's atmosphere is a challenging event for the vehicle and the occupants. While the vehicle grapples with re-entering Earth's atmosphere, the occupants are experiencing g-forces in excess of 1 g depending on the configuration of the seating and angle of re-entry. The occupants are in suits designed for re-entry as protection from a rapid decompression, restrained by harnesses, and in various seated configurations (recumbent, conformal, upright, etc.). The neuro-vestibular and sensorimotor systems (NV/SM) experience the return to gravity in many ways based on the re-entry scenario and landing site, land or ocean, make the immediate postlanding period challenging to evaluate given the variety of provocative stimuli. Immediate postlanding evaluations of crewmembers from US and Russian space vehicles have included basic clinical neurological evaluations and structured multisystem research protocols. The postlanding evaluations have enabled the operational and research community to understand the timeline and challenges associated with the re-adaptation phase. While there is significant inter-individual variability, crewmembers experience both NV/SM symptoms (dizziness, nausea, vomiting, unstable gate, nystagmus, diminished visual target acquisition, reduced fine motor and gross motor function: [9–12]) postlanding that can be disabling for the first 24 to 48 hours with an asymptotic recovery (Figure 1.2e). It is important to know that recovery is facilitated by regular yet limited provocative movements in the new environment. Therefore, it is important for crewmembers to experience movement in 1g in a variety of axes, but it is equally important to accommodate rest and recovery to allow re-adaptation of the sensitive sensory organs of the inner ear [13, 14]. The postlanding data are crucial information to understand what the vehicle and the human are capable of during Earth landings with assistance and what they will need to be capable of when humans land on Mars for the first time completely unaided. The immediate postflight decrements in performance are informing current operations and research with respect to pre-flight conditioning, in-flight countermeasures, and postflight aides to enable a safe landing environment and an operational approach that is reasonable given the expected change in human capability during the transit portion of the mission to Mars.

The occupational health model for spaceflight indicates that the best course is to remove the exposure that causes the undesirable outcome and in the case of spaceflight associated neuro-vestibular and sensorimotor disruption that exposure is microgravity. One of the proposed solutions is the implementation of artificial gravity (AG) as a multi-system countermeasure [15]. The suggestions are for the whole vehicle to rotate (long-arm centrifugation) providing a living and working environment that has approximately a 1 g exposure or to have a module or portion of a module rotate (short-arm centrifugation) providing intermittent exposure to a gravity gradient for prescribed amounts of time. There have been research studies (e.g. Bedrest) of artificial gravity as a countermeasure, continuous (rotating room) or intermittent (short-arm radius), designed to describe and quantify the potential benefits and risks to humans [16, 17]. There have also been some preliminary engineering studies of AG estimating the cost and risk of implementing a rotating vehicle, module, or exercise hardware with a centripetal component. The low-fidelity nature of the engineering studies and human research of the physiological and functional benefit of AG have left the issue in a chronic state of discussion and speculation. A properly funded collaborative effort could provide a comprehensive set of objective data that would facilitate the ability to make a

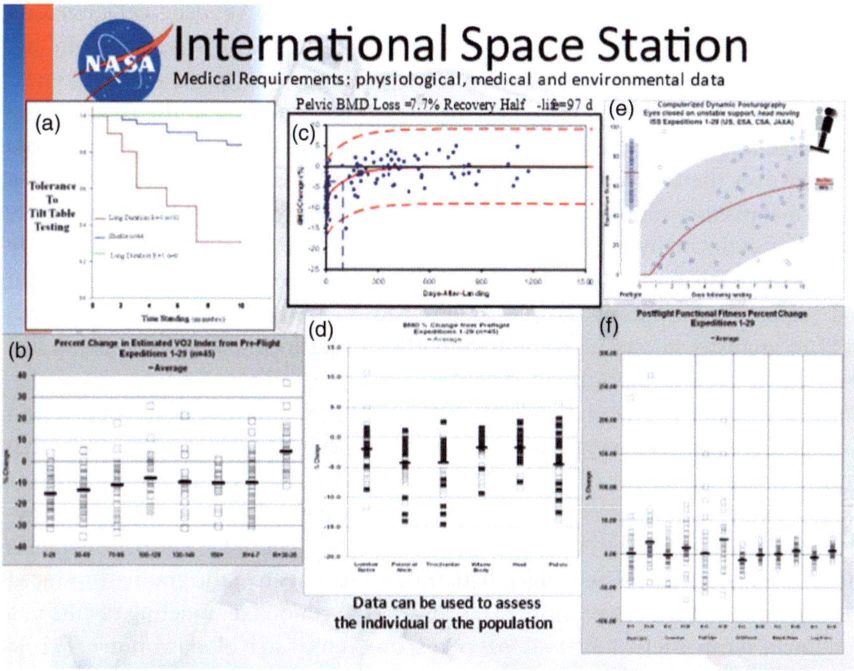

Figure 1.2 Pre- to Post-flight results from Multisystem Review of Inter-Individual Variability Among Astronauts, Fogarty 2016: https://ntrs.nasa.gov/citations/20160010832. Source: (a) NASA.

decision about the implementation of AG based on the risk mitigations benefits compared to risks it presents with respect to the engineering complexity and human experience.

1.5 Cardiovascular Adaptation

Much like the neurovestibular and sensorimotor systems, the cardiovascular system begins adjusting during launch as the body is exposed to an extended period in a recumbent seated position (crew on their back with legs up), various g-levels as the vehicle progresses through stages of flight, and the near absence of gravity within minutes of launch. The body begins adjusting the circulating amount of blood, plasma volume, due to sensors located in the upper chest perceiving an increased volume as a result of the headward (cephalad) shift of blood volume. Most of us have experienced this acutely when hanging upside down and even to a lesser degree when experiencing neutral buoyancy while swimming or diving. Without gravity, blood volume shifts toward the head, less volume is sequestered in the lower body and more easily returns to the heart while blood volume in the head experiences outflow resistance due to venous congestions on the way to the heart. This process leads to the puffy faces that you see in astronaut pictures since the additional volume causes an increase in pressure that exceeds the limits of the capillaries and plasma leaks into the tissue becoming trapped within and between cells creating swelling known as edema. Eventually some of that fluid will be removed by the lymphatics, a drainage system that parallels the venous system, and processed out by the kidneys with the residual remaining predominantly in the tissue of the upper torso and head. This fluid shift results in an absolute plasma volume loss of 10–20% over the first four to six weeks of spaceflight as crewmembers achieve a new normal plasma volume, normovolemia [19, 20]. This is possible because there is no gravity-induced hydrostatic gradient causing blood to collect in the lower body as it does on Earth and the cardiovascular system is capable of maintaining normal blood pressure accommodating metabolic needs of tissues at rest. However, this results in changes in vascular compartments and performance under stress like maximal exercise and exposure to gravity.

Components of the vascular system must adjust to the new plasma volume and the chronically reduced workload associated with living in the microgravity environment. Red blood cell (RBC) volume is acutely concentrated due to the rapid loss of plasma volume and RBCs are reduced to better match the spaceflight normovolemia maintaining proper viscosity. The venous system of the lower body begins a remodeling process that is associated with the lack of a hydrostatic gradient-induced column of blood in the lower limbs. This lower limb venous remodeling results in a reduction of sympathetic tone and loss of responsiveness to a blood volume [19]. So as with the absolute plasma volume loss, the relative RBC reduction and lower limb venous remodeling are not an issue during the microgravity portion of the mission unless there is a high aerobic capacity demand which has not occurred as a normal part of microgravity operations, but the demand is artificially created during exercise sessions to quantify a crewmember's aerobic capacity [25].

Aerobic capacity is the physiological metric of interest when assessing cardiovascular fitness that includes a role for metabolically active skeletal muscle. Changes to maximal aerobic capacity (VO2max; uptake of oxygen and production of carbon dioxide at a maximal workload) are often tracked to assess training, de-training, and fitness for duty. During long duration spaceflight estimated VO2max (based on heart rate versus workload as opposed to direct gas exchange) decreases 15–20% within the first month and recovers 5–10% during the mission, with plateau during months 2–6 [25]. This has not been an operationally relevant decrease since living and working in microgravity does not demand high workload, even microgravity EVAs are not a high total body workload, but it reflects the lost plasma volume and is very relevant for performance capability at landing especially in the scenario of self-egress and emergency evacuation of the vehicle. As expected in addition to neuro-vestibular disruption, aerobic capacity is significantly impacted immediately postlanding as the body struggles to adjust to gravity and plasma volume becomes sequestered in the lower body, both veins and tissues, and represents a state of dehydration. For long duration spaceflight, 80% of crewmembers experience orthostatic intolerance (OI) on landing day (Figure 1.2a), the inability to stand upright and maintain consciousness. Since dizziness, syncope, unsteady gate, nausea, and vomiting are symptoms of both NV/SM and OI, it is often difficult to identify the dominate cause of the postlanding inability to stand, walk, or run [30]. The current mitigation strategies, pre-landing fluid loading and lower body compression garment, have limited effectiveness and these symptoms significantly abate by 24 hours postlanding with OI having a quicker recovery than NV/SM.

Another critical cardiovascular component that remodels due to microgravity exposure is the heart. Without gravity exerting a force on the tissues of the body, blood flow to and from the heart is relatively unimpeded, and the heart does not have to produce the same contractile force as it does on Earth to maintain proper cardiac output. In addition, the heart is exposed to slightly lower intrathoracic pressures compared to that experienced on Earth [23]. Resistive and aerobic exercise can create an acute stimulus by redistributing blood volume to activated skeletal muscle and creating increased metabolic demand for oxygen supply and carbon dioxide removal. This exercise-induced demand may blunt the microgravity induced cardiac remodeling as indicated by spaceflight analog data but spaceflight missions have always included various forms and durations of exercise prescriptions so there is no spaceflight data to inform how extensive the remodeling could be. This cardiac remodeling is also central to another concern and that is the potential of arrhythmia or an abnormal heart rhythm.

Throughout spaceflight missions one of the main biomedical monitoring activities has been the electrocardiogram (ECG) to monitor the heart's electrical activity since it is a coordinated event necessary to elicit a proper cardiac contraction or heartbeat. Although crewmembers should be thoroughly screened upon selection and routinely monitored during their careers and training especially during the pre-launch period, abnormal heart rhythms can spontaneously develop with age or as a consequence of illness or even dehydration. Not all irregular heart rhythms are inherently dangerous and require context to understand their genesis as well as their potential to cause sudden cardiac death [21]. The concern has been that spaceflight as a unique physiological stressor may be a pro-arrhythmogenic environment. Although, irregular heart rhythms have been detected during early NASA missions there is no evidence that it has been caused by the complex environment of spaceflight but most likely by previously undetected presence of disease such as during the Apollo missions [24]. Subsequent missions have not demonstrated that crewmembers are more susceptible to arrhythmias during spaceflight compared to pre and postflight. The ongoing effort has been to establish a comprehensive baseline of ECG data for each crewmember at rest and during stress to compare over time to ECGs done during the mission for surveillance purposes. This process will provide context and allow interpretation of emerging cardiac rhythm issues, assessment of risk of a pathological process, and inform any treatment including something as straight forward as electrolyte replacement that is imperative before landing.

Immediately at landing the loss of plasma volume, changes to the vascular compartments, and (NV/SM) disruption in combination create an environment where the crewmember will have a measurable deficit compared to pre-flight in capabilities such as standing, walking, running, descending and ascending ladders, and operating vehicles. The magnitude and persistence of the deficit somewhat depends on the duration of the exposure to microgravity with longer spaceflights resulting in a greater degree of deficit and a longer persistence of the deficit (Figure 1.2a). While the effect is measurable at landing on Earth, there are questions about how this physiological adaptation will affect performance when crewmembers are exposed to partial gravity during planetary landings and EVAs. Of course we learned a great deal from the Apollo missions and have confidence that the short transit time will limit adaptation to microgravity and the exposure to lunar partial gravity will not be very provocative. Crewmembers will retain a significant amount of their pre-flight fitness and capability during short sortie-like missions. As lunar orbit and lunar surface missions increase in duration, as envisioned for the Artemis Program and the Moon to Mars initiative, it is unclear if partial gravity will have a mitigating effect on the adaptation to an extended duration microgravity exposure. Artemis missions, especially those later in the Program enabling long duration microgravity exposure followed by surface operations, will be invaluable to understanding the progression of adaptation and effectiveness of countermeasures for human system risks associated with exploration class missions like those proposed to Mars.

1.6 Musculoskeletal Adaptation a Tale of Disuse Atrophy and Remodeling

The human body is an amazingly energy efficient machine. Once a demand or stressor is removed the body will adjust supply. During spaceflight, the absence of gravity leads to biological processes that reduce the unnecessary muscle volume and bone density as it does in most Earth-relevant instances of disuse. The operational and research communities have been assessing changes in muscle volume and strength since the Apollo missions to understand the rate and persistence of the change to skeletal muscle and bone, but it has been challenging to relate any changes, particularly deficits, to a functional consequence. Astronauts and cosmonauts have been capable of completing mission objectives including EVAs, however, rarely are crewmembers required to self-egress upon landing and the off-nominal cases of self-egress in a crew of three have produced anecdotal evidence that only one of the three are capable of standing and often all three resort to crawling out of the vehicle and taking long periods of rest between activities until recovered. Therefore, the realistic concern is for future missions in which the disuse atrophy of both skeletal muscle and bone may progress given the extended duration of microgravity exposure and the workload demands on planetary surfaces may exceed the capability of the musculoskeletal system resulting in LOM or LOC.

While crewmembers' bodies adapt primarily to the absence of load in microgravity, new risks are created when the body is rapidly exposed to gravitational forces during landing and egress. Changes in muscle strength and volume have been studied during spaceflight for decades and not until recently have the changes been put into the context of the mission or expected tasks. The functional task test (FTT) has been a multi-disciplinary research effort to systematically assess and quantify changes in crewmember's performance starting immediately at landing and the recovery progression over several weeks [10, 11, 30]. The data analysis relates changes to the contribution of the various subsystems causing the pre to post deficits including NV/SM, cardiovascular, and musculoskeletal. Given the plasticity of the musculoskeletal system none of the changes appear permanent and crewmembers can recover both volume and strength with adequate training within weeks to months of return [27] (Figure 1.2f). This measurable change in muscle volume and strength can be mitigated with countermeasures such as resistance training and AG during the transit and at landing by having proper occupant protection to prevent landing injury and egress aids available to guide the crew. While skeletal muscle strength can be assessed and addressed in flight, it has been very challenging to get insight into bone density and only pre to post mission changes have been acquired to reflect back on how the mission has impacted that system.

Changes in bone mineral density (BMD) may put crewmembers at risk for fracture during deep space exploration missions but changes to BMD are monitored using pre and postflight dual-energy X-ray absorptiometry (DXA). Without insight into BMD loss during the mission it is difficult to know the true rate of loss. The current rate of bone loss is a simple calculation of the (Pre BMD – Post BMD)/duration of mission. This oversimplification has led many to believe that the loss rate is linear and would continue indefinitely for the duration of an exploration class mission of three years or more. This is not a reasonable assumption since we have observed crewmembers both female and male who have experienced varying mission durations such as extended ISS missions ranging from 8 to 12 months that have not demonstrated double the BMD loss as those who have had missions 4 to 6 months in duration. Unfortunately, this data has not been published due to the small "*n*" and inability to protect the identity of the subjects. It is more reasonable to hypothesize that the rate of BMD loss is asymptotic and does plateau at some point based on load exposure via countermeasures and genetics. It is still important to assess the projected loss of BMD and evaluate the mission objectives to understand the potential forces that will be encountered and the types of scenarios in which the forces might be applied to the skeleton (Figure 1.3).

To date regions across the entire skeleton are assessed and those with the most significant reductions in BMD are the trochanter, femoral head, and pelvis [26]

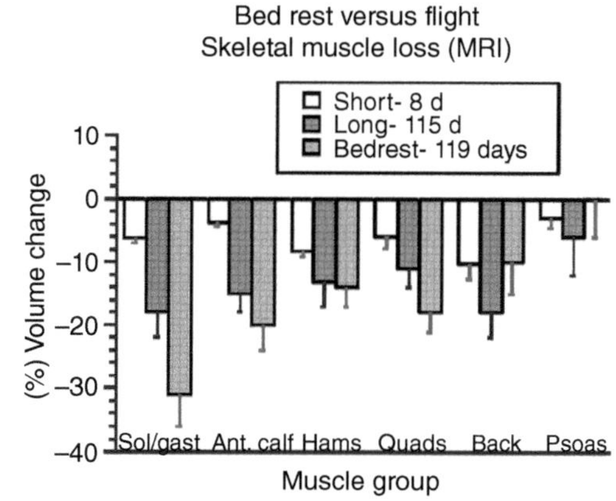

Figure 1.3 Percent change in selected muscle groups during short-duration (8 days; $n = 8$) and long-duration (115 days; $n = 3$) spaceflight (Mir 18) compared with long-duration bed rest (119 days). Data are from (LeBlanc et al. 1995; LeBlanc et al. 1992) and the Shuttle/Mir Final Report. [28].

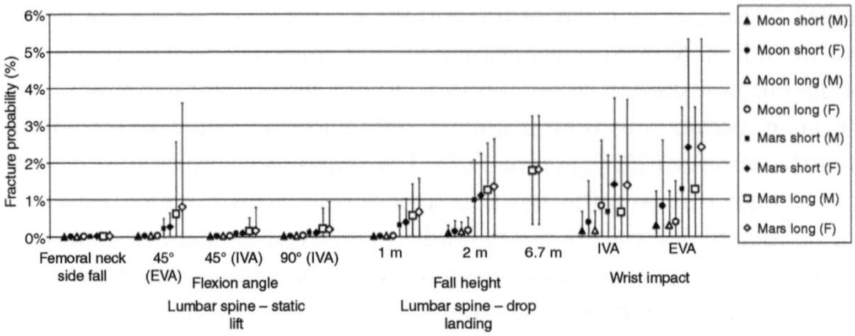

Figure 1.4 Fracture probability for astronauts during reference missions to the moon and Mars. Source: Nelson et al. [29], permission conveyed through Copyright Clearance Center, Inc. Republished with permission of Springer Nature.

(Figure 1.2c,d). Change in BMD for an individual overtime is a clinical surveillance tool and the results per bone region are compared to population norms for risk of fracture that include traumatic and atraumatic (or fragility) to be assessed. A complicating factor for crewmembers is that population norms for those younger than 65 are not robust and those comprising the population under 65 often have comorbidities. A unique effort sponsored by NASA [29] developed an astronaut fracture risk model and calculated the fracture risk for various missions (Figure 1.4). The results indicate that fracture risk varies by bone region based on the application of force and the mission duration aligning with greater BMD reductions.

One of the areas not specifically addressed by this model but still a significant area of research is the significance of the two different compartments of bone that are affected differentially by the BMD loss. Both the cortical and the trabecular compartments lose BMD, but the trabecular bone is disproportionately affected and also does not appear to fully recover [28]. Total BMD returns to pre-flight density due to the cortical bone exceeding its original volume to compensate for the loss of trabecular, however the trabecular bone plays a role in overall strength of the bone and the microarchitecture for the bone marrow. This differential response of the compartments has been challenging to assess due to applications of technologies that can elucidate these compartments and follow up over very long post mission time frames. Another complexity associated with the loss of BMD during spaceflight is the increased circulating calcium that increases the risk of renal stone formation [33]. If not prevented with hydration and potential manipulation of the urine pH, renal stones are a very serious and debilitating medical event that could be a contributor to LOM and, if left untreated due to an underprepared medical system, LOC [34]. Overall, in-mission countermeasures are necessary to maintain musculoskeletal health and fitness for duty given the expected rigors of planetary surface operations in addition to preventing unnecessary medical risks such as renal stone and fracture.

1.7 Multisystem Effects

Given that the human body is a system of systems, it is reasonable to treat all functions and outcomes as the product of multi-system interactions and with each action there will be opportunities and risks. While managing a complex endeavor like human spaceflight it is reasonable to articulate the known problems that drive requirements and factors of safety, the known unknowns that represent gaps in knowledge or technology and drive targeted research and development efforts, and the open space of the unknown unknowns that drive surveillance efforts and non-targeted research and data analysis. The first two categories drive formal programs that create bodies of work with budgets and schedules designed to culminate in quantifiable products that are delivered via requirements that can be debated and negotiated based on constraints such as time, money, launch mass capability, risk tolerance, etc. The third category is all about keeping your eyes and ears open and having some latitude to investigate or at least interrogate anomalies. It requires the ability to tolerate speculation and a rational approach to know if there is enough data available to constitute a real zero or a new gap. There are also many ways that unknown unknowns come to light and in the medical community they are often known as the sentinel case and often identified in hindsight.

Space Associated Neuro-Ocular Syndrome (SANS) is one of those non-targeted findings due to a sentinel case and now the most significant human finding of human spaceflight thus far this century. SANS represents how the body experiences the stressors of spaceflight and attempts to adapt using many compensatory mechanisms [35]. There is the pushing and pulling of fluid that drives conflict with organs and tissues that are very sensitive to pressure that attempt to autoregulate and at first succeed but overtime the mechanical forces prevail and the tissues succumb and remodel. The fluid shift that was mentioned earlier happens in all fluid compartments: blood volume (BV) and cerebral spinal fluid (CSF). The CSF and the brain shift up during flight, crowding the top of the skull where there are important veins to allow BV outflow. The CSF inundates the optic nerve stretching it and impacting the optic cup where it meets the eye and pressures in the tissues of the back of the eye are disrupted. The pressures exceed the tolerance of the tissues causing edema in the delicate choroid and retina. The choroid and retina begin to fold on themselves creating wrinkles that can affect vision but currently have not on orbit. The globe of the eye may flatten due to the CSF pressure in the optic nerve and/or the swelling of the choroid and retina. This flattening does affect vision to varying degrees and requires correction during the mission. One crewmember may have several prescriptions over the mission to address the evolving changes in eye shape. The venous outflow obstruction does not just exist at the top of the brain in the skull but is also functionally present in the neck and has been associated with blood clot-like formations found on ultrasound during a research study [36]. All of these findings have been unnerving at times and the community continues to find new ways to surveil for related changes as well understanding of the mechanisms. There has been extremely large inter-individual variability across the crewmembers and a large spectrum of presentations. Some aspects of the remodeling persist postflight and will forever be a scar on the individual who flew in space, and some aspects slowly return to the pre-flight state and show no long-term memory of the circumstances endured. There are also components of SANS that are still emerging given that the variety of case presentations has made understanding sensitivity and specificity challenging. A new emphasis has been placed on understanding the subtle anatomical brain changes in concert with functional outcomes. We are only now at the beginning of the long-term surveillance that is required to determine if spaceflight has forever changed the future of the health of crewmembers.

1.8 Vehicle-Induced Effects

The environmental control and life support system (ECLSS) is responsible for providing clean, breathable air at the appropriate temperature and pressure for human spaceflight. The provision of that environment and management of the air quality continues to evolve and be refined as new technologies are developed and tested. Selection and proper containment of materials used in the design of the vehicle and supplies are an important part of the process to prevent and control toxic exposures. Monitoring and surveillance are also present to alert the crew of a toxic release and indicate donning of proper protective gear while remediation occurs. Humans contribute carbon dioxide (CO_2) as a metabolic by product of respiration. The rate and quantity of production has challenged the current ECLSS system's capability to remove that CO_2 from the vehicle environment and that has challenged the human system with chronically elevated levels that contribute to physiological stressors of spaceflight. It has been difficult to quantify the contribution and studies using spaceflight analogs such as bed rest have not produced results that clearly identify the role CO_2 may be playing in spaceflight adaptation but there are several possible pathways that elevated CO_2 exposure may contribute that may worsen outcomes for crewmembers.

1.9 Conclusions

The abrupt g-transitions during and immediately following launch and landing induce some of the most provocative and short-lived disruptions to human health and performance and must be considered during the design of the vehicle and mission operations to enable not only survival but functionality during and after the transition is complete. As the duration of spaceflight has been extended such as that performed on the International Space Station, more substantial adaptations have been documented that persist after return to Earth and have various consequences as well recovery profiles.

Together the baseline health of the individual, the individual's response to spaceflight stressors, the ability to mitigate undesirable levels of adaptation, and the expected mission design and objectives are used to create a matrix of necessary information that can be used to assess the acceptable health status of the individual spaceflight participant. This chapter has provided the reader a limited overview of the human health findings in the context of vehicle and mission design from the historical short-duration spaceflight missions and the contemporary long duration missions as well as a perspective on the human health implications associated with the near future lunar orbit and surface missions. There are many more adaptations to discuss and elaborate upon including immunological, psychological, and psychosocial as well as more system level issues to debate such as food and nutrition, augmented and virtual reality, bioregenerative ECLSS with plants, and a well-informed crew medical system that can perform in-mission surveillance, diagnostics, crew medial decision support. Given the complexity of human exploration of space, the design and development transcends the concept of an enabling vehicle and approaches the concept of an exoplanet. To be successful, we must understand and respect the ecosystem required to support the most complex exploration endeavor attempted by humans.

References

1 NASA-STD-3001 VOL 1 I NASA Technical Standards System (NTSS).
2 NASA-STD-3001 VOL 2 I NASA Technical Standards System (NTSS).
3 Human Spaceflight Standards I NASA.
4 HRR – Evidence (nasa.gov); HRR – HRP Architecture (nasa.gov).
5 Nicogossian, Huntoon, and Pool (ed.) (1989). *Space Physiology and Medicine*, 2nde.
6 Heer, M. and Paloski, W.H. (2006). Space motion sickness: incidence, etiology, and countermeasures. *Autonomic Neuroscience* 129 (1–2): 77–79.
7 Lackner, J.R. and DiZio, P. (2006). Space motion sickness. *Experimental Brain Research* 175: 377–399.
8 Jones, E.M. and Glover, K. (2014). *Apollo Lunar Surface Journal* http://www.history.nasa.gov/alsj.
9 Clément, G. and Wood, S.J. (2013). Eye movements and motion perception during off-vertical axis rotation after space flight. *Journal of Vestibular Research* 23: 13–22.
10 Bloomberg, J.J., Feiveson, A.H., Laurie, S. et al. NASA's functional task test: providing information for an integrated countermeasure system. In: *Proceedings of the 20th IAA Humans in Space Symposium*, Prague, 28 June–3 July 2015, 1–2. International Academy of Astronautics.
11 Reschke, M.F., Kozlovskaya, I.B., Kofman, I.S. et al. (2015). Initial sensorimotor and cardiovascular data acquired from Soyuz landings: establishing functional performance recovery time constant. In: *Proceedings of the 20th IAA Humans in Space Symposium*, Prague, 28 June-3 July 2015, 1–2. International Academy of Astronautics.
12 Moore, S.T., Dilda, V., Morris, T.R. et al. (2019). Long-duration spaceflight adversely affects post-landing operator proficiency. *Scientific Reports* 9 (1): 2677.
13 Mulavara, A.P. et al. (2018). Physiological and functional alterations after spaceflight and bed rest. *Medicine and Science in Sports and Exercise* 50 (9): 1961–1980.
14 Reschke, M.F., Good, E.F., and Clément, G.R. (2017). Neurovestibular symptoms in astronauts immediately after space shuttle and International Space Station Missions. *OTO Open* 1 (4): 1–8.

15 Scott, W.B. (2005). Artificial gravity. *Aviat Week Space Technology* 162 (17): 62–64.
16 Vogt, T., Abeln, V., Strüder, H.K., and Schneider, S. (2014). Artificial gravity exposure impairs exercise-related neurophysiological benefits. *Physiology & Behavior* 123: 156–161.
17 Young, L.R. (1999). Artificial gravity considerations for a mars exploration mission. *Annals of the New York Academy of Sciences* 871: 367–378.
18 Bukley, A., Paloski, W., and Clement, G. (2006). Physics of Artificial Gravity, Chap. 2 , NASA Technical Reports Server (NTRS).
19 Meck, J.V., Waters, W.W., Ziegler, M.G. et al. (2004). Mechanisms of postspaceflight orthostatic hypotension: low alpha1-adrenergic receptor responses before flight and central autonomic dysregulation postflight. *American Journal of Physiology. Heart and Circulatory Physiology* 286 (4): H1486–H1495.
20 Waters, W.W., Ziegler, M.G., and Meck, J.V. (1985). Postspaceflight orthostatic hypotension occurs mostly in women and is predicted by low vascular resistance. *Journal of Applied Physiology* 92 (2): 586–594.
21 Khine, H.W., Steding-Ehrenborg, K., Hastings, J.L. et al. (2018). Effects of prolonged spaceflight on atrial size, atrial electrophysiology, and risk of atrial fibrillation. *Circulation. Arrhythmia and Electrophysiology* 11 (5): e005959.
22 Vernice, N.A., Meydan, C., Afshinnekoo, E., and Mason, C.E. (2020). Long-term spaceflight and the cardiovascular system. *Precis Clinical Medical* 3 (4): 284–291.
23 Norsk, P., Asmar, A., Damgaard, M., and Christensen, N.J. (2015). Fluid shifts, vasodilatation and ambulatory blood pressure reduction during long duration spaceflight. *The Journal of Physiology* 593 (3): 573–584.
24 Biomedical Results of Apollo (Section II Chapter 4 p. 115-128). Johnston, R. S., Dietlein, L. F., Berry, C. A., Parker, James F., West, Vita. , 1975. https://ntrs.nasa.gov/citations/19760005580.
25 Moore, A.D. Jr., Downs, M.E., Lee, S.M. et al. (1985). Peak exercise oxygen uptake during and following long-duration spaceflight. *Journal of Applied Physiology* 117 (3): 231–238. https://doi.org/10.1152/japplphysiol.01251.2013.
26 LeBlanc, A.D., Spector, E.R., Evans, H.J., and Sibonga, J.D. (2007). Skeletal responses to space flight and the bed rest analog: a review. *Journal of Musculoskeletal & Neuronal Interactions* 7: 33–47.
27 Ploutz-Snyder, L. (2015). Risk of Impaired Performance Due to Reduced Muscle Mass, Strength, and Endurance. NASA Johnson Space Center Human Research Program Evidence Report, Approved for Public Release: March 9, 2015, Muscle_for HRR clean (nasa.gov).
28 Sibonga, J. (2017). Risk of Accelerated Osteoporosis. NASA Johnson Space Center. Human Research Program Evidence Report May 9, 2017, available at: Microsoft Word - Fracture for IOM review_clean.docx (nasa.gov).
29 Nelson, E.S., Lewandowski, B., Licata, A., and Myers, J.G. (2009). Development and validation of a predictive bone fracture risk model for astronauts. *Annals of Biomedical Engineering* 37: 2337–2359.
30 Bloomberg, J. et al. (2017). The Functional Task Test: Results from the One-Year Mission. NASA Human Research Program Investigators Workshop (HRP IWS 2017), NASA publication. pp. 1-26. https://ntrs.nasa.gov/citations/20170001370

31 Ginsburg, G.S. and Phillips, K.A. (2018). Precision medicine: from science to value. *Health Aff (Millwood)* 37 (5): 694–701. https://doi.org/10.1377/hlthaff.2017.1624.

32 Blue, R., Nusbaum, D., and Antonsen, E.L. (2019). *Development of an Accepted Medical Condition List for Exploration Medical Capability Scoping NASA/TM-2019-220299*. NASA Johnson Space Center Available from: https://ntrs.nasa.gov/archive/nasa/casi.ntrs.nasa.gov/20190027540.

33 Pietrzyk, R.A., Jones, J.A., Sams, C.F., and Whitson, P.A. (2007). Renal stone formation among astronauts. *Aviation, Space, and Environmental Medicine* 78: A9–A13.

34 NASA (2017). *Evidence Report: Risk of Renal Stone Formation, Human Research Program Exploration Medical Capabilities Element*. Approved for Public Release: May 15,. Houston, Texas: NASA Johnson Space Center: Microsoft Word - Renal for IOM review.doc (https://ntrs.nasa.gov/citations/20170004709).

35 Lee, A.G., Mader, T.H., Gibson, C.R. et al. (2020). Spaceflight associated neuro-ocular syndrome (SANS) and the neuro-ophthalmologic effects of microgravity: a review and an update. *NPJ Microgravity* 6 (7): https://doi.org/10.1038/s41526-020-0097-9. Erratum in: *NPJ Microgravity*. (2020) 6:23.

36 NASA (2017). *Evidence Report: Risk of Spaceflight Associated Neuro-ocular Syndrome (SANS), Human Research Program Human Health Countermeasures Element*. Approved for Public Release: November 30,. Houston, Texas: NASA Johnson Space Center: SANS.pdf (nasa.gov).

2

Space Medicine and Countermeasures

Philip M. Williams, Tejasvi Shivakumar, and Valentine Anyanwu

University of Nottingham, School of Pharmacy, University Park, Nottingham NG7 2RD, UK

2.1 Introduction

The microgravity environment of orbital spaceflight creates new opportunities for the manufacture of medicines. And with these opportunities come significant risks and challenges to the health and wellbeing of crew. Understanding, restoring, and maintaining the health of astronauts in the extreme environment of space through the development of new diagnostic tools, drug manufacture processes, medicines formulations, and medicines management have immediate and long-term benefit for pharmaceutical and medical care on Earth. The study of how mankind can live in space onboard the International Space Station (ISS), enabled with technology that reclaims potable water, detoxifies an atmosphere, and supports life, has significantly improved our understanding of nutrition, immunity, regenerative medicine, health, and disease, and hinted at what adaptions occur and are necessary to survive in an adverse environment. In addition, the long list of health-related innovations that have so far stemmed from the exploration of space – including neutron cancer therapy, foldable walkers and disability aids, personal alert systems, computerized axial tomography (CAT) and magnetic resonance imaging (MRI) scanners, diabetic foot mapping technology, orthopedic foam cushioning, kidney dialysis machines, implantable heart pacemakers and defibrillators, emergency medical services in ambulances – shows how impactful space-focused research can be in medicine (Crump, Levy, and Billica 1996; Kirkpatrick et al. 2007).

As humans dive deeper into space, the need for Earth-bound pharmacy increases, along with the necessary range of medication of a rising variety of dosage and delivery forms required, increasing up-mass (Blue et al. 2019a; Eyal and Derendorf 2019; Hodkinson et al. 2017; Snyder et al. 2019). It is estimated that the rate of significant illness or injury (such as stroke, myocardial infarction, intracerebral hemorrhage,

In-Space Manufacturing and Resources: Earth and Planetary Exploration Applications, First Edition.
Edited by Volker Hessel, Jana Stoudemire, Hideaki Miyamoto, and Ian D. Fisk.

appendicitis, and bone fractures) on submarines, Antarctic expeditions, military aviation, and spaceflight is 0.06 cases per person-year; indicating a 67% probability of at least one such significant event in every six-crew, three-year Mars mission (Hodkinson et al. 2017; Houtchens 1993; Komorowski, Fleming, and Kirkpatrick 2016). The acceptance that in a Mars-class mission there will be one major medical event requiring intervention of the type usually delivered by a medical practitioner creates the need to prepare and plan for its occurrence, but as the nature of the event is unpredictable such preparations must account for every reasonable eventuality. With increased mission length and complexity comes the increase in the variety and number of medications that need to be available (Carnell 2020; Genah, Monici, and Morbidelli 2021; McLaughlin, Donoviel, and Jones 2017; Patel et al. 2020; Simon et al. 2021).

The extension of distance and duration of spaceflight is driving new medicines manufacture. This chapter looks at the development of medicines used in spaceflight, the prospects and opportunities of medicines manufacture in space, and the benefits of spaceborne medicines research and manufacture for use on Earth.

2.2 Medicines for Space

Medicines use in flight has existed since NASA's pioneering Project Mercury, with Gordon Cooper taking a 5 mg dextroamphetamine (Dexedrin) stimulant tablet 80 minutes before the retrofire burn of the 22-orbit (34 hour) MA-9 *Faith 7* mission (Carpentier et al. 2018). Autoinjectors of pethidine (an opioid anticholinergic pain killer) and trimethobenzamide (a dopamine antagonist antiemetic) were carried on all Mercury flights, where the drug could be quickly self-administered through the space suit, although none were used (Figure 2.1a). Through Gemini, Apollo, Skylab, and the Space Transportation System ("Space Shuttle") STS missions, the repertoire of medicines available in flight increased rapidly. Between 1981 and 1998, 498 (98%) of the 508 astronauts of shuttle flights STS-1 to STS-89 reported a medical event (1867 events in total) or symptom other than space motion sickness. Records of 79 STS missions reveal that 94% included some medication being taken during flight; of that number, 47% were for space motion sickness, 45% for sleep disturbances (Barger et al. 2014), and smaller percentages for headache, backache, and sinus congestion. Similarly, over a ten-year period (2002–2012), ISS records of long-duration flight showed the most used medications were for (in decreasing order of frequency) sleep, body pain, congestion/allergy, skin rash, and space adaptation syndrome (SAS) (Blue et al. 2019a). Of these indications, all except for SAS included both acute (less than seven days) and prolonged (more than seven days) medication use (Wotring 2015).

The medicines available onboard the ISS are in a variety of forms that include tablets and capsules, applications (creams and ointments for ear, eye, and skin),

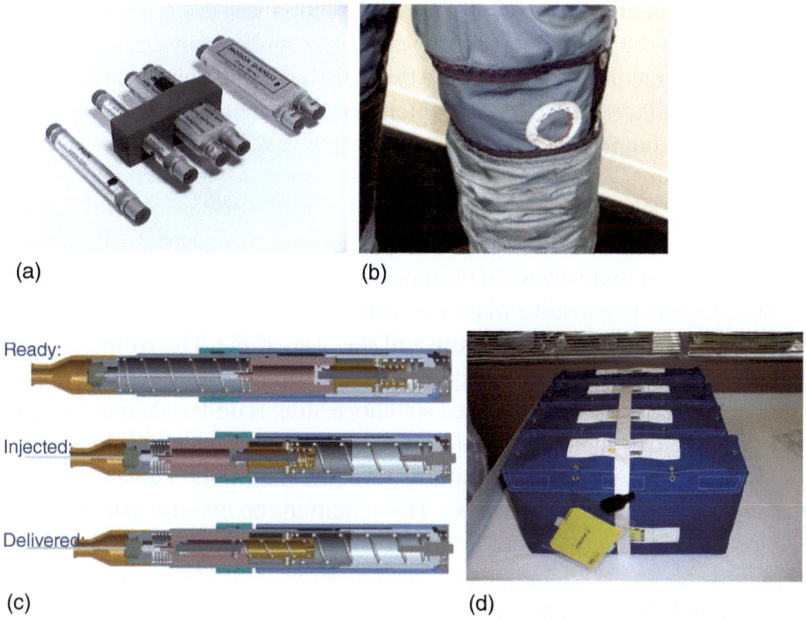

Figure 2.1 (a) Autoinjectors of medication for motion sickness and pain have been caried on all NASA crewed missions. (b) The injection rubber self-sealing patch on the suit worn by Apollo astronauts (with outer white thermal TMG layer removed). (c) Designs for the In-suit Injection System. (d) Packs of medication ready for flight in the medicines stability experiment. Source: NASA.

liquids (eye "drops," nasal sprays[1]), inhalers, and injectables. Most of the injectables are intramuscular (IM), and the emergency treatment medications for Advanced Life Support in cardiac arrest of atropine, epinephrine, and lidocaine given using an intraosseous (into the bone marrow of the tibia leg bone) injection device. Epinephrine (EpiPen) is available for treatment of anaphylaxis emergency by IM injection, and other IM and subcutaneous (SC) medicines available in single- and multiple-dose vials, some requiring reconstitution. Several liters of IV saline solution are carried, and experiments have been undertaken on the manufacture of such solutions in flight (see later). Considerations must be made for the use of injectables in emergency, such as the decompression of the environment causing reduced temperature and pressure (raising the propensity for bubble formation).

1 Oxymetazoline (Afrin) decongestant nasal spray was in the medication kit onboard Apollo. Problems with the packaging were reported by the astronauts of Apollo 11, where the contents "*bubbled out when the cap was removed and was therefore unusable,*" due to the lower atmospheric pressure in the spacecraft. The end of the spray bottle was filled with cotton in future flights. Harrison Jack Schmitt, the geologist that flew on Apollo 17, used three bottles of Afrin, plus took a pseudoephedrine/triprolidine (Actifed) tablet, after suffering from an allergic reaction to Moon dust. Afrin nasal spray and pseudoephedrine (Sudafed) tablets are stocked onboard the ISS.

In 2009, the *in suit* Injection System (ISIS) for IM injection was designed with four capabilities: to provide the astronauts with an easily accessible, easily administrable injection of necessary medications; to enable the injection of medicines through an EVA suit; to work in a low-pressure and low-temperature environment; to work in a microgravity environment (Figure 2.1c).

2.2.1 Stability and Shelf-life

The medicines carried in flight are all FDA-approved commercial products that have often been repackaged[2] to conserve mass and volume, and their quality is assured through the usual standards of manufacture and stated shelf-life. The United States Pharmacopoeia (USP) sets a medicine's quality, purity, strength, and identity standards. In product development, accelerated stability testing is undertaken to specify the shelf-life of a medicine, defining the time under specified storage conditions the medicine is expected to meet the specification. Beyond the expiration date the manufacturer cannot guarantee the medication. The conditions on board a spacecraft in flight are, of course, different to those under which product stability and shelf-life has been assured. The repackaging of the medicines may also affect a medicine's shelf-life or stability (Mehta and Bhayani 2017).

Nearly 90% of the medications flown onboard the ISS have shelf-lives of less than 24 months, and current practice ensures that the medications are restocked before their expiration date. Wotring undertook an opportunistic study of the stability of nine tablet medications returned from the ISS after 550 days of flight (Wotring 2016). At the time of testing, three medications (ibuprofen 400 mg, loperamide 2 mg, and modafinil 200 mg) were within their shelf-life, the other six (aspirin 325 mg, loratadine 10 mg, melatonin 3 mg, paracetamol 325 mg, pseudoephedrine 120 mg, zolpidem 10 mg) had passed their expiration date; three by nine months (aspirin, pseudoephedrine, zolpidem), and one (melatonin) by 11 months. Only the melatonin tablets failed to meet the USP specification for both active pharmaceutical ingredient (API) and degradants assay, although the paracetamol, loratadine, and the non-expired loperamide would have failed under the later-revised USP guidelines. A study between 2006 and 2008 compared APIs in 35 formulations flown in space for 14, 353, 596, and 880 days to ground-based controls, and found that flown tablets of amoxicillin/clavulanate, levofloxacin, sulfamethoxazole/trimethoprim, furosemide, and levothyroxine changed significantly compared to their ground control (Figure 2.1d) (Du et al. 2011). This study suggested that the rate of degradation for light-sensitive pharmaceuticals may be accelerated during spaceflight, which might be related to the radiation environment of the spacecraft. The study also revealed the percentage of degradation of clavulanate was higher on

2 The tablets and capsules in the medical kit on Apollo 11 had been repackaged in individually sealed plastic or foil containers. When the kit was unstowed in the Command Module during the flight, these containers expanded due to the lower pressure inside the spacecraft. This ballooning of the packages increased the volume of the medical kit contents so much so that part of the kit had to be cut away before it could be restowed.

the ground than it was in flight. This was attributed mainly to the higher relative humidity (RH) (36%–44%) recorded on the ground than in flight (23%–25%) since clavulanate is very sensitive to RH differences. Since the fragmentation of molecules through radiation in spaceflight is different to the degradation products seen on the ground, there is also the need to study and understand the effect and toxicity of such degradants (Blue et al. 2019b). The pharmaceutical formulation is also known to affect the susceptibility of the API to degradation in spaceflight, with liquid formulations degrading faster than solid. Hessel, for example, is studying the effect of spaceflight on the stability of APIs in tablets stored both inside and outside of the ISS in a Materials International Space Station Experiment (MISSE-FF) (Figure 2.2c), and the ability of heavy elements in silica, magnesium silicate, and calcium phosphate to protect against radiation-induced degradation (see "Cosmic-ray stable medicines").

MISSE is a series of experiments that are mounted externally on the ISS to study the effects of long-term exposure of materials to the harsh space environment. MISSE-6 included a study, led by Kaplan of Tufts University, of silk and collagen films (Figure 2.2a,b) (Hu et al. 2013). After 18 months of flight, approximately 15%

Figure 2.2 (a) The MISSE 6A and MISSE 6B tray on the European Columbus Laboratory as imaged during the fly-around of STS-123 Space Shuttle Endeavor. Source: NASA. (b) Post-flight analysis comparing the effects of long-term exposure to the space environment in LEO to ground controls of silk and silk-based composites. Source: Hu et al. 2013 / Springer Nature / CC BY NC-ND 3.0. (c) The Materials ISS Experiment Flight Facility (MISSE-FF) that replaced MISSE and is controlled robotically from the ground. Source: NASA.

of the film surfaces were etched by ionizing particles such as atomic oxygen, the major component of the low-Earth orbit (LEO) space environment. Furthermore, more than 80% of the silk and collagen materials were chemically crosslinked by the space radiation. In ground-based studies, Kaplan later researched the ability of such silk-protein-based biomaterials made as composites with silicon dioxide (SiO_2) and other compounds to protect encapsulated drug from radiation-induced degradation. Two drugs were studied: ampicillin that is known to be resistant to radiation damage, and clavulanate that is unstable. Humidity again had a significant impact, with the greater crystallinity of the silk at higher humidity impeding its dissolution and drug release. Drug release was further impacted through its interaction with SiO_2 particles. Under low humidity, however, under simulated Galactic Cosmic Radiation exposure at NASA's national radiation lab (Brookhaven, NY), 100% of the ampicillin was recovered when embedded in silk films, and increased recovery of clavulanate was also obtained compared to the free powder.

The fact that the efficacy and safety of medicines used in spaceflight is based on their use on the ground raises important considerations where the mission duration, propensity of underlying health conditions, severity of illness, or probability of incident may increase, such as in Mars-class and tourist spaceflight.

2.2.2 Changes in Pharmacokinetics (PK) and Pharmacodynamics (PD)

While no change in the dissolution rate of APIs from space-flown tablets compared to the ground controls has been observed, there are data that show the rate at which APIs are absorbed and processed in an astronaut's body does change in flight (Gandia, Saivin, and Houin 2005; Kast et al. 2017). Studies onboard STS, for example, suggest that the absorption of both paracetamol and scopolamine is changed in flight, and that the change is both time and subject dependent (Figure 2.3). A later experiment onboard the ISS compared tablet and capsule formulations of paracetamol. In these studies, the rate of drug absorption from tablets decreased significantly, whereas the bioavailability increased substantially under microgravity conditions. However, with the capsule form, the time of absorption decreased with no statistically significant change in bioavailability. The elimination time, the average retention time, and the distribution volume increased considerably (Eyal 2020). Although the experimental sample numbers are very limited, as it is with most experiments involving spaceflight, varying pharmacokinetics and pharmacodynamics is expected from the known and visible effects of microgravity, with fluid distribution changes followed by volume reductions. Astronauts' reports suggest that medications may not be as effective in managing their medical conditions as expected, although without sample measurement data to accompany the reports, any link to changes in PK/PD remains anecdotal.

Research is needed and new methods employed to manufacture and package medicines that have extended shelf-life stability in flight, produce medicines on-site and on-demand at the time of need, and understand and measure the PK/PD of the drug and the individual at the time and point of need.

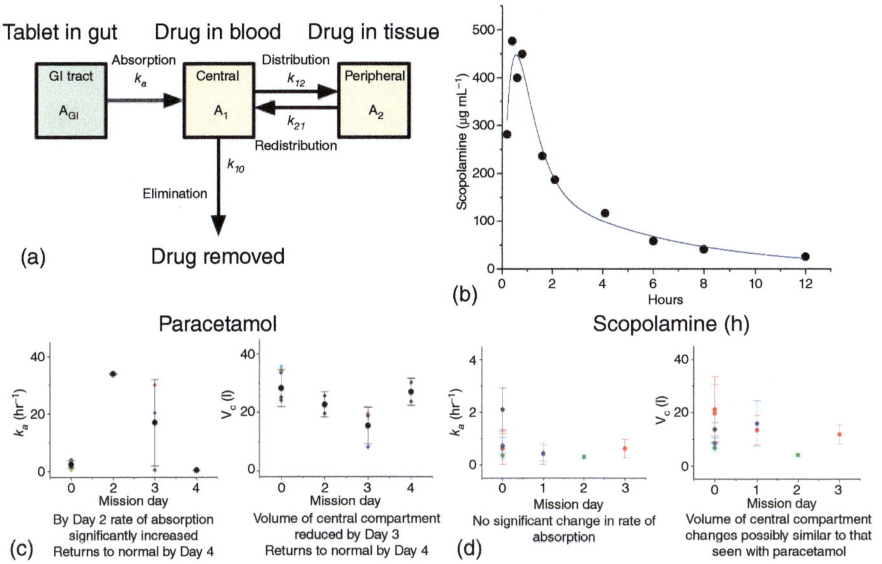

Figure 2.3 (a) A multi-compartment model can be used to model the pharmacokinetics of drugs. (b) Post-flight analysis of a series of mouth-swab saliva samples taken after ingestion of a tablet containing scopolamine. A two-compartment model has been fitted revealing apparent rates of absorption, distribution, redistribution, and elimination of the drug. (c) Plots of the apparent rate of absorption and volume of distribution of paracetamol taken before flight and on Mission Days 2, 3, and 4 for a number of astronauts. The apparent rate of absorption is increased by Day 2 and returns to normal by Day 4. The volume of distribution apparently drops on Day 3, returning to normal by Day 4. (a) The change in the pharmacokinetics of scopolamine is different to that of paracetamol. Data source: NASA.

2.2.3 The Vascular Echo Study: The Case for On-site Manufacture and Telepharmacy

The Vascular Echo study is part of a larger series of experiments to investigate the changes in circulation caused by microgravity and the body's tissue and organ response. The Vascular Echo experiment followed from previous findings that showed the arteries of some astronauts were much stiffer when they returned to Earth in comparison to when they launched, and the astronaut's carotid artery stiffness had increased by that expected from 10 to 20 years of normal aging. In Vascular Echo, ultrasound imaging and blood samples were taken during and after flight. The ultrasound equipment onboard the ISS had been replaced in 2011 with Ultrasound 2, a modified GE Healthcare *Vivid q* device, launched with STS-135/ISS assembly flight ULF7, the final mission of the Space Shuttle program (Figure 2.4a). A teleoperated ultrasound system developed by Sonoscanner was later sent to the ISS as part of an ESA and CNES series of experiments (Figure 2.4b).

There were 11 astronauts involved in the study, which included measuring the structure and function of the internal jugular vein at scheduled times in different positions during long-duration missions (Marshall-Goebel et al. 2019). One of the

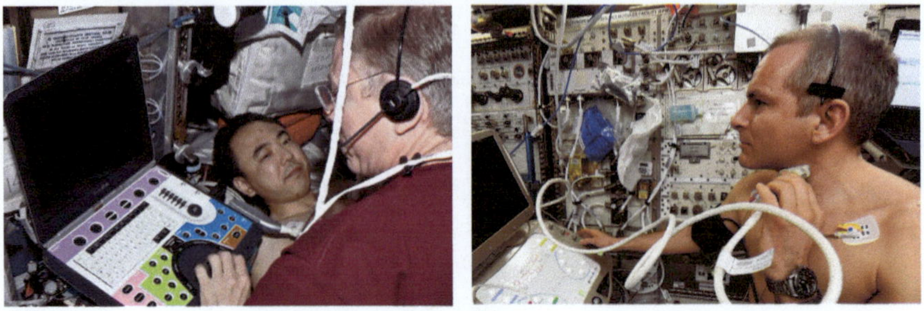

(a) (b)

Figure 2.4 (a) Commander Mike Fossum scans JAXA crew member Satoshi Furukawa using the Ultrasound 2 equipment on the ISS. (b) A teleoperated ultrasound scanner being operated by Thomas Pesquet. Source: NASA.

crewmembers developed a blood clot in the major vein in the neck, discovered in their Flight Day 50 ultrasound imaging (Aunon-Chancellor et al. 2020). The astronaut, guided by two radiologists on the ground to undertake further imaging and respiratory maneuvers, helped confirm the diagnosis of a venous thrombosis. Anticoagulant treatment was commenced with enoxaparin, a low-molecular-weight heparin, at a dose of 1.5 mg/kg body weight once daily. Twenty vials of 300 mg/3 mL enoxaparin each were on board the ISS, but no anticoagulation-reversal agent was available. These injections posed challenges – syringes are limited in number, and drawing liquids from vials was a significant challenge due to surface-tension effects. After 33 days, it was decided to reduce the dose to 1 mg /kg to extend therapy until an oral anticoagulant (apixaban tablets) and reversal agents (protamine and prothrombin complex concentrate) could be delivered to the ISS with a supply spaceflight. Although the size of the clot progressively shrank, spontaneous blood flow was still absent after 90 days of anticoagulation treatment. The astronaut took the apixaban until four days before the return to Earth. Spontaneous blood flow was seen on landing, and no further treatment was necessary.

The case highlights several significant issues that need to be addressed as we increase the range, time, and extent of manned spaceflight. Where early return to Earth or resupply is not possible, such as on the way to Mars, there would need to be either enough of any essential medicine to provide treatment for any life-threatening condition available for the whole of the mission duration, or the capability to manufacture the medicines on-site and on-demand. While the former presents further challenges of mass, volume, and stability as discussed, there is significant activity on researching the possibility of medicines manufacture in space.

2.3 Medicines in Space

In August 2020, the International Space Agencies Coordination Group, a forum of space agencies to advance the Global Exploration Strategy, reaffirmed their Global

Exploration Roadmap. NASA's Artemis program will enable human missions to the lunar surface beginning in 2024 and targets sustainable lunar exploration by 2028. The program objectives are based on the principle that human lunar surface exploration should focus on preparation for sustainable activities on the Moon and for human Mars missions. NASA highlighted in their Technology Roadmap TA 6 "Human Health, Life Support, and Habitation Systems" and in their 2009 Human Exploration of Mars Design Reference Architecture 5.0, that pharmaceutical provision could ultimately limit the acceptable duration of a mission.

Research to produce medicines in flight for use in flight started in 2010 with the IVGEN experiment, undertaken onboard the ISS during Expedition 23. The IntraVenous Fluid GENeration for Exploration Missions (IVGEN) experiment was designed to demonstrate a water purification and pharmaceutical mixing system that can make IV fluid in microgravity (Figure 2.5). The system used a deionizing column to remove contaminants from potable water to a purity level that could meet the USP standards. Six 1.5 L bags of purified water were produced. Two of these bags were mixed with USP-grade sodium chloride to make Normal Saline (0.9% NaCl) solution. Inline conductivity sensors quantified the input and output water purity, and normal saline mixing uniformity. The results indicated that the experimental success criteria were met except for salt concentration, highlighting the peculiar difficulties of working in microgravity. These two bags were returned to Earth for further testing. Problems with a large air bubble in

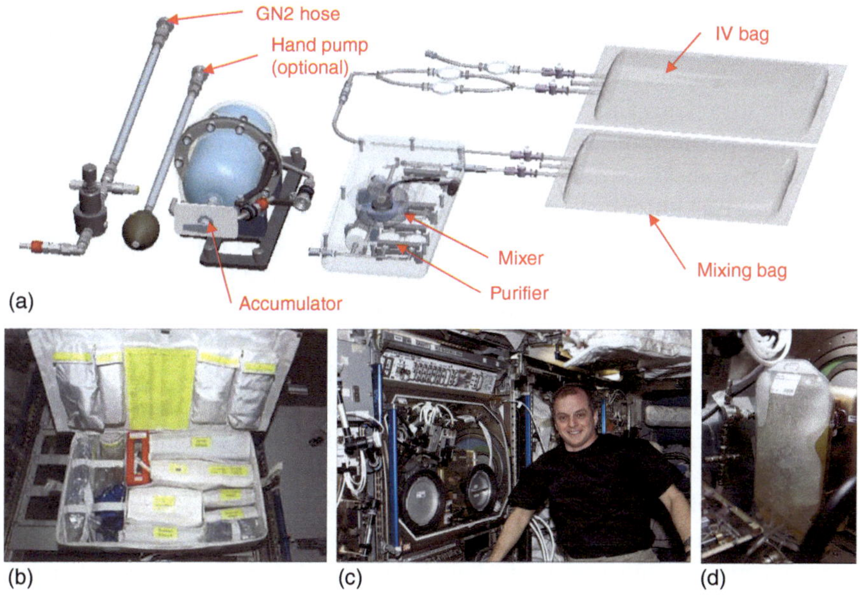

Figure 2.5 (a) Schematic diagram of the IVGEN equipment. (b) The Advanced Life Support Pack includes important items, such as saline solution. (c) Astronaut TJ Creamer with the IVGEN hardware in the Microgravity Science Glovebox onboard ISS. (d) An IVGEN Saline Bag. Source: NASA.

the first bag of purified water resulted in a slightly concentrated saline solution of 117%. The second bag had an inadequate amount of salt premeasured into the mixing bag resulting in a deficient salt concentration of 93.8% of the target value. The development of the next generation of this technology, IVGEN Mini, is underway with a focus on miniaturization, system improvements for reliability, ease of use, and compatibility with constraints of exploration missions (mass, volume, power, etc.). The project aims for a demonstration in-flight onboard the ISS in 2024, and deployment of a flight-ready unit on the Lunar Gateway in 2026/27.

2.3.1 Toward Chemical Synthesis in Space

To manufacture drugs in space is a significant challenge. Flow chemistry is cited as a route for molecular synthesis in microgravity since it benefits from the reactors having no headspace, the reaction zone is closed, often high yield, and a precise control of parameters and automation (Sipos et al. 2017). The *ISSpresso* coffee machine, first used on the ISS in 2015, is likened to the first example of a continuous-flow chemistry instrument used in space for the dissolution and extraction of material from complex solid mixtures using conditioned solvents. Here, water from a sealed pouch is aspirated and pressurized, heated, mixed with coffee, and collected into a receiving pouch.

Continuous-flow chemistry coupled with analytics and automation here on Earth is progressing toward the platform of "dial-a-molecule," where an automated system could be instructed, or programmed, to synthesize molecules on demand (Sans and Cronin 2016). Continuous manufacturing of nonsteroidal anti-inflammatory ibuprofen was demonstrated in 2009 (Bogdan et al. 2009), of an antimalarial drug candidate in 2015 (Lau et al. 2015), and the first end-to-end continuous manufacturing (synthesis, purification, and final dosage formation) of a medicine, tablets of the antihypertensive aliskiren, demonstrated by Novartis in 2013 (Mascia et al. 2013). In 2017, a team at Eli Lilly demonstrated the kilogram-scale synthesis of the anticancer drug prexasertib, under continuous-flow CGMP (current good manufacturing practices) conditions (Cole et al. 2017). These production-scale systems were large and massive – the Novartis system developed with the MIT Center for Continuous Manufacturing was more than 7 m long, and about 2.5 m high and deep – and thus not suitable for spaceflight. A smaller "refrigerator-sized" reconfigurable system was later developed at MIT, and demonstrated synthesizing four different drugs: diphenhydramine, lidocaine, diazepam, and fluoxetine (Adamo et al. 2016). Space-appropriate systems is a focus of SpacePharma, an Israeli start-up, who specializes in providing miniaturized laboratories for flight in CubeSats and on the ISS (Amselem 2019). In 2017, their DIDO2 platform ran an experiment designed by Cronin demonstrating aspects of "digital chemistry." A number of experiments were performed on the satellite, including the spectroscopic monitoring of the enzymatic oxidation of glucose by glucose oxidase, studying the self-assembly of phenylalanine and its crystallization, and imaging oil droplet coalescence in microgravity.

(a) (b)

Figure 2.6 (a) The Veggie plant growth system aboard the ISS has successful grown several plants. The color of the illumination is optimized to match that absorbed by the plants to grow. (b) Lettuce grown in Veggie. McDonald is engineering lettuce to produce drugs such as parathyroid hormone, which could be used to counteract bone density loss associated with spaceflight. Source: NASA.

2.3.2 Drugs from Plants in Space

Harnessing and reprogramming Nature's highly developed machinery to manufacture drugs on demand is an area of research undertaken by McDonald and her team at the University of California, Davis (He et al. 2021). With the lack of sunlight and gravity, growing plants in space is difficult but not impossible. The Vegetable Production System, known as Veggie, (discussed further in Chapter 12) is a space garden residing on the ISS, where a bank of light emitting diodes above the plants produces a spectrum of light optimized for and to guide growth (Figure 2.6). Russian kale, mizuna mustard, zinnia flowers, Chinese cabbage, and three types of lettuce have been grown successfully. To make drug molecules, McDonald and colleagues infected lettuce with *Agrobacterium tumefaciens* carrying the DNA sequence for parathyroid hormone (PTH), a protein that counteracts bone density loss associated with spaceflight. The bacterium transferred the DNA to the plants' genome, after which the plant produced the protein. After six days of growth the lettuce contained 10 mg/kg of PTH. This research is now being extended to produce two more drugs, granulocyte-colony-stimulating factor (G-CSF), and an anti-fungal peptide, as part of a Translational Research Institute for Space Health (TRISH) program on "Just in Time" medications.

The advent of such protein-based drugs – the biologics or biopharmaceuticals – has enabled new avenues for their creation on-site and on-demand (Crommelin et al. 2020; Crommelin et al. 2003). Of particular interest for application in space-based manufacture is the use of bacteria, and the use of cell-free protein synthesis, to make spaceflight-relevant protein therapeutics (Table 2.1).

2.3.3 Cell-Based Systems for Biopharmaceutical Manufacture

The Astropharmacy research of Rothschild (NASA AMES), Williams (Nottingham) and colleagues is developing on-demand drug manufacturing processes using both

Table 2.1 Example of biologics with potential use in a spaceborne emergency.

Drug	Indication
Alteplase (t-PA)	Thrombolytic used to treat ischemic stroke, myocardial infarction, pulmonary embolism,
Reteplase	Similar to alteplase but with sequence modifications to give a longer half-life of 13 to 16 minutes.
Tenecteplase	Similar to alteplase but with sequence modifications to provide higher fibrin specificity and greater resistance to inactivation. Glycosylated.
Bivalirudin	Anticoagulant (thrombin inhibitor)
Lepirudin	Anticoagulant (thrombin inhibitor)
Epoetin alfa	Human erythropoietin used for treating some anemias.
Salmon calcitonin	Severe hypercalcemia and osteoporosis
Teriparatide	Form of parathyroid hormone used to promote bone formation and in treatment of osteoporosis
Entolimod	Radioprotective agent. Unlike G-CSF entolimod would be given immediately before or after exposure, such as an un-shielded solar particle event.
Filgrastim (G-CSF)	Used for neutropenia following chemotherapy or radiation poisoning.

cell-based and cell-free systems. Our cell-based system uses bacteria, such as the *Vibrio natriegens* and *Bacillus subtilis*, which can be stabilized and stored for years in the dry state. *B. subtilis* has been shown to remain viable after several years in Earth orbit. In optimum conditions, *V. natriegens* can have a doubling time of less than 10 minutes, meaning proteins of sufficient quantity can be expressed quickly. For a mission, small stocks of bacteria would be transfected with the DNA encoding for the different biologic and then stored in the spacecraft, either lyophilized for the *V. natriegens* or in spore form for the *B. subtilis*, until use. When required, a sample of the appropriate bacteria would be cultured in media, during which time the drug is secreted and then purified through affinity chromatography and size exclusion. For the amounts of protein normally required (single person dose at a time) this can be all be performed "in-syringe."

Defense Advanced Research Projects Agency (DARPA), part of the United States Department of Defense, have selected on-demand biomanufacturing as part of their "Battlefield Medicine" program (Choi, Lewin, and Ling 2015). The program comprises two research areas: Pharmacy on Demand (PoD) and Biologically-derived Medicines on Demand (BioMOD) for terrestrial use on the battlefield. Successes in PoD are in the end-to-end manufacture discussed previously (Lewin, Choi, and Ling 2016). The BioMOD program aims to employ genetic engineering and cell-free protein synthesis to synthesize biologics on the battlefield and in other austere environments.

A team of DARPA-funded scientists have described a cell-based on-demand manufacturing system called InSCyT or Integrated Scalable Cyto Technology for terrestrial use (Crowell et al. 2018). InSCyT is an automated benchtop (total around 2 m in width) system capable of producing clinical-grade biologics in less than three days. The set-up involves three modules that are responsible for recombinant synthesis (using *Komagataella phaffii* yeast cells), purification and formulation, respectively. Expression of human growth hormone, G-CSF, and interferon α-2b was carried out to exhibit an end-to-end production system. Purity, potency, biochemical and biophysical traits of final products were tested to ensure similitude to marketed drug.

Cao and colleagues, driven by a similar vision, further described an integrated terrestrial platform for on-demand production of multiple therapeutics at one time (Cao et al. 2018). They demonstrate three strategies: inducible synthesis with control of drug expression ratios; consolidated bioprocessing (post-translational processing); co-expression/purification, for flexible and low-cost production of combinatorial therapies where resources are limited. Programmed expression of biologics was attained in portable micro-bioreactors, integrated with genetically engineered *Pichia pastoris* for on-demand applications (Perez-Pinera et al. 2016).

The significant interest of compartmentalization in synthetic biology is perhaps because of the need to control biochemical composition and structure. In their recent work, Johnson and colleagues present a compartmentalized microbial terrestrial platform, co-cultured and immobilized in hydrogels (Johnston et al. 2020). First, encapsulation of yeast and bacteria in hydrogels made of a temperature-responsive and shear-thinning polymer was shown. Next, mono-culture hydrogels were used to demonstrate on-demand production of four small molecule compounds and one antibiotic. Thirdly, it was shown that hydrogels provide spatial segregation of co-cultures, where microbial organization improved on-demand production of compounds on a mixed gel; microbial dynamics could be controlled by varying hydrogel ink. Lastly, the stability and lyophilized preservation of hydrogels were tested for repeated and long-term production of the compounds. The platform displayed long-term stability from three months (lyophilized) up to a year (continuous fermentation) and is an excellent example of cell-based platforms for on-demand production of therapeutics. On-demand terrestrial platforms like these, being developed for remote use, provide insight for consideration of systems that could have utilization for long-duration spaceflight once adapted to work in a microgravity environment.

2.3.3.1 Cell-Free Systems for Biopharmaceutical Manufacture

Cell-free protein synthesis (CFPS) refers to the *in vitro* production of proteins using molecular machinery derived from cellular extracts. CFPS is increasingly becoming the method of choice terrestrially over cell-based systems for expression of cytotoxic, synthetic, and complex proteins (Copeland et al. 2021; Dondapati et al. 2020; Silverman, Karim, and Jewett 2020). The numerous advantages of CFPS stem from its open, protean nature, and free of concern over viability (Carlson et al. 2012; Rasor et al. 2021).

An on-site and on-demand cell free system consists of three broad elements namely: cellular extract, template DNA, and reaction mixture. The extract strain is initially grown in media followed by lysis to obtain the cellular extract. The extract contains components such as RNA polymerases, ribosomes, translation factors, and metabolic enzymes that are generally responsible for transcription and translation in living cells. The reaction mixture consists of a cocktail of salts, nucleotides, amino acids, energy sources, transfer RNAs (tRNAs), and metabolic co-factors/additives that fuel expression. Lastly, linear or plasmid DNA serves as template for gene expression. Protein synthesis reactions occur upon combining the above components and incubation, on-site, at the desired temperature (typically between 30 °C and 37 °C), over a period from hours to days.

The successful deployment of CFPS for on-site and on-demand biomanufacturing of protein-based therapeutics for use in space will require integrated systems that are firstly portable for easy transportation and accessibility in extreme environments where gravity is absent and control of fluid dynamics and temperatures will be required, and secondly of high efficiency. The latter is veritable for sustainable production of target therapeutics of good-manufacturing-practice (GMP) quality within a reasonable time and at dose-able amount. Microfluidics and lab-on-a-chip devices are innovative technology platforms that assemble and operate biological processes in a mimetic, miniaturized and portable domain (Beneyton et al. 2014; Booth et al. 2017; Dittrich and Manz 2006; Dupin and Simmel 2019; Gerber, Maerkl, and Quake 2009; Haeberle and Zengerle 2007; Karzbrun et al. 2014; Melin and Quake 2007; Niederholtmeyer et al. 2013; Paegel and Joyce 2010; Shin and Noireaux 2012; Weiss et al. 2018). Detailed reviews on CFPS using microfluidics have been recently published (Damiati et al. 2018; Kamiya 2020; Supramaniam et al. 2019). Here, therefore, we focus on how microfluidics could and have been purposed for on-demand production of protein therapeutics in extreme and remote environments, such as in spaceflight.

Retterer and co-workers fabricated a microfluidic device for efficient and improved terrestrial production of single-dose quantity of therapeutic proteins (Timm et al. 2016). The device design through a nanofabricated membrane allowed its regulation of protein synthesis through the exchange of metabolites and reaction components between the cell-free protein reactor and feeder channels. Chang and co-workers developed a low-cost microfluidic reactor - "Therapeutics-On-a-Chip (TOC)," for on-site and on-demand CFPS-based therapeutic protein synthesis and purification through immunoprecipitation. To demonstrate the application of this platform, they first produced purified green fluorescent protein (GFP) using freeze-dried CFPS reagents. Thereafter, produced pharmaceutically active amount of antimicrobial peptide cecropin B (63 ng μ/L and 92% purity) within six hours and used a growth inhibition assay to validate its bactericidal activity on *Escherichia coli*, a well-known causative agent of biofilm-associated wound infections (Murphy et al. 2019). Similarly, taking advantage of the flexibility to automate and set-up a high-throughput cell-free protein production, Sullivan et al. developed a fluidic platform consisting of a eukaryotic or prokaryotic-based CFPS bioreactor and serial purification modules for a holistic production of protein therapeutics. Using yeast and bacterial extracts, they produced clinically active erythropoietin and G-CSF within 24 hours (Sullivan et al. 2016). Post-translational glycosylation, a limitation of CFPS (Zemella et al. 2015), of cell-free synthesized human erythropoietin has been demonstrated (Zemella et al. 2018).

Also, a group of researchers developed an automated and portable terrestrial on-demand platform for small-scale production of protein therapeutics. Using freeze-dried Chinese hamster ovary cell-free extracts, they produced pharmaceutical-grade histidine-tagged G-CSF, erythropoietin, glucose-binding protein, and diphtheria toxoid DT5 within an hour (Adiga et al. 2018). Recently, the same researchers reported the application of on-demand biologics manufacturing (Bio-MOD) platform that produced pharmaceutical-grade histidine-tagged G-CSF using cell-free protein synthesis (Adiga et al. 2020). As a proof of concept, they demonstrated the safety and efficacy of the produced G-CSF in a mouse model exposed to radiation at a dose estimated to be 30% lethal per 30 days postexposure. Overall, the results showed Bio-MOD consistently produced pharmaceutically certified on-demand biologics.

In our Astropharmacy program we are researching ways to miniaturize further the process of therapeutic protein production using stacks of paper each 3D-printed with reagents and CFPS components (Pardee 2018; Pardee et al. 2014). These stacks are capable of expressing and purifying protein (defined by whatever DNA is contained in one of the layers of the stack) and have an internal quality control/quality assurance mechanism. These programmable cellulose stacks may enable the delivery of in situ manufactured protein directly to the patient in an emergency, through a microneedle array for example, while the slower but more resilient techniques, such as the Bio-MOD or our cell-based systems, produce the drug required for the longer-term treatment.

2.3.4 On-site Formulation of the Drug

Almost all therapeutic biologics must be administered by injection and formulation, requires only the pure biologic and sterile, particle free water, such as could be provided by IVGEN Mini. For acute treatment delivery by injection is not a significant issue, and indeed desirable in case of emergency. As seen in the Vascular Echo study, long-term injection use can cause problems. Recently developed alternatives to injection through a needle, such as needle-free injection technology (NFIT), are of interest. These include liquid jet devices, particle "guns," and microneedle patches. We are currently exploring the use of both hollow and hydrogel-forming microneedles with the cell-free protein applied to the back of the patch and delivered through the skin as described by Donnelly et al.

3D printing provides a powerful route to the manufacture of solid drug dosage forms such as tablets and drug-eluting implants, in addition to manufacturing devices such as microneedles and injectors (Yang et al. 2020). Roberts and Wildman have published extensively on the use of ink-jet, extrusion-based, projection stereolithography, and multi-photon additive manufacturing techniques terrestrially to form tunable controlled-release tablets, "polypills," and drug-eluting devices (Clark et al. 2020; Liaskoni et al. 2021; Salimi et al. 2020). In 2015, the FDA approved Aprecia Pharmaceuticals' 3D-printed antiepileptic orodispersible Spritam (levetiracetam) tablets, and 2020 the commercial combined extrusion-based and direct powder extrusion M3DIMAKER 3D printer was launched by FabRx, a spin-out from UCL in the U.K., as the world's first pharmaceutical 3D printer for personalized medicines on Earth. A fused filament 3D printer was first sent to the ISS in 2014. This was later replaced by the Additive Manufacturing Facility (AMF), designed to allow immediate repair of essential components, upgrades for existing hardware, and provide manufacturing capability to support commercial interests on the ISS. This extrusion-based system can print a variety of materials, including those recycled on-site. The success of 3D printing onboard ISS indicates that the additive manufacture of pharmaceutical dosage forms and devices is possible in space and is something that should be further evaluated for proof-of-concept on ISS.

2.4 Medicines from Space

Space-based research has aided in the development of several medicines (Figure 2.7). The changes in physiological response to microgravity, which has been mentioned previously as complications in treating astronauts with medicines, can be used to an advantage in testing new drugs. The research on the change to the vasculature of astronauts, showing that six-month space flight caused changes similar to 10 to 20 years of normal aging, for example, shows that many processes related to aging, such as bone and muscle loss, are accelerated in flight. Any acceleration of the drug discovery and development process is welcome. The conditions of microgravity also facilitate the analysis of drugs and materials, for example by producing new formulations or higher quality samples.

Figure 2.7 (a) The Dragon cargo ship on SpaceX's CRS-22 resupply mission approaching ISS carrying several medicines-related experiments. (b) ESA astronaut Thomas Pesquet installs MME-2, an experiment that tests drugs to see whether they can improve health in space and possibly lead to new therapeutic targets to examine on Earth. (c) The Real-Time Protein Crystal Growth experiment uses on-station imaging to enable ground-based investigators to guide further experiments. (d) Lyophilization (freeze-drying) is an important and widely used method to protect delicate materials from degradation in storage. The Lyophilization-2 investigation examined gravity's effects on lyophilization. Source: NASA.

2.4.1 Accelerated Aging

Bone loss occurs at around 1.5% per month in microgravity (although varies considerably between individuals (Park et al. 2015; Ploutz-Snyder et al. 2014)), leading to the acceleration of age-related changes such as osteoporosis (Sibonga et al. 2019). In 2001, Amgen showed that osteoprotegerin (OPG), a protein that plays a role in the metabolism of bone by inhibiting its resorption, prevented bone loss in mice flown onboard STS-108. In 2007, Amgen showed that an experimental inhibitor of myostatin, a protein involved in muscle loss, also prevented bone loss in mice flown on STS-118. Amgen's research into osteoporosis and the role of OPG continued with human clinical trials, and in 2010 received FDA approval for their monoclonal antibody, denosumab, that mimics the action of OPG as a treatment for osteoporosis and bone cancer. Amgen and UCB also tested a second antibody, romosozumab, against another bone remodeling protein sclerostin, onboard STS-135, the last Space Shuttle flight. Upon completion of human clinical trials, this drug received FDA approval in 2019.

Similar investigations exploiting the accelerated processes in spaceflight include the Molecular Muscle Experiment (MME) and MME-2 of 2019 and 2021, respectively (Figure 2.7b). Here, the nematode *Caenorhabditis elegans*, a widely used proxy of human muscle, is used to investigate changes in protein expression in flight as the nematodes age more quickly and identify therapeutic interventions that may improve neuromuscular and mitochondrial health. Nematodes have been used as a research tool in spaceflight many times before. One experiment was flown on STS-107, the disastrous flight of Space Shuttle Columbia that broke up on re-entry on 1 February 2003. The passive experiment was housed in Biological Research in Canisters (BRIC) and was one of a number of experiments carried. Several of the experiments were successfully recovered from the debris field, including the worms, that were found to be alive when the canisters opened on 28 April (Szewczyk et al. 2005; Szewczyk and McLamb 2005). Other experiments that were recovered were several studies growing crystals, including Schering-Plough's investigation of interferon a-2b; a drug that they later developed for the treatment of hepatitis C and received FDA approval for in 2001.

Research into age-related muscle wasting aided by experiments onboard ISS has also led to the demonstration of new drug-delivery approaches. Formoterol, usually used as a long-acting bronchodilator in the management of asthma, is also known to attenuate muscle wasting and is being investigated as a potential treatment for sarcopenia. The Rodent Research-6 mission, launched to the ISS as part of SpaceX CRS-13 in December 2017, showed that formoterol, delivered slowly by an implanted nanofluidics device, attenuates muscle loss in both ground and flown mice. The success of the implantable drug-delivery platform suggests that such may be useful for many Earth-bound as well as spaceflight indications.

2.4.2 Crystal Studies

In a microgravity environment any process that involves fluids and mass transport, such as in crystal growth, is affected (Snell and Helliwell 2021). For crystallization to occur, molecules in solution first must come together, forming a nucleate, onto which further molecules assemble. As molecules come out of solution onto the crystal face, the solution at the surface becomes depleted. This solution usually has a lower density compared to the bulk, and in 1 g would rise (buoyancy-driven convection). In microgravity,[3] convection is all but stopped, and diffusion becomes the dominant method by which the molecules move. Combined with the reduced concertation of molecules at the surface of the crystal, crystal growth is slowed which leads to formation of generally larger crystals with fewer defects and of a more consistent size (Figure 2.8a,b) (McPherson and DeLucas 2015). Mosaicity, a measure of the spread of the crystal plane orientations, is also reduced. Overall, microgravity appears to be able to break the inverse relationship between crystal size and crystal quality.

A variety of different crystallization hardware and experiments have been flown, including 30 STS missions and on MIR and ISS resulting in the production of many crystals, and some of which have been used to gain structural information from their diffraction. Direct comparisons between the quality of crystals formed in microgravity compared to those on the ground are difficult, but well-controlled experiments have shown that microgravity-grown crystals have larger domains with a much-reduced mosaicity, and their diffraction peaks have a higher signal-to-noise with sometimes higher resolution. In 2017, Betzel and colleagues listed 29 different protein crystals that have been grown on the ISS. There is debate as to the value of spaceborne crystallography. In 2021, Snell and Helliwell updated their review of the field and commented that "there are surprisingly few studies where new biological information can be attributed solely to the growth of crystals in microgravity compared to those on the ground." (Snell and Helliwell 2021) Although crystals grown in microgravity tend to be larger and with fewer defects, advances in sample preparation, beamline, and detector instrumentation have largely nulled any benefit. For example, diffraction of crystals of the nucleosome core particle (NCP), a large multimolecular complex, grown in the Diffusion-Controlled Crystallization Apparatus for Microgravity onboard STS-73 resulted in an unprecedented 2.5 Å resolution

3 Newton's law of gravitation states that the gravitational force f between two objects of mass m_1 and m_2, separated by a distance between their centres r, is $f = \mathrm{G}\, m_1\, m_2 / r^2$, where G is the gravitational constant. The acceleration of an object of mass m_2, where $m_2 \ll \mathrm{m}_1$, at sea-level on Earth (m_1), where $r = 6.371 \times 10^6$ m, is 9.81 m s^{-2}, and termed 1 g. The acceleration a due to Earth's gravity given as a function of r is $a = 3.98 \times 10^{14} / r^2$. The ISS has mass of approximately 441 tonnes and is in LEO at approximately 400 km, where the acceleration due to Earth's gravity is 8.68 m s^{-2}, nearly 90% of that at sea level. The acceleration of an object traveling at velocity v toward the center of a circle radius r when cornering is $a = v^2 / r$. By traveling at nearly 7700 m s^{-1} ($\sqrt{6.771 \times 10^6 \cdot 8.68}$) the ISS is in constant free fall and orbits Earth once every 92 minutes. Objects are "weightless." This gravitational force of 8.68 m s^{-2} is at the center of mass of the ISS. An object situated in the ISS at a higher or lower position (equivalent to a slightly higher or lower orbit) will experience an ever-so-slightly different gravitational pull. The first derivative of the acceleration term w.r.t. r gives this variation in gravity with distance away from the CofG, as $da/dr = -7.96 \times 10^{14} / r^3$, which in orbit at 400 km is -2.6×10^{-6} s^{-2}, or 0.26 μg/m. This is microgravity. Objects in the ISS are also affected by "tidal" forces, where the gravitational pull is different between the ends of the object that are closest and furthest away from Earth, gravitational forces of other objects, motion of the astronauts imparting momentum to the ISS, gravitation of the Sun, etc.

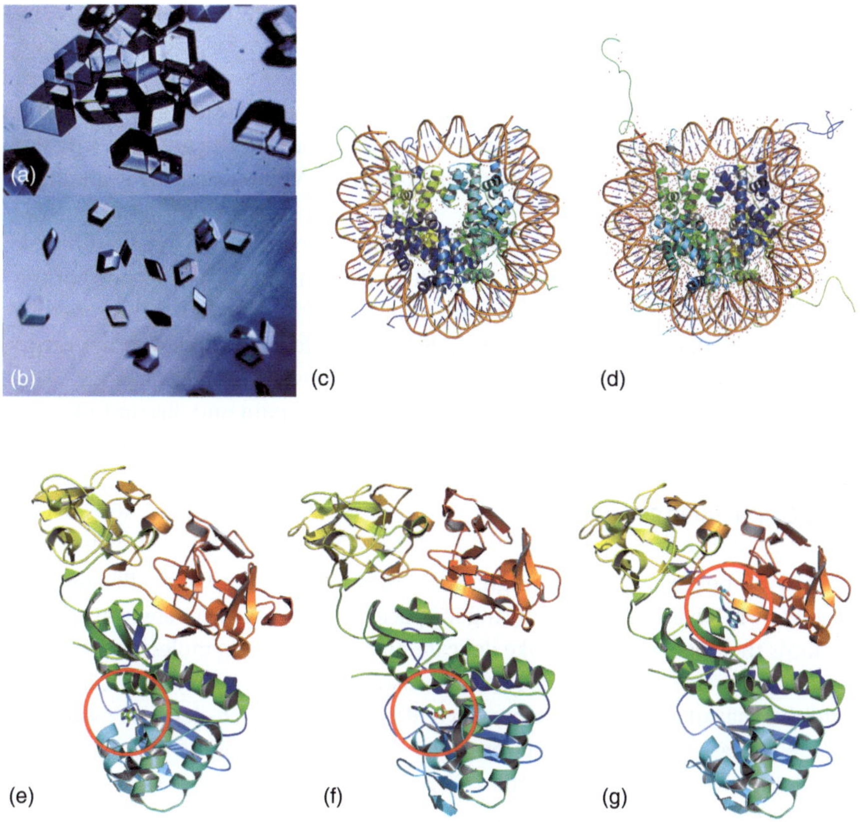

Figure 2.8 (a) Crystals grown in microgravity are often larger with fewer defects than those grown on Earth (b). (c) Microgravity enabled the first production of crystals of the large multi-molecular Nuclear Core Particle. The 2.5 Å structure revealed new, unexpected details of the structure. (d) Later ground-based studied were able to make crystals revealing the structure at 1.9 Å. (e) E (f) Representation of mistletoe lectin I (ML-I) with the nucleotide-binding site indicated in structures obtained from crystals grown in microgravity. The high water content of the crystal makes growing them on Earth difficult. E is at 1.9 Å resolution. (g) Further microgravity-derived studies of ML-I revealed new insight into the parasitic behavior of mistletoe. (PDB files C: 1EQZ; D: 1KX5; E: 1M2T; F: 3O5W; G: 3D7W). Source (a,b): NASA.

structure revealing significant additions to the understanding of the nucleosome and its unexpected structural asymmetry (Figure 2.8c) (Harp et al. 2000). Ground-based studies two years later produced a crystal structure at 1.9 Å (Figure 2.8d) (Davey et al. 2002).

There are other examples of crystallography of crystals grown in microgravity that have revealed new structure (Figure 2.8e–f). A study of the crystals of mistletoe lectin I soaked with the plant hormone zeatin grown onboard the Russian Service Module of the ISS revealed a specific binding site for this hormone (Meyer et al. 2008). Mistletoe is a parasite of trees from which it draws water. The tree hormone zeatin regulates the tree's transpiration. From this investigation, it is postulated that the mistletoe uses this lecitin protein to absorb the zeatin, thus maximizing its ability of to draw water from the host. Mistletoe is toxic, and ML-I causes cells to stop biosynthesis of proteins. ML-I has received interest, therefore, as an anticancer therapeutic. Due to the high water content of the crystals and the nature of the protein it has proven very difficult to grow crystals on the ground, whereas microgravity has allowed the collection of data with 1.9 Å resolution (Krauspenhaar et al. 2002; Malecki et al. 2012).

The ability to form crystals of greater uniformity in microgravity enabled Merck Sharp and Dome to produce uniform crystalline suspensions (1 mm–2 mm) of the immunotherapy drug pembrolizumab with acceptable viscosity (<12 cP) properties suitable for a SC injection (Reichert et al. 2019). SC reformulations of other drugs has revealed significant savings in time and staff effort over normal IV delivery, and that patients much preferred SC over IV with its reduced pain and discomfort.

2.5 Conclusions

In summary, long-duration space flight poses new pharmaceutical challenges that can be met by innovative manufacturing in space. The microgravity environment of LEO affords the opportunity to make delicate materials, such as molecular crystals, that is sometimes impossible in 1 g. Flow chemistry, molecular and synthetic biology, and additive manufacture can combine to realize the on-site, on-demand manufacture of medicines. Such an ability to provide pharmaceutical care to astronauts will have profound benefits for the support of people in extreme and remote environments on Earth, and coupled with advances in our understanding of pharmacology, physiology, aging, and diagnostics from experiments in LEO, will underpin truly personalized medicines.

References

Adamo, A., Beingessner, R.L., Behnam, M. et al. (2016). On-demand continuous-flow production of pharmaceuticals in a compact, reconfigurable system. *Science* 352 (6281): 61–67. https://doi.org/10.1126/science.aaf1337.

Adiga, R., Al-adhami, M., Andar, A. et al. (2018). Point-of-care production of therapeutic proteins of good-manufacturing-practice quality. *Nature Biomedical Engineering* 2 (9): 675–686. https://doi.org/10.1038/s41551-018-0259-1.

Adiga, R., Andar, A., Borhani, S. et al. (2020). Manufacturing biological medicines on demand: safety and efficacy of granulocyte colony-stimulating factor in a mouse model of total body irradiation. *Biotechnology Progress* 36 (3) https://doi.org/10.1002/btpr.2970: e2970.

Amselem, S. (2019). Remote controlled autonomous microgravity lab platforms for drug research in space. *Pharmaceutical Research* 36 (12) https://doi.org/10.1007/s11095-019-2703-7: 183.

Aunon-Chancellor, S.M., Pattarini, J.M., Moll, S., and Sargsyan, A. (2020). Venous thrombosis during spaceflight. *New England Journal of Medicine* 382 (1): 89–90. https://doi.org/10.1056/NEJMc1905875.

Barger, L.K., Flynn-Evans, E.E., Kubey, A. et al. (2014). Prevalence of sleep deficiency and use of hypnotic drugs in astronauts before, during, and after spaceflight: an observational study. *Lancet Neurology* 13 (9): 904–912. https://doi.org/10.1016/s1474-4422(14)70122-x.

Beneyton, T., Coldren, F., Baret, J.-C. et al. (2014). CotA laccase: high-throughput manipulation and analysis of recombinant enzyme libraries expressed in E. coli using droplet-based microfluidics. *Analyst* 139 (13): 3314–3323. https://doi.org/10.1039/c4an00228h.

Blue, R.S., Bayuse, T.M., Daniels, V.R. et al. (2019a). Supplying a pharmacy for NASA exploration spaceflight: challenges and current understanding. *NPJ Microgravity* 5: https://doi.org/10.1038/s41526-019-0075-2.

Blue, R.S., Chancellor, J.C., Antonsen, E.L. et al. (2019b). Limitations in predicting radiation-induced pharmaceutical instability during long-duration spaceflight. *NPJ Microgravity* 5. https://doi.org/10.1038/s41526-019-0076-1.

Bogdan, A.R., Poe, S.L., Kubis, D.C. et al. (2009). The Continuous-Flow Synthesis of Ibuprofen. *Angewandte Chemie, International Edition* 48 (45): 8547–8550. https://doi.org/10.1002/anie.200903055.

Booth, M.J., Schild, V.R., Box, S.J., and Bayley, H. (2017). Light-patterning of synthetic tissues with single droplet resolution. *Scientific Reports* 7. https://doi.org/10.1038/s41598-017-09394-9: 9315.

Cao, J., Perez-Pinera, P., Lowenhaupt, K. et al. (2018). Versatile and on-demand biologics co-production in yeast. *Nature Communications* 9. https://doi.org/10.1038/s41467-017-02587-w: 77.

Carlson, E.D., Gan, R., Hodgman, C.E., and Jewett, M.C. (2012). Cell-free protein synthesis: applications come of age. *Biotechnology Advances* 30 (5): 1185–1194. https://doi.org/10.1016/j.biotechadv.2011.09.016.

Carnell, L.S. (2020). Spaceflight medical countermeasures: a strategic approach for mitigating effects from solar particle events. *International Journal of Radiation Biology* https://doi.org/10.1080/09553002.2020.1820603 S125–S131.

Carpentier, W.R., Charles, J.B., Shelhamer, M. et al. (2018). Biomedical findings from NASA's Project Mercury: a case series. *NPJ Microgravity* 4. https://doi.org/10.1038/s41526-018-0040-5: 6.

Choi, E.J., Lewin, J.J., and Ling, G.S.F. (2015). Battlefield medicine: disrupting (bio)pharmaceutical production. *Pharmaceutical Bioprocessing* 3 (5): 361–369. https://doi.org/10.4155/pbp.15.17.

Clark, E.A., Alexander, M.R., Irvine, D.J. et al. (2020). Making tablets for delivery of poorly soluble drugs using photoinitiated 3D inkjet printing. *International Journal of Pharmaceutics* 578. https://doi.org/10.1016/j.ijpharm.2019.118805: 118805.

Cole, K.P., Groh, J.M., Johnson, M.D. et al. (2017). Kilogram-scale prexasertib monolactate monohydrate synthesis under continuous-flow CGMP conditions. *Science* 356 (6343): 1144–1150. https://doi.org/10.1126/science.aan0745.

Copeland, C.E., Langlois, A., Kim, J., and Kwon, Y.-C. (2021). The cell-free system: a new apparatus for affordable, sensitive, and portable healthcare. *Biochemical Engineering Journal* 175. https://doi.org/10.1016/j.bej.2021.108124: 108124.

Crommelin, D.J.A., Storm, G., Verrijk, R. et al. (2003). Shifting paradigms: biopharmaceuticals versus low molecular weight drugs. *International Journal of Pharmaceutics* 266 (1-2): 3–16. https://doi.org/10.1016/s0378-5173(03)00376-4.

Crommelin, D.J.A., Mastrobattista, E., Hawe, A. et al. (2020). Shifting paradigms revisited: biotechnology and the pharmaceutical sciences. *Journal of Pharmaceutical Sciences* 109 (1): 30–43. https://doi.org/10.1016/j.xphs.2019.08.010.

Crowell, L.E., Lu, A.E., Love, K.R. et al. (2018). On-demand manufacturing of clinical-quality biopharmaceuticals. *Nature Biotechnology* 36 (10): 988. https://doi .org/10.1038/nbt.4262.

Crump, W.J., Levy, B.J., and Billica, R.D. (1996). A field trial of the NASA telemedicine instrument pack in a family practice. *Aviation, Space, and Environmental Medicine* 67 (11): 1080–1085.

Damiati, S., Mhanna, R., Kodzius, R., and Ehmoser, E.-K. (2018). Cell-free approaches in synthetic biology utilizing microfluidics. *Genes* 9 (3) https://doi.org/10.3390/ genes9030144: 144.

Davey, C.A., Sargent, D.F., Luger, K. et al. (2002). Solvent mediated interactions in the structure of the nucleosome core particle at 1.9 angstrom resolution. *Journal of Molecular Biology* 319 (5): 1097–1113. https://doi.org/10.1016/s0022-2836(02)00386-8.

Dittrich, P.S. and Manz, A. (2006). Lab-on-a-chip: microfluidics in drug discovery. *Nature Reviews Drug Discovery* 5 (3): 210–218. https://doi.org/10.1038/nrd1985.

Dondapati, S.K., Stech, M., Zemella, A., and Kubick, S. (2020). Cell-free protein synthesis: a promising option for future drug development. *BioDrugs* 34 (3): 327–348. https://doi.org/10.1007/s40259-020-00417-y.

Du, B., Daniels, V.R., Vaksman, Z. et al. (2011). Evaluation of physical and chemical changes in pharmaceuticals flown on space missions. *The AAPS Journal* 13 (2): 299–308. https://doi.org/10.1208/s12248-011-9270-0.

Dupin, A. and Simmel, F.C. (2019). Signalling and differentiation in emulsion-based multi-compartmentalized in vitro gene circuits. *Nature Chemistry* 11 (1): 32–39. https://doi.org/10.1038/s41557-018-0174-9.

Eyal, S. (2020). How do the pharmacokinetics of drugs change in astronauts in space? *Expert Opinion on Drug Metabolism & Toxicology* 16 (5): 353–356. https://doi.org/10 .1080/17425255.2020.1746763.

Eyal, S. and Derendorf, H. (2019). Medications in space: in search of a pharmacologist's guide to the galaxy. *Pharmaceutical Research* 36 (10) https://doi.org/10.1007/s11095-019-2679-3: 148.

Gandia, P., Saivin, S., and Houin, G. (2005). The influence of weightlessness on pharmacokinetics. *Fundamental & Clinical Pharmacology* 19 (6): 625–636. https://doi .org/10.1111/j.1472-8206.2005.00374.x.

Genah, S., Monici, M., and Morbidelli, L. (2021). The effect of space travel on bone metabolism: considerations on today's major challenges and advances in pharmacology. *International Journal of Molecular Sciences* 22 (9) https://doi.org/10.3390/ijms22094585: 4585.

Gerber, D., Maerkl, S.J., and Quake, S.R. (2009). An in vitro microfluidic approach to generating protein-interaction networks. *Nature Methods* 6 (1): 71–74. https://doi.org/10.1038/nmeth.1289.

Haeberle, S. and Zengerle, R. (2007). Microfluidic platforms for lab-on-a-chip applications. *Lab on a Chip* 7 (9): 1094–1110. https://doi.org/10.1039/b706364b.

Harp, J.M., Hanson, B.L., Timm, D.E., and Bunick, G.J. (2000). Asymmetries in the nucleosome core particle at 2.5 angstrom resolution. *Acta Crystallographica Section D-Structural Biology* 56: 1513–1534. https://doi.org/10.1107/s0907444900011847.

He, W.S., Baysal, C., Gomez, M.L. et al. (2021). Contributions of the international plant science community to the fight against infectious diseases in humans-part 2: affordable drugs in edible plants for endemic and re-emerging diseases. *Plant Biotechnology Journal* https://doi.org/10.1111/pbi.13658 19 (10): 1921–1936.

Hodkinson, P.D., Anderton, R.A., Posselt, B.N., and Fong, K.J. (2017). An overview of space medicine. *British Journal of Anaesthesia* 119: I143–I153. https://doi.org/10.1093/bja/aex336.

Houtchens, B.A. (1993). Medical-care systems for long-duration space missions. *Clinical Chemistry* 39 (1): 13–21.

Hu, X., Raja, W.K., An, B. et al. (2013). Stability of silk and collagen protein materials in space. *Scientific Reports* 3. https://doi.org/10.1038/srep03428: 3428.

Johnston, T.G., Yuan, S.-F., Wagner, J.M. et al. (2020). Compartmentalized microbes and co-cultures in hydrogels for on-demand bioproduction and preservation. *Nature Communications* 11 (1) https://doi.org/10.1038/s41467-020-14371-4: 563.

Kamiya, K. (2020). Development of artificial cell models using microfluidic technology and synthetic biology. *Micromachines* 11 (6) https://doi.org/10.3390/mi11060559: 559.

Karzbrun, E., Tayar, A.M., Noireaux, V., and Bar-Ziv, R.H. (2014). Programmable on-chip DNA compartments as artificial cells. *Science* 345 (6198): 829–832. https://doi.org/10.1126/science.1255550.

Kast, J., Yu, Y., Seubert, C.N. et al. (2017). Drugs in space: pharmacokinetics and pharmacodynamics in astronauts. *European Journal of Pharmaceutical Sciences* 109: S2–S8. https://doi.org/10.1016/j.ejps.2017.05.025.

Kirkpatrick, A.W., Jones, J.A., Sargsyan, A. et al. (2007). Trauma sonography for use in microgravity. *Aviation, Space, and Environmental Medicine* 78 (4): A38–A42.

Komorowski, M., Fleming, S., and Kirkpatrick, A.W. (2016). Fundamentals of anesthesiology for spaceflight. *Journal of Cardiothoracic and Vascular Anesthesia* 30 (3): 781–790. https://doi.org/10.1053/j.jvca.2016.01.007.

Krauspenhaar, R., Rypniewski, W., Kalkura, N. et al. (2002). Crystallisation under microgravity of mistletoe lectin I from Viscum album with adenine monophosphate and the crystal structure at 1.9 angstrom resolution. *Acta Crystallographica Section D-Structural Biology* 58: 1704–1707. https://doi.org/10.1107/s0907444902014270.

Lau, S.-H., Galvan, A., Merchant, R.R. et al. (2015). Machines vs malaria: a flow-based preparation of the drug candidate OZ439. *Organic Letters* 17 (13): 3218–3221. https://doi.org/10.1021/acs.orglett.5b01307.

Lewin, J.J. III,, Choi, E.J., and Ling, G. (2016). Pharmacy on demand: new technologies to enable miniaturized and mobile drug manufacturing. *American Journal of Health-System Pharmacy* 73 (2): 45–54. https://doi.org/10.2146/ajhp150639.

Liaskoni, A., Wildman, R.D., and Roberts, C.J. (2021). 3D printed polymeric drug-eluting implants. *International Journal of Pharmaceutics* 597. https://doi.org/10.1016/j.ijpharm.2021.120330: 120330.

Malecki, P.H., Rypniewski, W., Szymanski, M. et al. (2012). Binding of the plant hormone kinetin in the active site of Mistletoe Lectin I from Viscum album. *Biochimica Et Biophysica Acta-Proteins and Proteomics* 1824 (2): 334–338. https://doi.org/10.1016/j.bbapap.2011.10.013.

Marshall-Goebel, K., Laurie, S.S., Alferova, I.V. et al. (2019). Assessment of jugular venous blood flowstasis and thrombosis during spaceflight. *JAMA Network Open* 2 (11) https://doi.org/10.1001/jamanetworkopen.2019.15011: e1915011.

Mascia, S., Heider, P.L., Zhang, H. et al. (2013). End-to-end continuous manufacturing of pharmaceuticals: integrated synthesis, purification, and final dosage formation. *Angewandte Chemie, International Edition* 52 (47): 12359–12363. https://doi.org/10.1002/anie.201305429.

McLaughlin, M.F., Donoviel, D.B., and Jones, J.A. (2017). Novel indications for commonly used medications as radiation protectants in spaceflight. *Aerospace Medicine and Human Performance* 88 (7): 665–676. https://doi.org/10.3357/amhp.4735.2017.

McPherson, A. and DeLucas, L.J. (2015). Microgravity protein crystallization. *NPJ Microgravity* 1. https://doi.org/10.1038/npjmgrav.2015.10: 15010.

Mehta, P. and Bhayani, D. (2017). Impact of space environment on stability of medicines: challenges and prospects. *Journal of Pharmaceutical and Biomedical Analysis* 136: 111–119. https://doi.org/10.1016/j.jpba.2016.12.040.

Melin, J. and Quake, S.R. (2007). Microfluidic large-scale integration: the evolution of design rules for biological automation. *Annual Review of Biophysics and Biomolecular Structure* 36: 213–231. https://doi.org/10.1146/annurev.biophys.36.040306.132646.

Meyer, A., Rypniewski, W., Szymanski, M. et al. (2008). Structure of mistletoe lectin I from *Viscum album* in complex with the phytohormone zeatin. *Biochimica Et Biophysica Acta-Proteins and Proteomics* 1784 (11): 1590–1595. https://doi.org/10.1016/j.bbapap.2008.07.010.

Murphy, T.W., Sheng, J., Naler, L.B. et al. (2019). On-chip manufacturing of synthetic proteins for point-of-care therapeutics. *Microsystems & Nanoengineering* 5. https://doi.org/10.1038/s41378-019-0051-8: 13.

Niederholtmeyer, H., Stepanova, V., and Maerkl, S.J. (2013). Implementation of cell-free biological networks at steady state. *Proceedings of the National Academy of Sciences of the United States of America* 110 (40): 15985–15990. https://doi.org/10.1073/pnas.1311166110.

Paegel, B.M. and Joyce, G.F. (2010). Microfluidic compartmentalized directed evolution. *Chemistry & Biology* 17 (7): 717–724. https://doi.org/10.1016/j.chembiol.2010.05.021.

Pardee, K. (2018). Perspective: solidifying the impact of cell-free synthetic biology through lyophilization. *Biochemical Engineering Journal* 138: 91–97. https://doi.org/10.1016/j.bej.2018.07.008.

Pardee, K., Green, A.A., Ferrante, T. et al. (2014). Paper-based synthetic gene networks. *Cell* 159 (4): 940–954. https://doi.org/10.1016/j.cell.2014.10.004.

Park, J.H., Seo, D.-H., Cho, S. et al. (2015). Effects of partial vibration on morphological changes in bone and surrounding muscle of rats under microgravity condition: comparative study by gender. *Microgravity Science and Technology* 27 (5): 361–368. https://doi.org/10.1007/s12217-015-9425-1.

Patel, Z.S., Brunstetter, T.J., Tarver, W.J. et al. (2020). Red risks for a journey to the red planet: the highest priority human health risks for a mission to Mars. *NPJ Microgravity* 6 (1) https://doi.org/10.1038/s41526-020-00124-6: 33.

Perez-Pinera, P., Han, N., Cleto, S. et al. (2016). Synthetic biology and microbioreactor platforms for programmable production of biologics at the point-of-care. *Nature Communications* 7: https://doi.org/10.1038/ncomms12211.

Ploutz-Snyder, L., Bloomfield, S., Smith, S.M. et al. (2014). Effects of sex and gender on adaptation to space: musculoskeletal health. *Journal of Womens Health* 23 (11): 963–966. https://doi.org/10.1089/jwh.2014.4910.

Rasor, B.J., Vogeli, B., Landwehr, G.M. et al. (2021). Toward sustainable, cell-free biomanufacturing. *Current Opinion in Biotechnology* 69: 136–144. https://doi.org/10.1016/j.copbio.2020.12.012.

Reichert, P., Prosise, W., Fischmann, T.O. et al. (2019). Pembrolizumab microgravity crystallization experimentation. *NPJ Microgravity* 5. https://doi.org/10.1038/s41526-019-0090-3: 28.

Salimi, S., Wu, Y., Barreiros, M.I.E. et al. (2020). A 3D printed drug delivery implant formed from a dynamic supramolecular polyurethane formulation. *Polymer Chemistry* 11 (20): 3453–3464. https://doi.org/10.1039/d0py00068j.

Sans, V. and Cronin, L. (2016). Towards dial-a-molecule by integrating continuous flow, analytics and self-optimisation. *Chemical Society Reviews* 45 (8): 2032–2043. https://doi.org/10.1039/c5cs00793c.

Shin, J. and Noireaux, V. (2012). An *E. coli* cell-free expression toolbox: application to synthetic gene circuits and artificial cells. *ACS Synthetic Biology* 1 (1): 29–41. https://doi.org/10.1021/sb200016s.

Sibonga, J., Matsumoto, T., Jones, J. et al. (2019). Resistive exercise in astronauts on prolonged spaceflights provides partial protection against spaceflight-induced bone loss. *Bone* 128. https://doi.org/10.1016/j.bone.2019.07.013, 112037.

Silverman, A.D., Karim, A.S., and Jewett, M.C. (2020). Cell-free gene expression: an expanded repertoire of applications. *Nature Reviews Genetics* 21 (3): 151–170. https://doi.org/10.1038/s41576-019-0186-3.

Simon, A., Smarandache, A., Iancu, V., and Pascu, M.L. (2021). Stability of antimicrobial drug molecules in different gravitational and radiation conditions in view of applications during outer space missions. *Molecules* 26 (8) https://doi.org/10.3390/molecules26082221: 2221.

Sipos, G., Bihari, T., Milankovich, D., and Darvas, F. (2017). Flow chemistry in space-A unique opportunity to perform extraterrestrial research. *Journal of Flow Chemistry* 7 (3-4): 151–156. https://doi.org/10.1556/1846.2017.00033.

Snell, E.H. and Helliwell, J.R. (2021). Microgravity as an environment for macromolecular crystallization - an outlook in the era of space stations and commercial space flight. *Crystallography Reviews* 27 (1): 3–46. https://doi.org/10.1080/0889311x.2021.1900833.

Snyder, J.E., Walsh, D., Carr, P.A., and Rothschild, L.J. (2019). A makerspace for life support systems in space. *Trends in Biotechnology* 37 (11): 1164–1174. https://doi.org/10.1016/j.tibtech.2019.05.003.

Sullivan, C.J., Pendleton, E.D., Sasmor, H.H. et al. (2016). A cell-free expression and purification process for rapid production of protein biologics. *Biotechnology Journal* 11 (2): 238–248. https://doi.org/10.1002/biot.201500214.

Supramaniam, P., Ces, O., and Salehi-Reyhani, A. (2019). Microfluidics for artificial life: techniques for bottom-up synthetic biology. *Micromachines* 10 (5) https://doi.org/10.3390/mi10050299: 299.

Szewczyk, N.J. and McLamb, W. (2005). Surviving atmospheric spacecraft breakup. *Wilderness & Environmental Medicine* 16 (1): 27–32. https://doi.org/10.1580/pr21-03.1.

Szewczyk, N.J., Mancinelli, R.L., McLamb, W. et al. (2005). *Caenorhabditis elegans* survives atmospheric breakup of STS-107, space shuttle Columbia. *Astrobiology* 5 (6): 690–705. https://doi.org/10.1089/ast.2005.5.690.

Timm, A.C., Shankles, P.G., Foster, C.M. et al. (2016). Toward microfluidic reactors for cell-free protein synthesis at the point-of-care. *Small* 12 (6): 810–817. https://doi.org/10.1002/smll.201502764.

Weiss, M., Frohnmayer, J.P., Benk, L.T. et al. (2018). Sequential bottom-up assembly of mechanically stabilized synthetic cells by microfluidics. *Nature Materials* 17 (1): 89. https://doi.org/10.1038/nmat5005.

Wotring, V.E. (2015). Medication use by US crewmembers on the International Space Station. *FASEB Journal* 29 (11): 4417–4423. https://doi.org/10.1096/fj.14-264838.

Wotring, V.E. (2016). Chemical potency and degradation products of medications stored over 550 earth days at the International Space Station. *The AAPS Journal* 18 (1): 210–216. https://doi.org/10.1208/s12248-015-9834-5.

Yang, R., Zhou, J., Yang, C. et al. (2020). Recent progress in 3D printing of 2D material-based macrostructures. *Advanced Materials Technologies* 5 (9) https://doi.org/10.1002/admt.201901066: 1901066.

Zemella, A., Thoring, L., Hoffmeister, C., and Kubick, S. (2015). Cell-free protein synthesis: pros and cons of prokaryotic and eukaryotic systems. *Chembiochem* 16 (17): 2420–2431. https://doi.org/10.1002/cbic.201500340.

Zemella, A., Thoring, L., Hoffmeister, C. et al. (2018). Cell-free protein synthesis as a novel tool for directed glycoengineering of active erythropoietin. *Scientific Reports* 8. https://doi.org/10.1038/s41598-018-26936-x: 8514.

3

Frontier Medical Technologies to Support Space Exploration

Quy D. Tran, Nam N. Tran, and Volker Hessel

University of Adelaide, School of Chemical Engineering and Advanced Materials, North Terrace, SA 5005, Adelaide, Australia

3.1 Introduction

Space is an ever-young field of frontier-science studies. For decades, we have benefitted from technology innovation derived from space research, from communication, engineering, to medical–biological research. As we move forward on exploring space and using it to its fullest potential, new questions are being raised every day. The exploration of using space to manufacture could create a significant opportunity for business, launching Earth's economy to its next great leap. In the old space context, the research was trying to understand space effects on human health better to foster longer duration exploration missions to other planets. The transition from the old to the new space economy follows the evolution of space technologies. The new space context focuses on how to effectively use space resources through advanced manufacturing, R&D, space construction and services (Figure 3.1). Research is now being focused on in the context of space travel and human space exploration, and, maintaining the health of people who will one-day live and work off the planet.

Yet, space offers another set of living conditions, the lack of gravity for instance. As we take gravity for granted, we are unaware of what happens when we do not have it. What we commonly know on Earth changes in space, whether it is the way plants grow, liquids flow, or even how fire burns. Microgravity is only one of the factors contributing to the space environment. Others factors are strong ionizing radiation, high level of vacuum, and extreme heating–cooling cycles [1]. This requires measures of protection against those threats, and intensive research and development is a way to provide solutions.

In comparison with research in space environments, Earth experiments simulating space conditions can only provide a proxy to true space conditions, and there may be a remarkable difference in some cases. The current single opportunity to conduct human health space experiments is provided by the International Space Station (ISS), located in the low-Earth orbit (LEO). The ISS provides a living space

In-Space Manufacturing and Resources: Earth and Planetary Exploration Applications, First Edition.
Edited by Volker Hessel, Jana Stoudemire, Hideaki Miyamoto, and Ian D. Fisk.

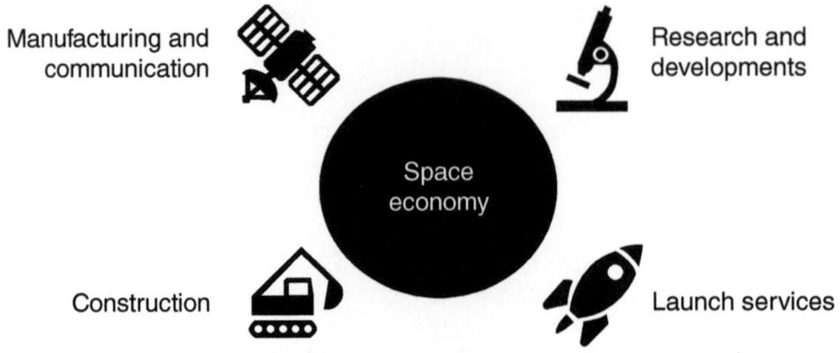

Figure 3.1 The expanding space economy.

for humans in space, allowing medical personnel and researchers to monitor their well-being and physiological/psychological changes, while its laboratories provide a unique platform to conduct experiments not possible on Earth. The outcomes of these space research efforts can be leveraged both on Earth and as support for future exploration of space. Indeed, many scientific achievements have been achieved and reported from work conducted on the ISS over the past few decades. Thousands of experiments have been carried out on ISS, of which more than 300 are related to human health research (https://www.nature.com/articles/d41586-020-03085-8). Up to the present, it has been challenging and limiting for researchers to conduct research on the ISS because of the limited access and high costs, as well as limitations in the scope of experimentation given existing capabilities.

Technology is the answer to both limitations. In particular, disruptive technologies and innovative ways of experimental automation have delivered innovative approaches, compensating for crew time limitations, and reducing expenses (or, at least, increase the ratio of gain to expenses).

In this chapter, we present a review of frontier technologies, address the future of space health and outline the potential opportunities for future assessment of heath measures and research tools. A short analysis of astronaut health risks (Section 3.2), analytical diagnostic measures (Section 3.3), analytical medical microchip devices as human analog models (Section 3.4), and computational-based telemedicine technologies (Section 3.5) are introduced.

3.2 Health Risks of Space Flights

Space exploration is challenging for astronauts both in a positive and negative way; the latter meaning a risk to human health. As well as exposure to elevated radiation levels and altered gravity fields, space travel poses other risks to well-being and mental health, due to the confined and isolated living in a closed environment, altered day-night cycles, and generally the stress of being so far from the "home" Earth [2, 3]. The derived risks associated with each of these risk factors must be properly identified and sufficiently mitigated for the well-being of crew members [3]. For a mission to succeed, ensuring the health and well-being of the crew, including their physical and mental safety, is imperative. With the increase in the length of the mission,

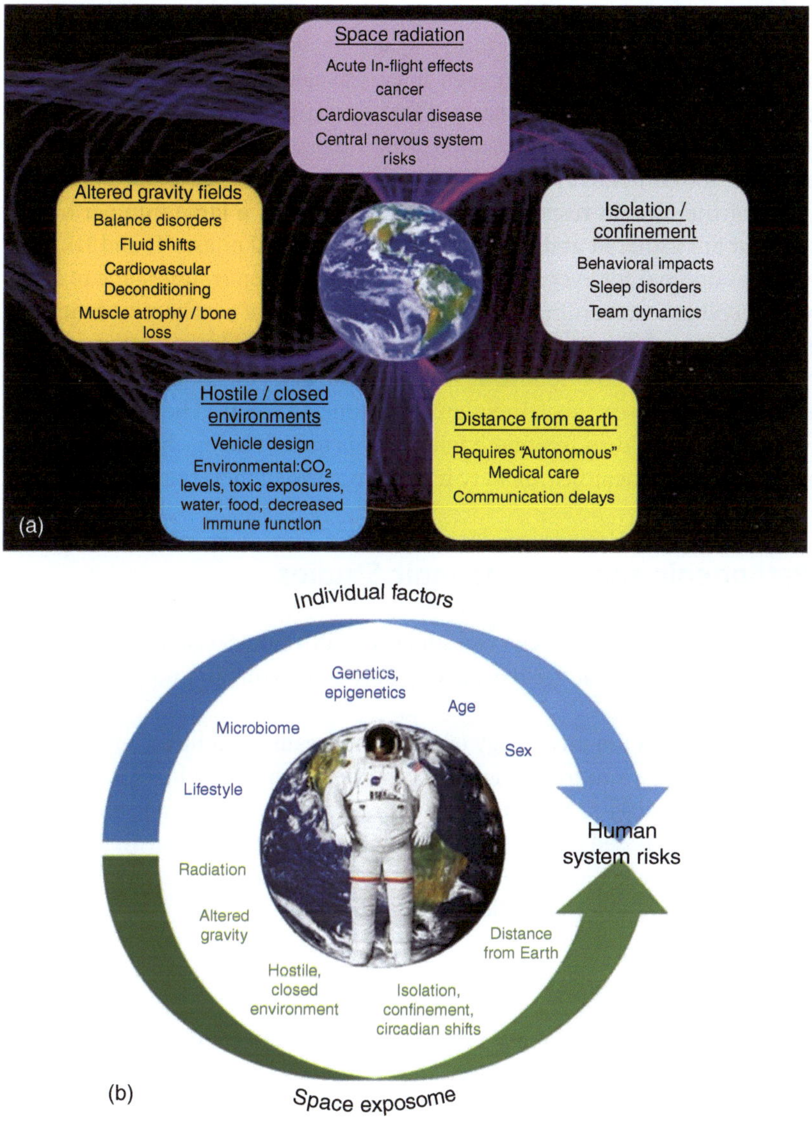

Figure 3.2 The key risks to human health and performance. Source: Reprinted from Petel et al. [3] with permission from Springer Nature.

the greater time delay in communication, and the more limited medical resources, the requirements for real-time medical assistance increase. Advanced medical technologies need to be employed in order to provide sufficient health care resources for flight crews (Figure 3.2).

Using state-of-the-art technologies, miniaturization of devices, and innovative space health care may help in reducing invasive methods, expanding telemedicine options, and implementing on-orbit analytics. Astronauts must learn how to use new technologies to support a stringent healthy lifestyle and take active part in health care decisions; just as patients on Earth are becoming increasingly authorized in their own care. The future space healthcare will change from a directed-type to self-directed care between space travelers and remote doctors. The concept of "digital healthcare" has been put into operation together with disruptive technologies [2].

In the course of traditional drug development, medicines are developed based on average data of the human response to the medicine from a large number of test persons, which are gathered and documented in the clinical phases II and III. This approach is not optimal for every individual and may be even troublesome to some (sensitive) individuals. Accordingly there is a future vision of a customized medical treatment in the sense of precision medications, practices, and interventions, which are tailored to the predicted response or risk of failure of a patient using individualized medicine [4]. The use of such individually optimized solution helps prevent the risk of medical errors and/or treatment failure. This benefit being shown on Earth, likely can also improve healthcare safety in remote areas such as space.

3.3 Pathogenic and Immunogenic Studies

Pathogenic and immunogenic assays can support health care decisions, in the sense of the above-mentioned proactive partnership between astronaut and remote space medical personnel.

Microgravity changes the physiology of pathogens, causing cellular and molecular changes directly relevant to contagious properties of the microbial [5]. This can add to our understanding of how infectious diseases are caused and transmitted. There are health risks posed to astronauts due to their immunological system being dysregulated during space flight. The following section will discuss the recent developments in R&D on pathogenic studies.

3.3.1 Genome Sequencing

The exploration of new environments urges an understanding of the causes and risk factors associated with disease so that we can better protect astronauts' health in the future. That is why work has been put into developing quick scanning devices that give a real-time assessment. A colorimetric loop-mediated isothermal amplification or rDNA nanopore sequencing has been studied to provide real-time analysis during a space mission, allowing for real-time decisions and remediation strategies [6, 7]. In as little as 30 minutes, the detection system can detect the repetitive telomeric deoxyribonucleic acid (DNA) sequence at the ISS through a simple color change in the reaction mixture [7].

3.3.2 Microbiology Studies of Pathogens

The health of a crew in space needs to be protected against the virulence-enhancing effects of microbes, and this information can also be used to guide antimicrobial strategies on Earth. Microbiology studies conducted on ISS can also help to develop preventative measures against microbes that threaten human health. As an example, sample cultures of Salmonella typhimurium were compared in space with identical control cultures on the ground. A decrease in time to death, an increase in mortality, and a reduction in LD50 value were seen in mice infected with the bacteria from the flight cultures as compared to mice infected with ground controls. The space flight environment may provide insights into the mechanistic regulation of microbial responses, in particular, in relation to the role of Hfq as a conserved bacterial ribonucleic acid (RNA)-binding protein that plays a key role in post-transcriptional gene regulation [8] (Figure 3.3).

3.3.3 Vaccination in Space

Astronauts normally undergo vaccination before the flight to prevent disease. However, restrictions are known in the applicability of Earth medicines in the space environment. Some formulations require the medication to be re-taken several times at different intervals in order to be effective.

Restriction may be given also to the active medical component itself, e.g. the vaccine. A twin study has been conducted to evaluate the efficiency of vaccines in space compared to Earth. During this study, in order to determine the response to influenza immunizations in a space flight environment compared to response on Earth, samples were taken from three time periods: pre-flight, during performance, and post-flight. The purpose was to evaluate the presence of CD4 and CD8 T cells from frozen peripheral blood mononuclear cells. With this information, the data can be used to predict the degree of sensitivity of a human body to an infection. To better explain this, the depletion of T-cells corresponds to the decrease in immunological protection. And as a result, humans become susceptible to infection. Based on the results of the vaccine study on orbit, no significant differences have been found in the percentage of CD4 or CD8 T cells. This suggests that vaccination processes that occur on Earth and in space have similar effects [9].

3.3.4 New Formulations for Medication in Space

New formulations may assist the efficiency of vaccines in space, or more generally of pandemic drugs. While this has not been tested so far, the impact of a formulation on central analytical characteristics of a low-molecular drug has recently be reported. Remdesivir, the first anti-Covid 19 drug, has been tested in space to overcome limits known from its performance on Earth. The Earth environment limits the research on drug–cyclodextrin complexes [10]. In space, this is different. The absence of gravity actually makes perfect crystal formation possible, as it leaves diffusion as only driving force for seed molecular process as uniform mechanism for mass transfer. On the contrary and on Earth, natural and forced convection provides a complex means of achieving mass transfer. This means that mass transfer varies under gravity effects, leading to non-uniform crystal formation. From this theoretical perspective, the size distribution of space-made (ISS) Remdesivir drug–cyclodextrin complexes has been investigated. Results show that the expectations are met, meaning that space samples have smaller size and are more uniform in distribution. In addition, high performance liquid chromatography (HPLC) analysis indicates that one of the flight-Remdesivir samples exhibited less decomposition as compared to Earth samples. The carrier cyclodextrin appears to be intact [10].

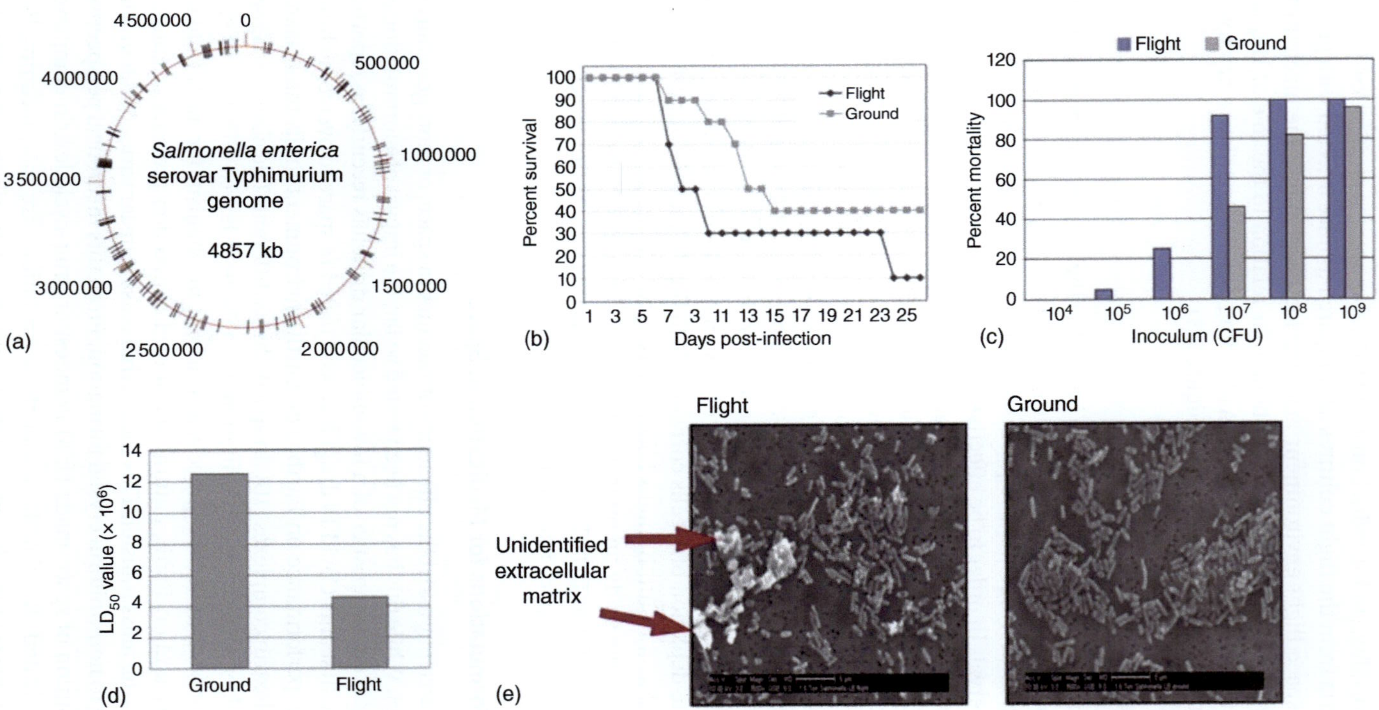

Figure 3.3 Comparison study from space samples to ground control subjects. Source: Reproduced with permission from Wilson et al. [8], © (2007), National Academy of Sciences.

3.4 Organ-on-Chip Microfluidics

Micro-engineered systems are known to offer compact footprints and low weight, as well as facilitating automation, which is necessary for space operations with no human personnel. However, due to the lack of commercially available systems such systems are not really available for use in space. Considerable R&D is required to develop such systems that are able to give accredited results. On a positive note, solving the problem of those space-tested micro-engineered systems should facilitate a further improvement in automation standards on Earth. And that may offer considerable potential for technology transfer.

Most scientists are now studying diseases and testing products without using animals to avoid the translational limitations associated non-human models. This gave a push to the development of tissue chip and organoid models. As a human analog, micro-physiological system (MPS), or organs-on-chips, are being used to mimic the physiological and mechanical conditions experienced in the human body. Their goal and capability are to investigate the mechanism of disease and test drugs on human tissue. The ability to study an individualized tissue isolating the effect of other organs, and combining multiple multiple tissues in these models is a significant step in health sciences. In addition, stem cells taken from astronauts can be used to create a unique MPS to facilitate development of customized countermeasures prior to a mission.

The main purpose of this section is to give an overview of MPS experiments being carried out recently on ISS and their supporting technological capability, as tested on Earth. These include studies on the cardiovascular system, the blood–brain barrier, the musculoskeletal system, lung tissue, and kidney tissue (Figure 3.4).

3.4.1 Cardiovascular System

Body fluids do not flow the same way as they do on Earth since there is no gravity to guide fluids. Fluid shifts in the body causes space traveling motion sickness, cognitive impairments, insomnia, and impacts on cardiovascular function that mimic accelerated aging. In this context, cardiovascular-analog chip has been developed to study the effect of short- or long-term exposure to microgravity on cardiac tissue. The

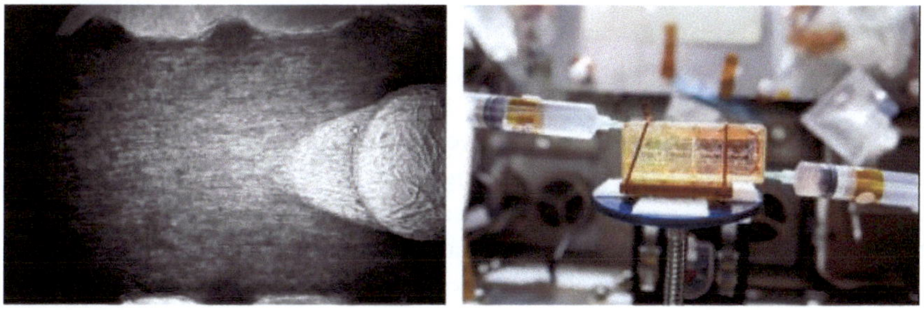

Figure 3.4 Examples of MPS systems used on ISS (muscle left, heart right). Source: Reprinted from Sharma et al. [11] with open access from Elsevier.

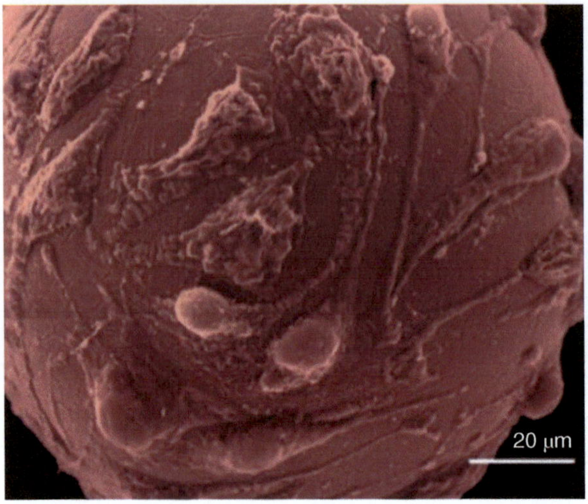

Figure 3.5 Electron microscopy image of muscle cell attached to a microbead carrier (not pictured) after three days of simulated microgravity conducted in conjunction with studies on ISS. Source: Reprinted from Low and Giulianotti [15] with permission from Springer Nature.

goal of this study was to provide insights into the progression of chronic heart disease and investigate therapeutic strategies that could benefit patient's here on Earth.

By developing an MPS chip, it was possible in microgravity to identify, monitor, and screen phenotypic changes similar to ischemic cardiomyopathy [12, 13]. Also, immunological senescence and it's impact on tissue stem cells and regeneration have been studied [14]. This research evaluating the effects of immunosenescence on bone healing and vascular regeneration, advances understanding the mechanisms of immune response in regeneration of specific stem cells and can be used to test potential future therapeutic strategies [14] (Figure 3.5).

3.4.2 Blood–Brain Barrier (BBB)

The integrity of epithelial cell layers play a major role in the organ's immune response to counteract external variations [16]. Skin, for example acts as a natural barrier preventing pathogen infection to the blood-brain barrier preventing toxic access to the central nervous system. Thus, an organ's immune response is based on the integrity of tissue-specific epithelial cells. It is thought that MPS studies will help to learn more about the mechanisms involved in neuro-degenerative diseases like Alzheimer's and Parkinson's disease where neuroinflammation plays a major role [17]. MPS studies are helpful in researching the mechanism of neurodegenerative diseases.

An ISS experiment was conducted to reveal the effect of space stressor on the functioning of the blood-brain-barrier in microgravity. Changes in tight junction proteins and epithelial layers have been seen in response to launch and simulated microgravity. In particular, this ISS-based research evaluated the effects of spaceflight-relevant stressors such as hypoxia on blood-brain-barrier integrity to provide insight into neurodegenerative disease [17]. While in-flight results are yet not published, the BBB organ-on-chip has been used to evaluate abundance of phosphoSer129-alpha-synuclein (phospho-aSyn129), a post-translational

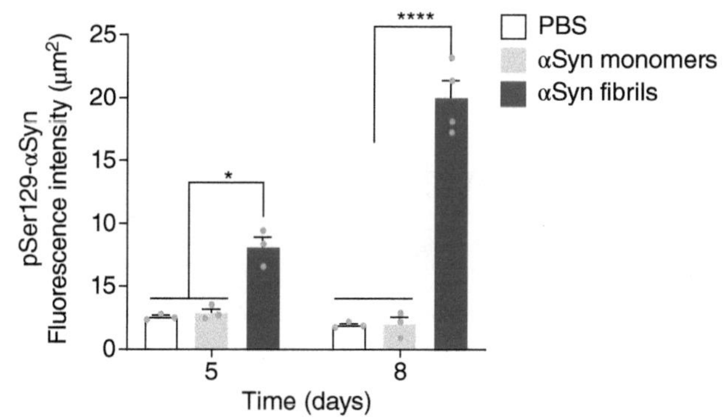

Figure 3.6 Quantitative analysis of fluorescence intensity in BBB chip. Source: Pediaditakis et al. [18], figure 3c/Springer Nature/Licensed under CC BY 4.0.

modification characteristic for the pathogenic aSyn species. Induction of phospho-aSyn129 was only evident following exposure to aSyn fibrils, as the same amount of aSyn monomer or phosphate-buffered saline (PBS) did not lead to the induction of detectable phospho-aSyn129 in the culture. This means genes associated with Parkinson's disease are altered with the presence of αSyn fibrils, suggesting the potential utility of this model for research into the dynamics of cell-cell interactions in human synucleinopathies associated with neurodegenerative diseases like Parkinson's disease, and a testing platform for target identification and validation of novel therapeutics [18] (Figure 3.6).

3.4.3 Musculoskeletal Organ

Studies have shown that a space traveler's bone density decreases and lose weight-bearing capacity after long-term space flight. This leads to osteopenia and muscular atrophy. Upon traumatic joint injury, an immediate upregulation of inflammatory cytokines is observed in the synovium fluid. Deterioration of cartilage as well as subchondral bone is often associated with this process. Upon returning to Earth's gravity field, this can increase the risk of injury and weakness [19]. The degradation of cartilage can be a source for damage in the future, and efforts have been made to investigate this further in spaceflight by Grodzinsky and Cirit [20]. The goal of this study is to develop strategies for bone resorption and solve the problem of chronic osteoarthritis pain on Earth and in long-term space flight.

While in-flight results are yet to be published, a novel cartilage degeneration MPS model has been used to predict the loss of fixed charge density (FCD) of proteoglycans by incorporating deviatoric and maximum shear strains, coupled with fluid velocity control algorithms [21]. Results show the degeneration algorithm driven by fluid velocity exhibited a uniform pattern around cartilage lesions, whereas discontinuous pattern were found around cracks (Figure 3.7).

Another study elucidated the effects of microgravity on muscle cells and tissue in real time. With the use of tiny electrodes and camera, tiny contractions of the MPS myocyte muscle can be monitored in real time. This provides directional information to investigate microgravity-related muscle atrophy in human specimens [22].

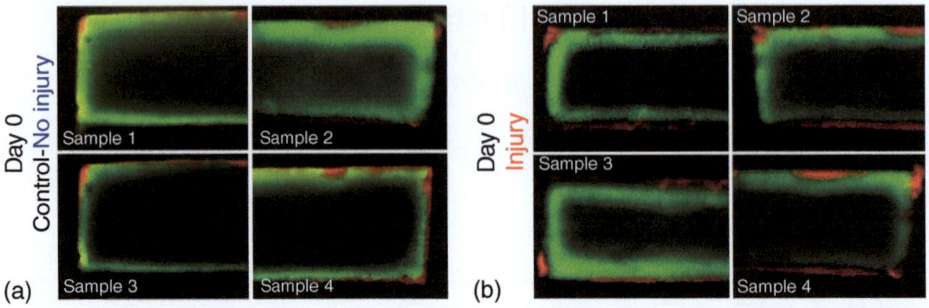

Figure 3.7 Cell viability in cartilage MPS after treatment; green are live cells and red are dead cells. Source: Reprinted from Orozco et al. [21] with open access from Springer Nature.

3.4.4 Lung

As with any virus infection, the human immune system responds to it in a complex way. Using a human lung model in microgravity, researchers studied immune response to better understand potential mechanisms for counteracting immunosuppression involved in infection. A three-layer chip containing both airways and vascular connections was used to examine mitigating factors for infectious airway diseases caused by pathogens. Under microgravity, lung-bone marrow cells immune response were studied on ISS [23].

While in-flight results are yet to be published, a case study on host–pathogenic lung MPS chip has been conducted terrestrially by Seo and Huh [24]. In this study, in order to allow for volatile contact between the pathogens and epithelial cells via open airway lumens, the authors developed a closed system containing separate chambers for A. fumigatus and Pseudomonas aeruginosa with three human bronchial MPS modules. Results show an increased epithelial cytokine production without direct exposure to pathogens. The information provided here could help understanding how host responses vary depending on the pathogen exposure, and provide insight on mechanisms that can enhance the early host response [24] (Figure 3.8).

3.4.5 Kidney

Microgravity makes astronauts more vulnerable to kidney dysfunction that can lead to proteinuria, osteoporosis, and kidney stones. The cellular function in the kidney cannot be reproduced by statics models. An MPS system has been designed to mimic the kidney's function by interacting with ions and bioactivating vitamin D. As part of the MPS study, oxalate microcrystals were grown in a laboratory to simulate the formation of kidney stones. The use of microgravity can accelerate the crystal formation process, thereby speeding up the lengthy process. This can help researchers understand (i) the function of kidney proximal and distal tubule epithelium in ion and solute transporters; and (ii) the mechanism of stone formation in humans on Earth [25].

While in-flight results are yet to be published, a similar MPS system has been tested on Earth for the purpose of assessing drug-related nephrotoxicity from

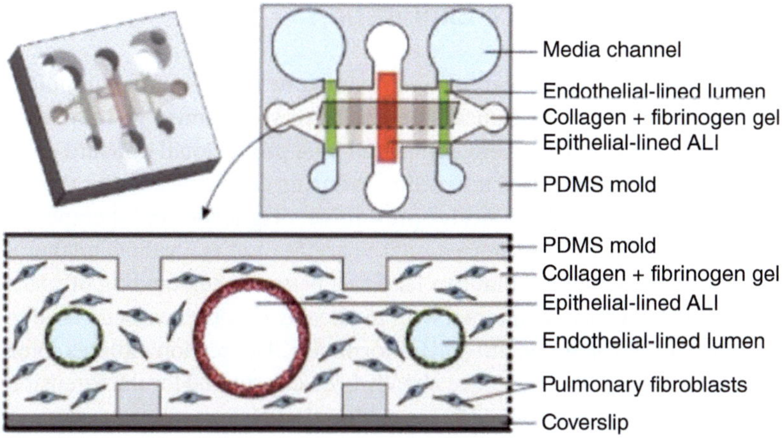

Figure 3.8 In vitro model of airway epithelial cells, lung fibroblasts, and endothelial cells. Source: Seo and Huh [24], figure 8 (p. 201)/Reproduced with permission of Elsevier.

treatment and predicting optimal dosing regimens. Results show that clinical dosing of rifampicin (600 mg/d) is effective and less likely to cause nephrotoxicity, whereas higher doses of greater than 600 mg/d or frequent dosing regimens would not improve efficacy any further. Pharmaceutical development for clinical treatment in the future is supported [26] (Figure 3.9).

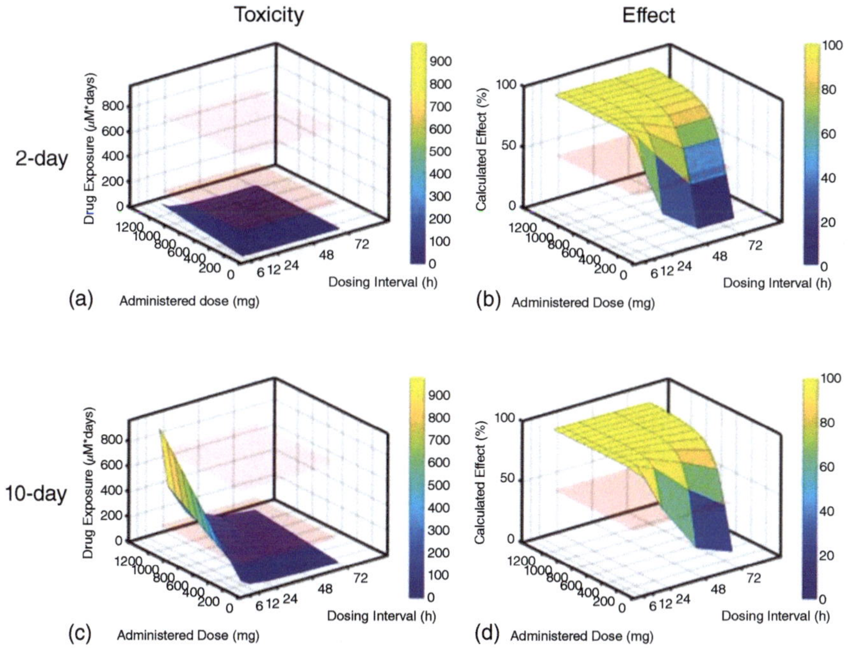

Figure 3.9 Using kidney MPS system to identify optimum dose regime. Source: Republished with permission of Maass et al. [26], John Wiley & Sons.

3.5 Telemedicine for Space Exploration

Longer missions aboard the space station are providing new data on the physiological effects of microgravity. Yet, astronauts on extended space travel will not have the option of having medical treatment completed by a professional onboard or in real time. Therefore, having access to telemedicine during deep-space exploration is vital. Telemedicine allows people to receive preventive, diagnostic, and therapeutic care when they are in space. The International Space Station is a valuable laboratory for telemedicine experiments that can lead to a better understanding of space health. Technological advances combined with medical expertise have been used to monitor and maintain astronaut health and performance. The section outlines some of the technological innovations that have already been considered for the extension of space exploration to enable extended space exploration.

3.5.1 Point-of-Care (POC) Diagnostic Device

Technologies for research and central laboratories to meet most diagnostic requirements use a variety of peripheral equipment, several sampling methods, flexible protocols, and a number of signal detection formats. In contrast, a point-of-care (POC) microfluidic device is optimized for a particular application. POC devices are portable and networked diagnostic devices. The development of POC diagnostics involves the choice of appropriate materials, surface treatment, sample processing, microfluidic elements (such as valves, pumps, and mixers), receptors, and analytes, with various biosensing concepts embedding all functions on chip such as for processing, analyzing, and reporting [27]. A microfluidic POC device is uses a small, minimally invasive volume of unprocessed sample taken directly from the patient, which may be as low at 1 μl, and thus much lower than a drop from a finger prick (typically 25 μl). POC devices can multiplex the analysis of up to 100 analytes such as proteins and nucleic acids, achieve quantitative results within minutes, and detect multiple biomarkers with high sensitivity [28] (Figure 3.10).

As an alternative to POC bio-analyzers, astronauts have to store their blood in a small freezer on the ISS. While freezing reduced sample quality, it remains a necessary step because on-board testing is not yet available for all astronaut health testing. POC diagnostics make it possible to analyze samples from space in as little as two to three hours, saving scientists the time and cost of freezing and returning them and their use is likely to expand in the future [29]. Through this on-board instrument, researchers will have much faster access to scientific data. The application of different microfluidic principles has resulted in the miniaturization of existing biomedical protocols as well as the development of entirely new diagnostic and research strategies [30].

In one study, a commercial test device capable of analyzing white blood cell (WBC) count and differential from a fingerstick sample within minutes, known as "Hemocue," has been tested on board the International Space Station to determine whether it can provide in-flight capabilities [31]. An in-flight experiment was conducted using standard WBC samples. The tests were classified into three

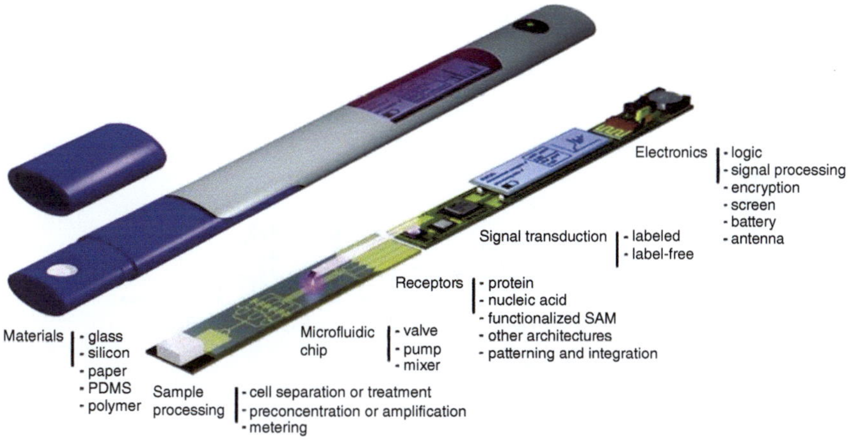

Figure 3.10 The ideal POC diagnostic device. Source: Reprinted from Gervais et al. [27] with permission from Wiley-VCH.

ranges, low, normal, and high, according to their WBC concentration: 2.7, 8, and 17 K cells/ul. Although there was a slight variation between ground and flight data, research suggests that this was due to the compromise of the standard solutions due to long storage periods before flight operations. Therefore, results from this study indicate that with improved storage of standard solutions, the miniaturized device can be used in microgravity for astronauts to monitor their wellbeing during missions [31] (Figure 3.11).

Considering from a technical side, blood has been the most widely used source of tissue for gene expression in biodosimetry studies. When considering less-invasive sampling methods, saliva is becoming more common. Sampling saliva is a non-invasive and convenient alternative to collecting blood. In recent years, saliva has been shown to contain RNA biomarkers for prediction and diagnosis of numerous diseases [32]. Vertical Integrated Flow Assay System Technology, or VerIFAST, is a method for assessing physiological effects by evaluating biomarkers of humans using blood or saliva. Multiplex molecular diagnostics can be performed to assess radiation exposure as well as other diagnostic biomarkers. An array of reagents is arranged on rows of nanoporous membranes with spots containing diagnostic markers. After introduction of sample and interaction with reagents, the spots change color providing results in minutes [32]. The method helps researchers develop an expeditious way of predicting the risk of radiation exposure and information on other diagnostic biomarkers of health on the International Space Station and future space missions.

3.5.2 Handheld Ultrasound Devices in Emergency Medicine

Ultrasonography offers a great deal of flexibility over other imaging therapies, including radiography, computed tomography, and magnetic resonance imaging. It is an essential tool for diagnostic and interventional procedures, and has been used

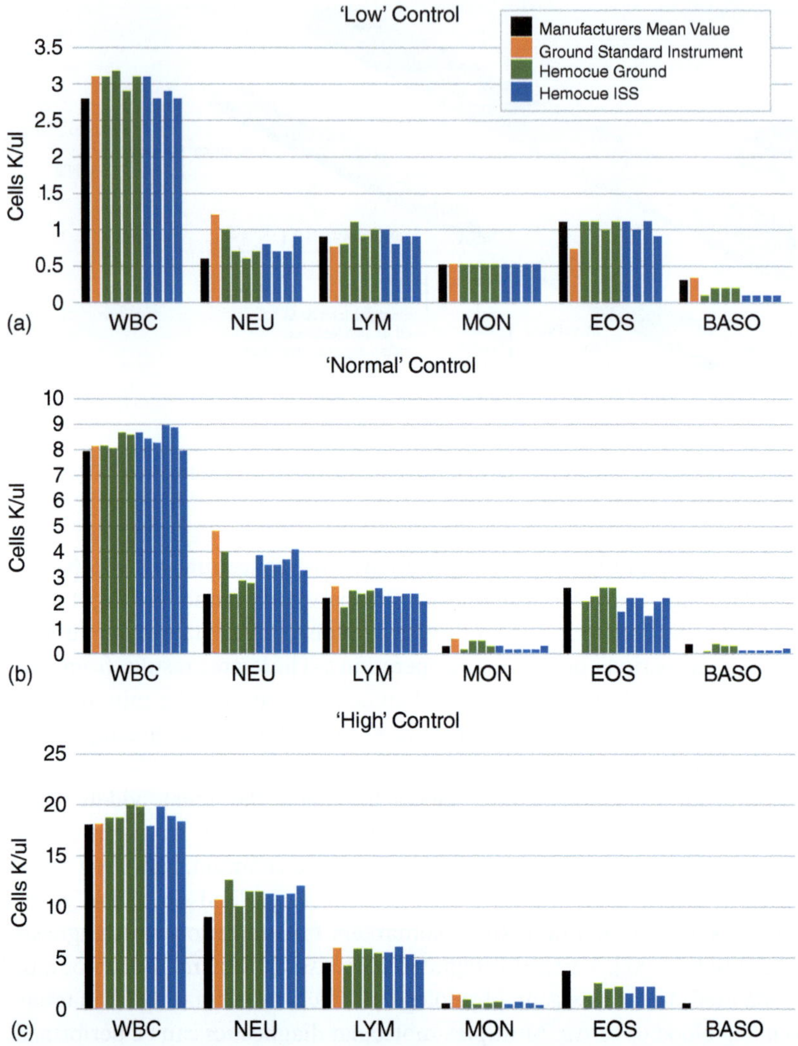

Figure 3.11 Comparison POC sample data collected on ground and aboard the International Space Station. Source: Crucian et al. [31], figure 2 (p. 32)/Reproduced with permission of Elsevier.

to study almost every organ system. With little training and a little assistance from a ground sonographer, the crew members are able to produce a variety of high-quality scans [33]. The significant role for ultrasound has been and will remain in the study of physiological changes that occur in space or are a result of space flight [34].

It has become more affordable and practical in recent years to use portable ultrasound devices, which previously only were available in specialized departments as bulky and expensive machines. In recent years, portable ultrasound devices offer rapid visual correlates of clinical findings and complement clinical examinations. Bulky devices cannot be used in spacecrafts and are troublesome

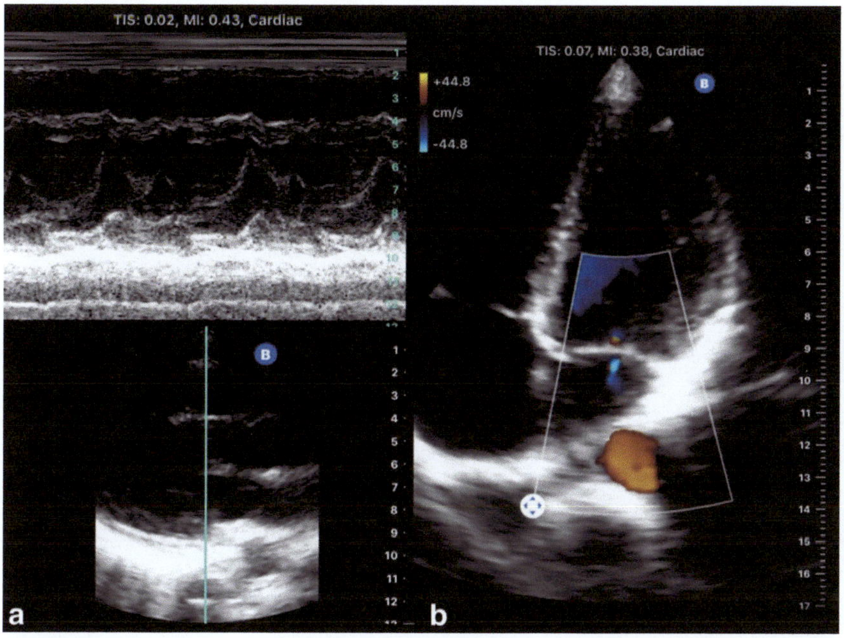

Figure 3.12 Cardiac ultrasound imaging by Butterfly IQ (a) M-Mode (b) color Doppler view. Source: Reprinted from European Society of Radiology [35] with open access from Springer Nature/CC BY 4.0.

even in extraterrestrial habitats; rather portable medical devices seem to be the right space solution for monitoring.

One of the most recent examples for portable ultrasound devices is the handheld ultrasound device, namely Butterfly iQ that connects to a mobile device such as a tablet or smartphone. As an alternative to piezoelectric crystals, the Butterfly iQ uses silicon chips to create ultrasound waves. With this, it is possible to scan 2–30 cm deep in M-, B-, or color Doppler modes mirroring the curved, linear, or phased transducers [35] (Figure 3.12). Astronauts have tested Butterfly iQ on ISS and imaging capabilities like this, which are critical items for a spaceflight medical kit given their reduced mass, and self-reliant usability by crew, would be well suited for future long-duration spaceflight, like NASA's Artemis missions.

3.5.3 VisualDx

VisualDx is a decision support tool designed to be a point-of-care reference for confirming diagnostics based on visual appearances and textual descriptions of diseases and symptoms of the skin. It is a clinical evidence technology that provides comprehensive understanding of a variety of skin conditions for the end user that is planned to be tested in future spaceflight [36].

VisualDx can be used as a way of guiding astronauts' medical decision-making on deep-space missions, where visiting a doctor is not an option. Astronauts can take a

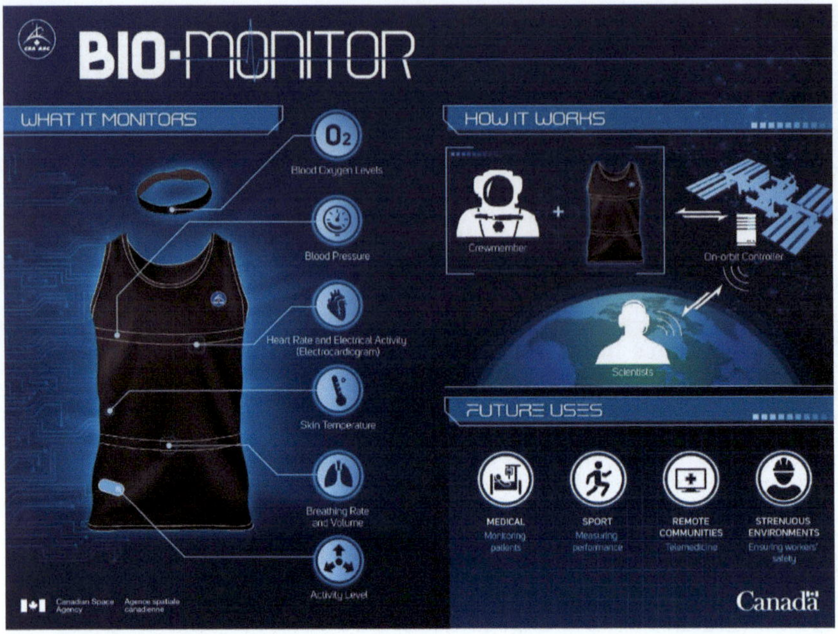

Figure 3.13 Bio-monitor system. Source: Asrar et al. [37].

photo of a skin condition for example, to guide their diagnosis of the condition using an onboard digital platform.

3.5.4 Bio-Monitor

The Bio-Monitor is a smart gadget designed to be worn by astronauts aboard the ISS to record vital signs and monitor them as part of their daily routine. Incorporating a smart shirt and tablet application, this new system allows astronauts to keep track of their health and collect new scientific data by continuously measuring their physiological states [37] (Figure 3.13).

3.6 Conclusion

In this chapter, a review of frontier medical technologies is presented, for future space health. Their potential is outlined by experiments on the International Space Station (ISS), where accessible, and derived from Earth experiments, when space results have not (yet) been reported. The chapter starts with a problem definition by giving a short analysis of astronaut health risks (Section 3.2; Introduction being the first). The conclusion is that based on astronaut data from studies on ISS, spaceflight has a variety of effects on the body both physically and psychologically. Using this data and the ISS to more fully understand how these effects change as we travel beyond low-Earth orbit will continue to be an important area of study for human health in space. This asks for providing space-specific solutions to the healthcare, which is underpinned by a few selected medical diagnosis methods, tested in space (Section 3.3). Conclusion here is that measures for human health in space, including more sophisticated transcriptomic, proteomic, and microbiome analysis of body fluids from astronauts on ISS, along with physiologic, cognitive and mental health assessments, are essential in the development of countermeasure development for future space travelers who will one day live and work in low-Earth orbit, or continue our exploration efforts to other planets. With much broader consideration, a survey on tissue-based microchip devices for medical organ-tissue studies (Section 3.4). Conclusion is here that, although much fundamental knowledge has been investigated on Earth, the publication of the outcome of recent in-flight results is highly awaited to uncover what these human analog models can tell us about spaceflight effects for astronauts, and improved understanding of mechanisms responsible for many of the significant health burdens here on Earth and potential novel pathways for treatments that have not yet been identified in terrestrial research. Those are the foremost current technology standards with regard to this chapter's objective. The chapter ends with introducing some modern computational-based telemedicine technologies (Section 3.5), which have been used in the space health research or are likely to be used in near future. We expect this branch of technology and research to continue to expand and to see technology advancements in the near future that will foster the creation of future cities in space and continued exploration well beyond low-Earth orbit.

References

1 Thirsk, R., Kuipers, A., Mukai, C., and Williams, D. (2009). The space-flight environment: the International Space Station and beyond. *CMAJ* 180 (12): 1216–1220.

2 Mesko, B. (2017). Digital health technologies to support human missions to Mars. *New Space* 6 (2): 109–116.

3 Patel, Z.S., Brunstetter, T.J., Tarver, W.J. et al. (2020). Red risks for a journey to the red planet: the highest priority human health risks for a mission to Mars. *NPJ Microgravity* 6 (1): 33.

4 Vogenberg, F.R., Isaacson Barash, C., and Pursel, M. (2010). Personalized medicine: part 1: evolution and development into theranostics. *P & T* 35 (10): 560–576.

5 Nickerson, C.A., Ott, C.M., Wilson, J.W. et al. (2004). Microbial responses to microgravity and other low-shear environments. *Microbiology and Molecular Biology Reviews* 68 (2): 345–361.

6 Boguraev, A.-S., Christensen, H.C., Bonneau, A.R. et al. (2017). Successful amplification of DNA aboard the International Space Station. *NPJ Microgravity* 3 (1): 26.

7 Rubinfien, J., Atabay, K.D., Nichols, N.M. et al. (2020). Nucleic acid detection aboard the International Space Station by colorimetric loop-mediated isothermal amplification (LAMP). *FASEB BioAdvances* 2 (3): 160–165.

8 Wilson, J.W., Ott, C.M., Höner zu Bentrup, K. et al. (2007). Space flight alters bacterial gene expression and virulence and reveals a role for global regulator Hfq. *Proceedings of the National Academy of Sciences of the United States of America* 104 (41): 16299–16304.

9 Garrett-Bakelman, F.E., Darshi, M., Green, S.J. et al. (2019). The NASA twins study: a multidimensional analysis of a year-long human spaceflight. *Science* 364 (6436), eaau8650.

10 Mezohegyi G., Buchholcz B., Sohajda T., et al. (eds.) (2021). Space chemistry and global health: drug development against Covid-19 in space. *58th Scientific and Technical Subcommittee*, Vienna, Austria.

11 Sharma, A., Clemens, R.A., Garcia, O. et al. (2022). Biomanufacturing in low Earth orbit for regenerative medicine. *Stem Cell Reports* 17 (1): 1–13.

12 Wu, J.C. and Pruitt, B.L. (n.d.). *Effect of Microgravity on Drug Responses Using Engineered Heart Tissues*. The National Institutes of Health (NIH). Available from: https://reporter.nih.gov/project-details/9644885.

13 Kim, D.-H. (n.d.). *A Human iPSC-based 3D Microphysiological System for Modeling Cardiac Dysfunction in Microgravity*. The National Institutes of Health (NIH) Available from: https://reporter.nih.gov/project-details/9644709.

14 Schrepfer, S. and Deuse, T. (n.d.). *Microgravity as Model for Immunological Senescence and Its Impact on Tissue Stem Cells and Regeneration*. The National Institutes of Health (NIH) Available from: https://reporter.nih.gov/search/ACWKH3GbGEeiqc7Pb4O56A/project-details/10005510.

15 Low, L.A. and Giulianotti, M.A. (2019). Tissue chips in space: modeling human diseases in microgravity. *Pharmaceutical Research* 37 (1): 8.

16 Moens, E. and Veldhoen, M. (2012). Epithelial barrier biology: good fences make good neighbours. *Immunology* 135 (1): 1–8.

17 Hinojosa, C.D. and Karalis, K. (2018). *Organs-on-Chips as a Platform for Studying Effects of Microgravity on Human Physiology: Blood-Brain Barrier-Chip in Health and Disease*. National Institutes of Health (NIH). Available from: https://reporter.nih.gov/project-details/9402972.

18 Pediaditakis, I., Kodella, K.R., Manatakis, D.V. et al. (2021). Modeling alpha-synuclein pathology in a human brain-chip to assess blood-brain barrier disruption. *Nature Communications* 12 (1): 5907.

19 Lang, T., Van Loon, J.J.W.A., Bloomfield, S. et al. (2017). Towards human exploration of space: the THESEUS review series on muscle and bone research priorities. *NPJ Microgravity* 3 (1): 8.

20 Grodzinsky, A. and Cirit, M. *Cartilage-Bone-Synovium MPS: Musculoskeletal Disease Biology in Space*. The National Institutes of Health (NIH). Available from: https://reporter.nih.gov/project-details/9402868#details.

21 Orozco, G.A., Tanska, P., Florea, C. et al. (2018). A novel mechanobiological model can predict how physiologically relevant dynamic loading causes proteoglycan loss in mechanically injured articular cartilage. *Scientific Reports* 8 (1): 15599.

22 University of Florida (2020). UF tissue chip experiment to study age-related muscle loss aboard International Space Station: University of Florida. https://

pharmacy.ufl.edu/2020/12/01/uf-tissue-chip-experiment-to-study-age-related-muscle-loss-aboard-international-space-station/.

23 Worthen, G.S. and Huh, D. (n.d.). *Lung Host Defense in Microgravity*. The National Institutes of Health (NIH). Available from: https://reporter.nih.gov/project-details/9403062.

24 Seo, J. and Huh, D. (2019). Microphysiological models of human organs: A case study on microengineered lung-on-a-chip systems, Chap. 7. In: *Microfluidic Cell Culture Systems*, 2nde (ed. J.T. Borenstein, V. Tandon, S.L. Tao and J.L. Charest), 187–208. Elsevier.

25 Himmelfarb, J. and Kelly, E.J. (n.d.). Effects of microgravity on the structure and function of proximal and distal tubule MPS.

26 Maass, C., Sorensen, N.B., Himmelfarb, J. et al. (2019). Translational assessment of drug-induced proximal tubule injury using a kidney microphysiological system. *CPT: Pharmacometrics & Systems Pharmacology* 8 (5): 316–325.

27 Gervais, L., de Rooij, N., and Delamarche, E. (2011). Microfluidic chips for point-of-care immunodiagnostics. *Advanced Materials* 23 (24): H151–H176.

28 Gomez, F.A. (2012). The future of microfluidic point-of-care diagnostic devices. *Bioanalysis* 5 (1): 1–3.

29 Canadian Space Agency (2019). Bio-Analyzer: Near-real-time biomedical results from space to Earth. https://www.asc-csa.gc.ca/eng/iss/bio-analyzer.asp.

30 Serra, M., Ferraro, D., Pereiro, I. et al. (2017). The power of solid supports in multiphase and droplet-based microfluidics: towards clinical applications. *Lab on a Chip* 17 (23): 3979–3999.

31 Crucian, B., Valentine, R., Calaway, K. et al. (2021). Spaceflight validation of technology for point-of-care monitoring of peripheral blood WBC and differential in astronauts during space missions. *Life Sciences and Space Research* 31: 29–33.

32 Ostheim, P., Amundson, S.A., Badie, C. et al. (2021). Gene expression for biodosimetry and effect prediction purposes: promises, pitfalls and future directions – key session ConRad. *International Journal of Radiation Biology* 2021: 1–12.

33 Law, J. and Macbeth, P.B. (2011). Ultrasound: from Earth to space. *McGill Journal of Medicine* 13 (2): 59.

34 Martin, D.S., South, D.A., Garcia, K.M., and Arbeille, P. (2003). Ultrasound in space. *Ultrasound in Medicine and Biology* 29 (1): 1–12.

35 European Society of Radiology (2019). ESR statement on portable ultrasound devices. *Insights Into Imaging* 10 (1): 89.

36 Burke, M. and Littenberg, B. (2019). Effect of a clinical evidence technology on patient skin disease outcomes in primary care: a cluster-randomized controlled trial. *Journal of the Medical Library Association* 107 (2): 151–162.

37 Asrar, F.M., Saint-Jacques, D., Chapman, H.J. et al. (2021). Can space-based technologies help manage and prevent pandemics? *Nature Medicine* 27 (9): 1489–1490.

Part II

Space Biology

4

Plant Biology and a New Approach to Space Farming

Natasha Haveman, Anna-Lisa Paul, and Robert Ferl

University of Florida, Department of Horticultural Sciences, 2550 Hull Road, Gainesville, FL 32611, USA

4.1 Introduction

Plants have been proposed to regenerate oxygen and water in bioregenerative life support systems while supplementing the astronaut's diet with fresh food, as well as providing psychological benefits for the crew. In addition to those life-supporting roles, plants can also provide a variety of natural and synthetic products that can be utilized by a wide array of manufacturing industries. Products such as terpenoids, resins, fiber, essential oils, and biopolymers are just some examples of the added resources that plants can provide as humans expand space manufacturing capabilities. The current approach to integrating plants into exploration scenarios is to move conventional crop production and standard plant varieties to vehicular, contained exploration environments, dealing with issues such as microgravity, to produce mission-valuable resources. However, this creates an intrinsic design problem in that mission plans have adopted varieties selected for growth in terrestrial fields or greenhouses, with food production as the major goal. These varieties may not perform well in closed environments, and their performance is a complete unknown for space-based applications where the specter of failure creates a barrier for application. In addition, this traditional approach limits plants as resources for non-food products that would greatly enable in space manufacturing. Plants have been adopted when they should have been adapted. This chapter will discuss how using traditionally bred or synthetically engineered plants for mission-specific objectives would revolutionize the way we think about integrating biology into exploration missions.

4.2 Historical Context of Plant Selections

Plants are vital forms of life as they are a major source of oxygen and food for all animals on Earth. Plants also shape the Earth's landscape as they reduce noise levels, wind speed, water runoff, and soil erosion. Throughout human history,

In-Space Manufacturing and Resources: Earth and Planetary Exploration Applications, First Edition.
Edited by Volker Hessel, Jana Stoudemire, Hideaki Miyamoto, and Ian D. Fisk.

Figure 4.1 Examples of domesticating ancient crops and artificially selecting desired phenotypes. The wild ancient crops are shown next to their modern domesticated versions. Source: Carrot – (Ellison et al. 2018). Strawberry – (Aharoni et al. 2003). Brassica – adapted from geneticliteracyproject.org: Unnatural natural history of some favorite green veggies. Gray / Adobe Stock; dasuwan / Adobe Stock; Andrew Scheck / Adobe Stock; Alexander Potapov / Adobe Stock.

humans have used, domesticated, and developed plants and fungi for an array of needs (i.e. food, medicine, materials, fuels, poisons, environmental uses, and social uses) (Kew 2016). The domestication of many ancient plants for food has resulted in the physiological and morphological change between these crops and their wild progenitors (Figure 4.1). A great example of how humans have dramatically modified plants through artificial selection can be illustrated by the evolution of the wild mustard plant.

The wild mustard plant belongs to a family of plants called *Brassicaceae*. Well-known members of this family make up many of the cruciferous vegetables we consume today. These variations are the result of centuries of human selection to diversify one species to produce separate cultivars with nutritious leaves, starchy roots, high protein flower heads, or oily seeds (Vaughan 1977).

Particularly, the species *Brassica oleracea* and *Brassica rapa* comprise hundreds of useful crop varieties used throughout the world and have been widely recognized for their value to human health (van Poppel et al. 1999). Cauliflower, Brussel sprouts, kale, radishes, cabbage, and kohlrabi are all derived brassicas, all developed for efficient cropping in the fields (Figure 4.1).

Figure 4.2 Examples of mutations that create dramatic, and potentially useful phenotypes. (a) Wild-type Arabidopsis compared to Arabidopsis mutant exhibiting larger, ruffled leaves (jaw-D/stip-D/cuc2g-m4). Source: Blein et al. 2013/CC BY 3.0. (b) The wild-type progenitor of "cole" crops, *Brassica oleracea*, compared to the cultivar vegetable kale. Source: Alexander Potapov/Adobe Stock. (c). Wild-type Arabidopsis flowers compared to a flower mutant resembling cauliflower (apetala1/cal). Source: Smyth 1995. (d) Wild-type *B. oleracea* flowers compared to the cultivar vegetable cauliflower. Source: Scisetti Alfio/Adobe Stock; Andrew Scheck/Adobe Stock.

Within the Brassicaceae family is the model organism laboratory plant *Arabidopsis thaliana* (Arabidopsis). Arabidopsis is widely used in genetic studies as a model plant organism, as its entire genome has been sequenced and well annotated. Studies in Arabidopsis have revealed insights that have led the way in dissecting the genetic origins of the familiar vegetable traits (Figure 4.2). A particularly good example is the *CAULIFLOWER* (*CAL*) gene in Arabidopsis that was identified from a mutation that resulted in a phenotype in Arabidopsis that looked like tiny cauliflowers, and it was discovered that the common vegetable was deficient in the same gene (Smyth 1995). Other mutations in the Arabidopsis genome such as *Jaw-D*, *Stip-D*, *Mir164a-4/CUC2g-m4* double and triple mutants also demonstrate how the leaf structure can be altered to resemble other known vegetables within the family (Blein et al. 2013). Thus, using Arabidopsis as a model lab organism, it is easy to relate how breeding for certain traits can eventually lead to parallel mutations in the progenitor of many terrestrial horticultural crop species.

In 1983, the first transgenic plant was constructed in tobacco through the insertion of a bacterial antibiotic resistance gene called neomycin phosphotransferase II (nptII) (de Framond et al. 1983). The addition of the NPTII enzyme in tobacco plants served as a selectable marker that enabled the quantification of transformation efficiency and gene expression (Curtis et al. 1995). Since then, the applications of transgenic plants have been growing exponentially. Here we highlight a few examples of the applications of transgenic plants worldwide.

4.3 Resistance to Biotic or Abiotic Stresses

Naturally occurring stresses by other living organisms (i.e. insects, bacteria, fungi, and viruses – biotic stresses) or the environment (i.e. temperature, water availability, soil composition, and sunlight availability) can have profound impacts on plant health, growth, and development. In the case of biotic stresses, a variety of crops have been engineered with disease resistance genes to confer resistance towards pathogenic diseases, which in turn reduces the usage of pesticides and herbicides in the field.

Corn is the most widely produced crop in the United States (Capehart and Proper 2019). In 2018, about 92% of all corn planted in the United States were genetically modified (US Food and Drug Adminstration 2020). The *Bacillus thuringiensis* (Bt) corn, which has been genetically modified to express one or more proteins from the bacterium *Bacillus thuringiensis*, was approved in 1996 as insect-resistant corn. These proteins are toxic to insects like the European corn borer that destroys corn crops by burrowing into the stems of the plant and causing them to lose the structural integrity necessary for plant growth and development. Instead of spraying the crops with insecticides that may not have a consistent effect, Bt corn produces a protein that is activated by the digestive system of a specific group of insects and ruptures the insect's intestinal lining within days. These crops can better withstand the impact of European corn borer outbreaks without dire consequences to the industry.

4.4 Improving Crop Yield and Nutritional Value

Vitamin A deficiency is a public health problem in many developing countries that can cause blindness. A solution to address this problem was to fortify a major staple food, rice, with provitamin A. The provitamin A (β-carotene) biosynthetic pathway was introduced into rice endosperm by genetic engineering and Golden Rice was created (Beyer et al. 2002). It has been projected that Golden rice can provide 60% of the vitamin A dose recommended to a child's daily needs and can diminish mortalities and diseases related to vitamin A at a lower cost compared to other approaches (Shahbaz et al. 2020).

Another example that has spaceflight relevance is the genetically engineered plum trees (*Prunus domestica L*) that over-expresses the poplar (*Populus trichocarpa*) Flowering Locus T-1 (FT1) gene (Srinivasan et al. 2012). These transgenic plum trees now have several key features that allow them to be considered as candidate crops in bioregenerative life support systems for human space exploration missions (Graham et al. 2015). These features include dwarf growth habits, early flowering, and continuous fruit production, all of which have eliminated obvious constraints of growing fruit trees in confined conditions such as the ISS. In addition to these optimal traits of the FT1 plums, studies have also demonstrated the beneficial effects of dried plums (*P. domestica L.* "Improved French") on bone health. This osteoprotective dietary supplement is of added value to crewmembers as NASA scientists have identified bone loss as one of the greatest challenges to crewmembers on long-duration space missions.

4.5 Transgenic Plants as Bioreactors for Recombinant Proteins

Biopharmaceuticals have traditionally been produced in mammalian cell culture systems; however, plants can carry out all of the complex post-translational modifications necessary for the function of many biopharmaceuticals (Chen and Davis 2016). By removing the endogenous plant-specific β1,2-linked xylose and α1,3-linked fucose, a variety of plant-made antibodies have demonstrated improved immune receptor binding and greater potency compared to commercially available antibodies produced in mammalian cell (Zeitlin et al. 2011; Marusic et al. 2018; Diamos et al. 2020). In the last few years, plant-made antibody therapies have been developed for the dengue virus (Dent et al. 2016), West Nile virus (Sun et al. 2018), and chikungunya virus (Hurtado et al. 2020). In a recent study aimed at producing high amounts of monoclonal antibodies using an optimized plant expression system, up to four fluorescent proteins could be co-expressed simultaneously, reaching high yields of approximately 3–5 g/kg leaf fresh weight or ~50% total soluble protein (Diamos et al. 2020).

In the time between now and when our ancestors first started domesticating plants, humans have dramatically changed the way plants have been utilized to provide a variety of products under various terrestrial conditions. The current emphasis on using more renewable resources has led to increased development in technologies that not only diversify manufacturing power sources but also in the way raw materials are synthesized, harvested, and utilized for a large spectrum of applications.

4.6 Plant Products for Manufacturing

For centuries humans have been using the plant kingdom as a major source of chemical products. According to the United States Department of Agriculture (USDA) 2020 Plant Protection report, agricultural trade in plants and plant products is currently worth nearly $1.7 trillion annually (USDA 2020). Rice, wheat, and maize are the top three most heavily planted, popular, and profitable crops in the world (https://www.visualcapitalist.com/the-worlds-most-valuable-cash-crop/) as these food crops are a staple source of carbohydrates for more than half the world's population. Apart from human consumption, these crops are also harvested for animal feed and biofuel (ethanol). Cotton is another cash crop, one that is harvested for its fiber which is widely used to make textile for clothing. Sunflower, soybean, rapeseed, cotton, and peanut are plants that produce oils with low levels of saturated fats and linolenic acids. These oil crops, especially soybean, make up a large proportion of all oil crops produced worldwide (Sharma et al. 2012).

Although industrial crops are largely known to produce food and feed, many of them are also grown to supply raw materials, such as biofuels, specialty chemicals, pharmaceuticals, fibers, and biopolymers, for a variety of manufacturing industries (Lin and Eudes 2020). Many of the aforementioned raw materials used as building blocks in a myriad of industrial applications are metabolic products synthesized by plants. Over the course of evolutionary time, plants have evolved to accumulate and synthesize many metabolic compounds to combat an array of ecological forces. Some of these forces include pests, diseases, dealing with changing environmental conditions and enhancing reproductive success (Wu and Chappell 2008).

The complexity of harnessing these raw materials for commercial production should not be underestimated. The vast number of these biologically unique metabolites produced by plants is usually associated with a particular plant family grown under a specific geographical locale and climatic condition. In addition, these raw materials often comprise mixtures of different compounds that accumulate in the plants over long growth periods (Wu and Chappell 2008; Lin and Eudes 2020) (Figure 4.3). For example, most plant resins are mixtures of organic compounds like terpenoids and/or phenolics (Dilworth et al. 2017). Many plant species secrete these substances in response to injury from insects, herbivores, and pathogens (Langenheim 1990). Another similar plant product is latex. Like resins, latex is produced by plants for defense and comprises mixtures of secondary metabolites such as rubber, alkaloids, phenolics, and terpenoids, to name just a few (Agrawal and Konno 2009). Interestingly, research has shown that latex production is not only species-dependent, but both the biotic and abiotic environment can have a strong influence on latex exudation (Agrawal and Konno 2009).

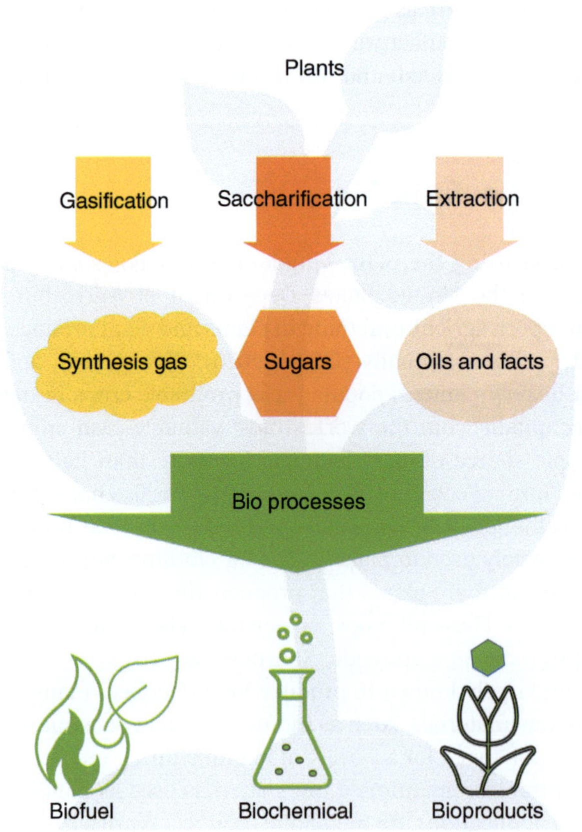

Figure 4.3 Schematic of various processes used to refine plant biomass for renewable raw materials. Source: Schematic adapted from ViaSpace.com: Biorefinery: Biofuels, Biochemicals and Bioplastics.

To utilize plants as a source of these organic compounds, the products must accumulate to levels that justify the costs associated with growing and processing the plant material. Traditional selective plant breeding for farming has aptly demonstrated the enormous gains in productivity that can be achieved when plants are actively selected for specific traits and specific production scenarios. However, selective plant breeding also has its challenges. The process is slow, unspecific – certain traits are impossible to select for, and may result in unintentional selection for deleterious recessive genes. With the development of high-throughput transformation protocols and synthetic biology molecular tools, the technology now exists to redirect, in a much more rapid and directed fashion, the genetics of plant structure and metabolism toward desired goals (Dutt et al. 2014; Pennisi 2016).

The concept of using plant compounds to replace petroleum products in the manufacturing sector has been around for decades. Although using natural oils in place of petroleum is currently not cost effective and the performance of these products tends to be lower than that of the conventional petrochemical nature, technologies have been developed to obtain photocurable polymers, vinyl monomers, biobased oxygen-containing monomers, and olefins of various chain lengths from plant oils (Sharma and Biresaw 2016). These products have been used in various industrial markets such as lubricants, paints, textiles, coatings, resins, adhesives, linoleum, plasticizers, and cosmetics (Langenheim 1990; Wagner et al. 2001; Lazzeri et al. 2006; Wu and Chappell 2008; Agrawal and Konno 2009; Xia and Larock 2010; Miao et al. 2014). As humans embark on extended journeys off planet, such plant-based polymers and monomers may become the only means of obtaining these materials for our manufacturing goals away from Earth. Using synthetic plant biology, plants can be designed to produce desired amounts of natural oils and fats in various space-relevant environments so that these products will be available for industrial needs. Synthetic biology also offers the opportunity to engineer entire pathways, normally found in one plant species, into plants designed for applications such as spaceflight environments.

4.7 Synthetic Plant Biology

Selective breeding, which is a traditional form of genetic engineering, has progressed with the advancement of modern biotechnology. Genetic engineering and gene editing can be employed to accelerate the development of desirable traits, (Liu et al. 2015) and create biochemical "factories" (Taylor 1998; Borrelli and Trono 2016; Wurtzel and Kutchan 2016). These capabilities have changed the approaches that took humans thousands of years to attain, to now only taking a few years to accomplish. Genetically engineered plants, also known as synthetic plants or transgenic plants, are constructed to produce plants that have ideal agronomic traits, product quality, and high product yield. These synthetic plants will have their genome modified either by adding a new beneficial gene/gene cassette or by the removal of a specific detrimental gene. Gene editing, and the targeted control of those edits throughout the plant's life cycle, is key to the production of desired compounds.

Transgenic plants for the production of natural products such as biopolymers (i.e. polyethylene and polypropylene), fibrous proteins (i.e. silk, collagen, and keratin), polyhydroxyalkanoates (PHAs), and terpenoids have been successfully developed in the past few decades (Van Beilen and Poirier 2008). In some instances, the production of these natural products in transgenic plants has not yet reached the commercial stage as it is not as cost effective when compared to production from other sources available terrestrially. However, *in planta* production of chemicals and biopolymers may take center stage as humans venture farther away from earth and its terrestrial-bound resources.

Silks are categorized as a broad class of polymers that are loosely defined as externally spun fibrous protein secretions largely from insects and spiders (Van Beilen and Poirier 2008). Synthetic silk genes have been successfully expressed in transgenic plants such as tobacco, potato, and *Arabidopsis thaliana* (Arabidopsis), which have been cultivated in greenhouses and field trials (Menassa et al. 2004; Scheller et al. 2004). In the case of synthetic silk production in tobacco and potato leaves, targeting synthesis to the endoplasmic reticulum yielded 2% of total soluble protein (TSP) (Scheller et al. 2001). On the other hand, expression of synthetic silk in leaf apoplasts of Arabidopsis yielded 8.5% TSP, while targeting to seed endoplasmic reticulum yielded 18% TSP (Yang et al. 2005). It has been suggested that improvements in fibrous protein synthesis in plants may require several approaches: (i) optimization of the amino acid and tRNA pools for the amino acids that are over-represented in these proteins (such as glycine and alanine in spider silk), and (ii) co-expression of several fibrous proteins as found in natural silk, and, (iii) improved targeting to subcellular compartments and tissues that are optimal for protein synthesis and storage (Van Beilen and Poirier 2008).

PHAs are a group of biological polyesters that are produced by a wide variety of bacteria as osmotically inert carbon and energy storage compounds that accumulate in the form of granules (Van Beilen and Poirier 2008). Characteristics such as impermeability to water and air make PHAs very suitable for consumer products such as bottles, films, and fibers. Although commercially viable PHA production in plants has not gained traction, which has been attributed largely to competing production methods. The production of PHA polymer such as 3-hydroxybutyrate with a fraction of longer-chain monomers (C6 or higher) was tested in plants. Arabidopsis expressing the PHA synthases in the peroxisomes produced a maximum of 0.02–0.04% plant dry weight (Arai et al. 2002; Matsumoto et al. 2006). The same process in bacteria (*Pseudomonas oleovorans*) resulted in over 50% PHA accumulation (Lageveen et al. 1988). Thus, efforts are currently on the way to improve PHA production in plants, by further investigating various types of PHA production in different plant species, subcellular compartment sites, and innovative ways to enhance the synthesis (Van Beilen and Poirier 2008).

Sorbitol is another versatile chemical that is on the list of the top 12 most useful chemical building blocks (Werpy and Petersen 2004). Sorbitol is a polyol that is used as a sweetener and as a monomer for PET-like polyesters such as polyethylene isosorbide. Sorbitol could be produced in sugar cane at levels up to 61% of soluble sugars in the leaves or 12% dry weight, but only 1% was found in the stalk pith (Chong et al. 2007). Although biomass yields (30–40% less) and plant health were affected, this study demonstrates that metabolic intermediates of the sucrose biosynthetic pathway can be effectively diverted for the synthesis of alternative valuable compounds (Van Beilen and Poirier 2008).

The list of raw materials that can be harvested from plants is consistently growing and evolving as the push for more renewable manufacturing practice has advanced technologies in plant genetic engineering. In concert, technologies to utilize these renewable raw materials in space are also developing. Made in Space, Inc., a US-based company that specializes in engineering and manufacturing microgravity compatible 3D printers, has worked with NASA to develop the capability of 3D printing an array of products using many different types of materials in microgravity. Modern additive manufacturing allows 3D printing to be performed onboard the ISS or on future space craft which, will allow crew members to produce spare parts for spaceships on demand, equipment for expeditions, and food supplies. Made in Space's additive manufacturing facility is designed as a modular device that can be upgraded to increase functionality, and is compatible with over 30 different polymers, including space-rated, high-performance thermoplastics. For example, the Vulcan Advanced Hybrid Manufacturing (VULCAN) System combines several manufacturing technologies into one seamless system to produce precision metallic, polymer, and hybrid parts (https://madeinspace.us/capabilities-and-technology/vulcan/). Different extruders and machine tools can be exchanged on demand to 3D print products using multiple types of materials - some of these materials could be made by plants. Together, the development of genetically engineered plants and space-ready hardware will open up the possibilities for mission-dependent applications in space.

4.8 Engineering Plants for Spaceflight Application

Plants have long been held as important to missions of exploration (Halstead and Dutcher 1987; Ferl et al. 2002; Wolverton and Kiss 2009; Wheeler 2010; Ruyters and Braun 2014), and plants as components of spaceflight and planetary life support have been specifically called out for more than a decade of NASA imperatives. First among these is the Advanced Life Support Project Plan (Russo and Henninger 2002), then the Mars Exploration Program Advisory Group (MEPAG – http://mepag.nasa.gov/) (Taylor 2004), and in the last Decadal Survey (Decadal-Survey-Committee 2011).

Transcriptomic studies have revealed how plants respond to the spaceflight environment and have also begun to reveal how some genes have more important roles in the physiological adaptation to that environment (Paul et al. 2013; Ferl et al. 2015) (Figure 4.4). Certain cultivars are better at physiologically adapting to the spaceflight environment, and altering a single gene could improve spaceflight adaptation (Paul et al. 2017). Similarly, analog environment studies have revealed

Figure 4.4 Scientists Dr. Rob Ferl and Dr. Anna-Lisa Paul at the University of Florida have been conducting research on how plants respond to the spaceflight environment for the past 30+ years. They have flown plants (Mostly *Arabidopsis thaliana* and others) on the ISS, Zero G parabolic flights, Starliner flighter jets, and other remote environments on Earth, like Antarctica and the Arctic. By using telemetric hardware and genetically engineered plants they can investigate how plants regulate their genes in response to these environments. Source: UF Space Plants Lab.

which genes in Arabidopsis are important for successful adaptation to the types of closed habitats that are envisioned for Mars and lunar greenhouses (Paul et al. 2004; Paul and Ferl 2006), (Figure 4.5) and the possible impact of in situ resources, such as regolith (Visscher et al. 2010).

The concept of developing new environmentally tailored crops for spaceflight application has not been given much consideration when compared to the many ways humans use this approach in the terrestrial context. Engineering plants to thrive under various spaceflight mission-relevant scenarios will have the potential to enhance mission success and to also revolutionize terrestrial horticulture particularly within spaces shared with humans – green living spaces. Primarily plants would be designed to thrive well in a vehicular microgravity environment such as the ISS, cis-lunar habitats, and transit to Mars, where metabolically demanding, gravity-defying structures can be removed. Spaceflight vehicles and habitats for the coming decades will not employ large-scale artificial gravity, placing microgravity as the single major element for redefining plant structures and metabolism. In these scenarios, metabolically demanding and gravity-defying structures can be removed. Microgravity-engineered plants could perhaps allow the elimination of unneeded components such as lignin and morphologies that normally direct photosynthetic carbon to gravity-defying compounds and structures that are wasteful or otherwise difficult to process postharvest. One focus might be to redirect carbon away from structures useless in microgravity, and toward fruits and vegetables that are readily edible (highlighted in the FT1 plum example). The other longer-term scenario under consideration involves planetary habitats where production volumes and in situ resources are limited and specialized, but where gravity again plays a role. In addition, surface habitat deployment designs change focus to highly defined plant growth, requiring that all structures that intercept light be maximized, photosynthetic organelle position be optimized to intercept light, and that production in the limited volumes be maximized. Here, light and nutrient gathering segments can be separated from the plant and from the fruits and edible tissues that can be harvested without disrupting the production facilities of the plant. Within the contexts of the mission scenarios, the biology of the plants will be altered by gene editing, transformation, and breeding (based on the informatics toolboxes of DNA regulatory sequences), and habitat light regimes will be tailored to precisely control metabolism, development, and morphology.

Apart from optimizing plants to grow in novel microgravity and planetary habitats, plants can be synthetically engineered to produce raw materials required for in-space manufacturing. Manufacturing in space allows us to progress toward a sustainable presence both in transit and on planetary surfaces. To date, many space manufacturing companies are investing in testing their capacity to develop or enhance production in the microgravity environment https://www.factoriesinspace.com/manufacturing-companies. Efforts in this list, include 3D printing food, material processing, and growing organs from stem cells. Being able to not only perform these processes under novel environments but also harvest the raw materials needed will allow for a more self-sustaining manufacturing process that is needed as humans strive to be independent of Earth's resources.

Figure 4.5 *Arabidopsis thaliana* is the model plant organism that allows scientists to investigate the molecular responses of plants under novel environments. Due to their small size, they can be grown in an array of configurations that enable scientists to investigate their response to various environmental conditions. In addition, since their genomes are well annotated, investigations of their transcriptome and genetically altering specific genes for further investigations can be easily performed. Source: UF Space Plants Lab.

Many plant-based compounds have huge applications for space manufacturing. PLA (polylactic acid) plastics that can be used as material for 3D printers are manufactured by fermentation of polysaccharides out of plant-based resources such as corn, potatoes, or sugar canes/beets (Murariu and Dubois 2016). In addition, the company Made in Space, also uses Green PE (polyethylene) in their AMF onboard the ISS (https://madeinspace.us/news-product/amf/). Green PE is a renewable plastic produced from sugarcane and is used in a wide array of applications from building tools to replacement parts on station. The presence of synthetic plants producing the raw materials for these applications will be beneficial for advancing space manufacturing goals.

Advances in 3D printing technology have provided other beneficial applications such as printing food for the crew on long-duration space missions. NASA has funded the development of a prototype 3D food printer so that astronauts may one day enjoy a 3D-printed pizza on Mars (https://www.space.com/21250-nasa-3d-food-printer-pizza.html). By extruding sugars, starch, and/or proteins it is possible to use various concoctions to design and print food for space missions (Terfansky and Thangavelu 2013). Moisture from these food components is removed and the food powder is stored in prepackaged cartridges that are expected to have shelf lives of more than 30 years. In addition, the food printed can be customized with essential nutrients and specific dietary requirements for each crew member on their long-duration missions (Lin 2015). Zero G Kitchen is a company that has developed and demonstrated the Space Oven, which is a cylinder-shaped insulated container designed to hold and bake food samples in the microgravity environment of the International Space Station (https://www.factoriesinspace.com/zero-g-kitchen). Once the pizza is 3D printed, it could be cooked in the Space Oven without the need for lasers to cook the pizza as it is printing. Continuous innovations such as these would elevate the in-space food manufacturing industry which will increase in demand as humans embark on longer and further space missions.

4.9 Discussion

Plants will play a prominent role in human space exploration agendas. However, the current state of the art in plant production concepts for exploration life support is completely embedded within the horticulture of plants that have been in production for generations of human use in terrestrial contexts. This creates a conundrum for exploration utilization as the plants are not optimized for the conditions of their production and engineering struggles to keep plants healthy and productive. In addition, there is a niche in space exploration to utilize plants in a multipurpose fashion. For example, while plants are contributing to bioregenerative life support systems during their development, once mature, the plants could then be harvested for renewable raw materials which could be used in an array of in-space manufacturing processes. This ensures that limited resources used to grow plants are maximized with high efficiency.

Genetically engineering plants to enable them to thrive and produce raw materials needed for various mission objectives will be the best approach to ensure successful and efficient cultivation. Space exploration will have wholly new plants optimized for space, thereby creating entirely new equations in the life support decision. In addition to the potential impacts for the exploration mission, such concepts will also create an avenue for Earth benefits generating enthusiasm that draws upon this exploration innovation – many modern plant applications, such as living walls, indoor gardens, and nutraceutical production, all currently struggle to make traditional field plants thrive in specialized production circumstances where those plants simply were never meant to grow. Currently available are the genetic tools and genomic insights necessary for a credible chance at truly revolutionizing the way preparations for spaceflight missions are conducted.

The concept is familiar and virtually universally embraced in science fiction; science fiction writers take it for granted that plant genetic engineering will be part of the human exploration equation. And their readers do as well. Successes on this front will have as huge an impact on public space exploration dreams as the development of the rocket to take us there, especially as genetic editing is more widely applied.

The immediate future goals will be to genetically engineer a foundation plant for spaceflight missions while researching the mechanisms to fulfill the complete concept and engineering principles that can include planetary surface mission scenarios. Perhaps altering genes associated with spaceflight stress responses that may not be needed for the physiological adaptation to that environment would help define the foundation plant (Paul et al. 2012a, b, 2013; Ferl et al. 2015). Chief among these are genes associated with secondary cell wall formation. Altering the cell wall composition to reduce the amount of carbon that is used to maintain secondary structures that are not necessary for support in microgravity may result in enhanced growth responses in space. It is necessary to strike a balance that maintains sufficient structure to keep the plants healthy while recognizing that altering the cell wall can also compromise the resistance to pathogen infection (e.g. Zhao and Dixon 2014). Once a successfully adapted plant is created, incorporating gene cassettes with raw material production pathways can then be inserted. Over time, a synthetic plant that thrives and produces renewable materials in the spaceflight environment will be established. The next steps will then be to address other important facets of growing plants extraterrestrially; these include topics such as protecting plants from space radiation, pollination methods, and prevention of genetic drifts. All these considerations will be necessary to ensure reliable and continuous availability of raw materials for in-space manufacturing.

4.10 Conclusion

In these early days of expanding sustained human presences in space, there is still an immense amount of leg work that needs to be done before a self-sustaining plant-reliant manufacturing facility can be established in low-earth orbit, cis-lunar orbit, or on a planetary base. However, efforts to establish these spaceflight adapted plants should begin as such procedures can be arduous and requires a lot of trial and errors. Numerous spaceflight missions with plants have been conducted and familiarity with the effects of microgravity and the unique constraints of mission structures and architectures have been acquired. Thus, a list of candidate genes for spaceflight-adapted engineered plants can be easily obtained. Starting the process now and testing plants adaptability onboard the ISS would ensure that engineered plants are available when needed. It is also possible that with new plants optimized for spaceflight scenarios, hardware designs can be freed from terrestrial heritages and better designs can be established to maximize the available resources.

Genetically engineered plants developed for enhanced vigor in the various spaceflight environments will be the first piece of the puzzle, hardware development and mission-directed production of raw materials will then follow. It will be enlightening to see the specialization of plants for each of the various spaceflight environments; however, being able to produce resources required as humanity establishes our home away from home takes the cake.

References

Aharoni, A., Giri, A.P., Deuerlein, S. et al. (2003). Terpenoid metabolism in wild-type and transgenic arabidopsis plants. *The Plant Cell* 15 (12): 2866–2884.

Agrawal, A.A. and Konno, K. (2009). Latex: a model for understanding mechanisms, ecology, and evolution of plant defense against herbivory. *Annual Review of Ecology, Evolution, and Systematics* 40: 311–331.

Arai, Y., Nakashita, H., Suzuki, Y. et al. (2002). Synthesis of a novel class of polyhydroxyalkanoates in Arabidopsis peroxisomes, and their use in monitoring short-chain-length intermediates of β-oxidation. *Plant and Cell Physiology* 43: 555–562.

Beyer, P., Al-Babili, S., Ye, X. et al. (2002). Golden rice: introducing the β-carotene biosynthesis pathway into rice endosperm by genetic engineering to defeat vitamin A deficiency. *The Journal of Nutrition* 132: 506S–510S.

Blein, T., Pautot, V., and Laufs, P. (2013). Combinations of mutations sufficient to alter Arabidopsis leaf dissection. *Plants (Basel)* 2: 230–247.

Borrelli, G.M. and Trono, D. (2016). Molecular approaches to genetically improve the accumulation of health-promoting secondary metabolites in staple crops – a case study: the lipoxygenase-B1 genes and regulation of the carotenoid content in pasta products. *International Journal of Molecular Sciences* 17: 1177.

Capehart, T. and Proper, S. (2019). USDA – Corn is America's Largest Crop in 2019. https://www.usda.gov/media/blog/2019/07/29/corn-americas-largest-crop-2019 (accessed 14 April 2021).

Chen, Q. and Davis, K.R. (2016). The potential of plants as a system for the development and production of human biologics. *F1000Research* 5: F1000 Faculty Rev-912. https://doi.org/10.12688/f1000research.8010.1.

Chong, B.F., Bonnett, G.D., Glassop, D. et al. (2007). Growth and metabolism in sugarcane are altered by the creation of a new hexose-phosphate sink. *Plant Biotechnology Journal* 5: 240–253.

Curtis, I.S., Power, J.B., and Davey, M.R. (1995). NPTII assays for measuring gene expression and enzyme activity in transgenic plants. In: *Plant Gene Transfer and Expression Protocols* (ed. H. Jones), 149–159. Totowa, NJ: Springer.

De Framond, A.J., Bevan, M.W., Barton, K.A., Flavell, R., and Chilton, M.-D. (1983). Mini-Ti plasmid and a chimeric gene construct: New approaches to plant gene vector construction. *MIAMI Winter Symposium 1983.*

Decadal-Survey-Committee (2011). *The Decadal Survey on Biological and Physical Sciences in Space – National Research Council: Recapturing a Future for Space Exploration: Life and Physical Sciences Research for a New Era*, National Research Council. The National Academies Press http://www.nap.edu/catalog.php?record_id=13048.

Dent, M., Hurtado, J., Paul, A.M. et al. (2016). Plant-produced anti-dengue virus monoclonal antibodies exhibit reduced antibody-dependent enhancement of infection activity. *The Journal of General Virology* 97: 3280–3290.

Diamos, A.G., Hunter, J.G.L., Pardhe, M.D. et al. (2020). High level production of monoclonal antibodies using an optimized plant expression system. *Frontiers in Bioengineering and Biotechnology* 7: 472, 1–15.

Dilworth, L.L., Riley, C.K., and Stennett, D.K. (2017). Chapter 5 – Plant constituents: carbohydrates, oils, resins, balsams, and plant hormones. In: *Pharmacognosy* (ed. S. Badal and R. Delgoda), 61–80. Boston, MA: Academic Press.

Dutt, M., Dhekney, S.A., Soriano, L. et al. (2014). Temporal and spatial control of gene expression in horticultural crops. *Horticulture Research* 1: 14047.

Ellison, S.L., Luby, C.H., Corak, K.E. et al. (2018). Carotenoid presence is associated with the Or gene in domesticated carrot. *Genetics* 210: 1497–1508.

Ferl, R., Wheeler, R., Levine, H.G., and Paul, A.L. (2002). Plants in space. *Current Opinion in Plant Biology* 5: 258–263.

Ferl, R.J., Koh, J., Denison, F., and Paul, A.-L. (2015). Spaceflight induces specific alterations in the proteomes of Arabidopsis. *Astrobiology* 15: 32–56.

Graham, T., Scorza, R., Wheeler, R. et al. (2015). Over-expression of FT1 in Plum (*Prunus domestica*) results in phenotypes compatible with spaceflight: a potential new candidate crop for bio-regenerative life-support systems. *Gravitational and Space Research* 3 (1): 39–50.

Halstead, T.W. and Dutcher, F.R. (1987). Plants in space. *Annual Review of Plant Physiology* 38: 317–345.

Hurtado, J., Acharya, D., Lai, H. et al. (2020). In vitro and in vivo efficacy of anti-chikungunya virus monoclonal antibodies produced in wild-type and glycoengineered *Nicotiana benthamiana* plants. *Plant Biotechnology Journal* 18: 266–273.

Kew, R.B.G. (2016). *State of the World's Plants.* https://stateoftheworldsplants.org/2016/ (accessed 26 April 2021).

Lageveen, R.G., Huisman, G.W., Preusting, H. et al. (1988). Formation of polyesters by *Pseudomonas oleovorans*: effect of substrates on formation and composition of poly-(R)-3-hydroxyalkanoates and poly-(R)-3-hydroxyalkenoates. *Applied and Environmental Microbiology* 54: 2924–2932.

Langenheim, J.H. (1990). Plant resins. *American Scientist* 78: 16–24.

Lazzeri, L., Mazzoncini, M., Rossi, A. et al. (2006). Biolubricants for the textile and tannery industries as an alternative to conventional mineral oils: an application experience in the Tuscany province. *Industrial Crops and Products* 24: 280–291.

Lin, C.-Y. and Eudes, A. (2020). Strategies for the production of biochemicals in bioenergy crops. *Biotechnology for Biofuels* 13: 71.

Lin, C. (2015). 3D food printing: a taste of the future. *Journal of Food Science Education* 14: 86–87.

Liu, X.X., Lang, S.R., Su, L.Q. et al. (2015). Improved Agrobacterium-mediated transformation and high efficiency of root formation from hypocotyl meristem of spring *Brassica napus* 'Precocity' cultivar. *Genetics and Molecular Research* 14: 16840–16855.

Marusic, C., Pioli, C., Stelter, S. et al. (2018). *N*-glycan engineering of a plant-produced anti-CD20-hIL-2 immunocytokine significantly enhances its effector functions. *Biotechnology and Bioengineering* 115: 565–576.

Matsumoto, K.I., Arai, Y., Nagao, R. et al. (2006). Synthesis of short-chain-length/medium-chain-length polyhydroxyalkanoate (PHA) copolymers in peroxisome of the transgenic *Arabidopsis thaliana* harboring the PHA synthase gene from *Pseudomonas* sp. 61-3. *Journal of Polymers and the Environment* 14: 369–374.

Menassa, R., Zhu, H., Karatzas, C.N. et al. (2004). Spider dragline silk proteins in transgenic tobacco leaves: accumulation and field production. *Plant Biotechnology Journal* 2: 431–438.

Miao, S., Wang, P., Su, Z., and Zhang, S. (2014). Vegetable-oil-based polymers as future polymeric biomaterials. *Acta Biomaterialia* 10: 1692–1704.

Murariu, M. and Dubois, P. (2016). PLA composites: from production to properties. *Advanced Drug Delivery Reviews* 107: 17–46.

Paul, A.-L., Amalfitano, C.E., and Ferl, R.J. (2012a). Plant growth strategies are remodeled by spaceflight. *BMC Plant Biology* 12: 232.

Paul, A.-L. and Ferl, R. (2006). The biology of low atmospheric pressure-implications for exploration mission design and advanced life support. *Gravitational and Space Biology* 19: 3–17.

Paul, A.-L., Schuerger, A.C., Popp, M.P. et al. (2004). Hypobaric biology: Arabidopsis gene expression at low atmospheric pressure. *Plant Physiology* 134: 215–223.

Paul, A.-L., Sng, N.J., Zupanska, A.K. et al. (2017). Genetic dissection of the Arabidopsis spaceflight transcriptome: are some responses dispensable for the physiological adaptation of plants to spaceflight? *PLoS One* 12: e0180186.

Paul, A.-L., Zupanska, A.K., Schultz, E.R., and Ferl, R.J. (2013). Organ-specific remodeling of the Arabidopsis transcriptome in response to spaceflight. *BMC Plant Biology* 13: 112.

Paul, A.L., Zupanska, A.K., Ostrow, D.T. et al. (2012b). Spaceflight transcriptomes: unique responses to a novel environment. *Astrobiology* 12: 40–56.

Pennisi, E. (2016). The plant engineer. *Science* 353: 1220–1224.

Russo, D.M. and Henninger, D.L. (2002). Advanced life support project plan. *NASA-JSC CTSD-ADV-348 Rev C – JSC 39168*.

Ruyters, G. and Braun, M. (2014). Plant biology in space: recent accomplishments and recommendations for future research. *Plant Biology* 16: 4–11.

Scheller, J., Gührs, K.-H., Grosse, F., and Conrad, U. (2001). Production of spider silk proteins in tobacco and potato. *Nature Biotechnology* 19: 573–577.

Scheller, J., Henggeler, D., Viviani, A., and Conrad, U. (2004). Purification of spider silk-elastin from transgenic plants and application for human chondrocyte proliferation. *Transgenic Research* 13: 51–57.

Shahbaz, U., Yu, X.-B., Akhtar, W. et al. (2020). Golden rice to eradicate the Vitamin A deficiency in the developing countries. *European Journal of Nutrition & Food Safety* 53–63.

Sharma, B.K. and Biresaw, G. (2016). *Environmentally Friendly and Biobased Lubricants*, 1e (ed. Kenneth M. Doll, Bryan R. Moser, Zengshe Liu, and Rex E. Murray), CRC Press https://doi.org/10.1201/9781315373256.

Sharma, M., Gupta, S., and Mondal, A. (2012). Production and trade of major world oil crops. In: *Technological Innovations in Major World Oil Crops*, vol. Vol. 1, 1–15. Springer.

Smyth, D.R. (1995). Flower development: origin of the cauliflower. *Current Biology* 5: 361–363.

Srinivasan, C., Dardick, C., Callahan, A., and Scorza, R. (2012). Plum (*Prunus domestica*) trees transformed with poplar FT1 result in altered architecture, dormancy requirement, and continuous flowering. *PLoS One* 7: e40715.

Sun, H., Chen, Q., and Lai, H. (2018). Development of antibody therapeutics against flaviviruses. *International Journal of Molecular Sciences* 19: 54.

Taylor, C.B. (1998). Factories of the future? Metabolic engineering in plant cells. *The Plant Cell* 10: 641–644.

Taylor, G.J. (2004). MEPAG (Mars Exploration Program Analysis Group) – scientific goals, objectives, investigations, and priorities. http://mepag.jpl.nasa.gov/reports .cfm.

Terfansky, M.L. and Thangavelu, M. (2013). 3D printing of food for space missions. *AIAA SPACE 2013 Conference and Exposition*.

US Food and Drug Administration. (2020). GMO crops, animal food, and beyond. https://www.fda.gov/food/agricultural-biotechnology/gmo-crops-animal-food-and-beyond (accessed 14 May 2021).

USDA. (2020). USDA APHIS|2020: The Year Plant Health Awareness Goes Global! https://www.aphis.usda.gov/aphis/ourfocus/planthealth/ppq-program-overview/plant-protection-today/articles/iyph (accessed 26 March 2021).

Van Beilen, J.B. and Poirier, Y. (2008). Production of renewable polymers from crop plants. *The Plant Journal* 54: 684–701.

Van Poppel, G., Verhoeven, D.T.H., Verhagen, H., and Goldbohm, R.A. (1999). Brassica vegetables and cancer prevention. In: *Advances in Nutrition and Cancer*, 2e (ed. V. Zappia, F.D. Ragione, A. Barbarisi, et al.), 159–168. Boston, MA: Springer US.

Vaughan, J.G. (1977). A multidisciplinary study of the taxonomy and origin of Brassica crops. *Bioscience* 27: 35–40.

Visscher, A.M., Paul, A.L., Kirst, M. et al. (2010). Growth performance and root transcriptome remodeling of Arabidopsis in response to Mars-like levels of magnesium sulfate. *PLoS One* 5: e12348.

Wagner, H., Luther, R., and Mang, T. (2001). Lubricant base fluids based on renewable raw materials: their catalytic manufacture and modification. *Applied Catalysis A: General* 221: 429–442.

Werpy, T. and Petersen, G. (2004). *Top Value Added Chemicals from Biomass: Volume I – Results of Screening for Potential Candidates from Sugars and Synthesis Gas.* Golden, CO: National Renewable Energy Laboratory.

Wheeler, R.M. (2010). Plants for human life support in space: from Myers to Mars. *Gravitational and Space Biology* 23: 25–35.

Wolverton, C. and Kiss, J.Z. (2009). An update on plant space biology. *Gravitational and Space Biology* 22: 13–20.

Wu, S. and Chappell, J. (2008). Metabolic engineering of natural products in plants; tools of the trade and challenges for the future. *Current Opinion in Biotechnology* 19: 145–152.

Wurtzel, E.T. and Kutchan, T.M. (2016). Plant metabolism, the diverse chemistry set of the future. *Science* 353: 1232–1236.

Xia, Y. and Larock, R.C. (2010). Vegetable oil-based polymeric materials: synthesis, properties, and applications. *Green Chemistry* 12: 1893–1909.

Yang, J., Barr, L.A., Fahnestock, S.R., and Liu, Z.-B. (2005). High yield recombinant silk-like protein production in transgenic plants through protein targeting. *Transgenic Research* 14: 313–324.

Zeitlin, L., Pettitt, J., Scully, C. et al. (2011). Enhanced potency of a fucose-free monoclonal antibody being developed as an Ebola virus immunoprotectant. *Proceedings of the National Academy of Sciences* 108: 20690–20694.

Zhao, Q. and Dixon, R.A. (2014). Altering the cell wall and its impact on plant disease: from forage to bioenergy. *Annual Review of Phytopathology* 52: 69–91.

5

Stem Cell Biology and Tissue Engineering in Space

Madelyn Arzt[1,2], *Amelia Jenkins*[1,2], *and Arun Sharma*[1,2,3]

[1] *Board of Governors Regenerative Medicine Institute, Cedars-Sinai Medical Center, 8700 Beverly Blvd, Los Angeles, CA 90048, USA*
[2] *Smidt Heart Institute, Cedars-Sinai Medical Center, 8700 Beverly Blvd, Los Angeles, CA 90048, USA*
[3] *Department of Biomedical Sciences, Cedars-Sinai Medical Center, 8700 Beverly Blvd, Los Angeles, CA, USA*

5.1 Historical Context

Over the past 15 years, stem cell research has become a major focus area in biomedical research (Zakrzewski et al. 2019). Stem cells and their derived cell types can model and alleviate human diseases, and improved utilization of these cells is an ongoing goal. Stem cells can transform, or differentiate, into different cell types of a specific adult tissue lineage (multipotent stem cells) or into nearly all somatic cell types (pluripotent stem cells) that are found within adult humans. Recently, human-induced pluripotent stem cells (hiPSCs), often made from a small sample of primary adult skin or blood tissue instead of derived from embryos, have alleviated some ethical concerns traditionally associated with embryonic stem cell research (Rowe and Daley 2019). Given the tremendous differentiation potency associated with both multipotent and pluripotent stem cells, academic and commercial entities have begun widely using stem cells for *in vitro* and *in vivo* applications in regenerative medicine.

During the 2000s, the United States Congress developed and implemented the International Space Station National Laboratory (ISSNL) to maximize ISS employment toward conducting scientific research aimed toward the benefit of humans on Earth. Ideally, this extraterrestrial laboratory environment can be used by commercial entities, academic institutions, or private enterprises. In collaboration with the United States and select international partners, the ISS provides a persistent microgravity environment in low-Earth orbit (LEO), and is a unique research laboratory for studying the impact of microgravity on biology and physiology. Stem cell biology has recently been investigated in the context of microgravity physiology and basic biology research to determine the potential roles of microgravity in altering properties such as stem cell proliferation, differentiation, culture, or potency (Grimm et al. 2018) (Figure 5.1). Various stem cell microgravity research studies have been conducted over the years, either using artificial microgravity simulators on Earth or,

In-Space Manufacturing and Resources: Earth and Planetary Exploration Applications, First Edition.
Edited by Volker Hessel, Jana Stoudemire, Hideaki Miyamoto, and Ian D. Fisk.

more recently, in LEO aboard the International Space Station. Future private space stations under development will also serve as LEO research laboratories. Microgravity research has contributed to augmenting the fundamental knowledge of stem cell function, regeneration, and differentiation (Vunjak-Novakovic et al. 2002). Bioengineering experiments in LEO, in combination with parallel ground-based studies on Earth, have shown that microgravity impacts stem cell proliferation and differentiation rates (Yuge et al. 2006). Printing biological tissues using stem cell-derived cell types in microgravity coupled with low-viscosity bioinks may enable the production of custom biomaterials. Both stem cell function and tissue engineering rely heavily on biomechanical properties impacted by gravity. A focus of such studies has been on stem cell populations associated with bone and muscle, since atrophy of these tissue types is a major physiological side-effect experienced after extended exposure to microgravity. There is much work that remains to be conducted in the realm of examining the impact of microgravity on fundamental stem cell properties, modeling human physiology during spaceflight, or enhancing tissue biomanufacturing. (Berthiaume et al. 2011).

5.2 Stem Cells and Space Flight

Stem cells and stem cell–derived tissues have potential as research tools and therapeutic products (Sharma et al. 2020). Recently, there has been much interest in using microgravity to potentially enhance stem cell properties such as pluripotency and multipotency, division, differentiation, and/or maturation of stem cell derivatives. One method toward achieving these goals is employing LEO platforms, such as the ISS, to cultivate stem cells in microgravity. The ability of microgravity to alter stem cell function is documented in peer-reviewed scientific literature showing that sustained microgravity influences the aforementioned stem cell properties (Table 5.1). Stem cells are in part defined by their potency, or the ability to give rise to derivative cells either within a particular lineage of interest (multipotency) or across multiple germ lineages (pluripotency). For example, multipotent hematopoietic stem cells can give rise to many lineages of blood cells, but cannot typically give rise to other tissue types. Similarly, pluripotent stem cells are able to transform into nearly all cells in the body except for the extraembryonic placental tissues, although recent developments in naïve state pluripotency enable the production of these extraembryonic tissues as well (Theunissen et al. 2016). Stem cell differentiation protocols have rapidly evolved over the past 3 decades, enabling efficient production of a variety of stem cell derivatives (Engel et al. 2016). However, a major common issue in the stem cell field is variability in stem cell potency and differentiation capacity between cell lines. This often detrimentally influences experimental reproducibility in academic and clinical applications. Thus, it is critical to identify novel methodologies that will either maintain or enhance the potency and differentiation capacities of multipotent and pluripotent stem cell lines. These improvements in stem cell properties will have an impact on biomanufacturing of stem cell-derived products from a research and development perspective but also for clinical and therapeutic applications.

Table 5.1 Relevant publications on the impacts of microgravity on stem and progenitor cell function.

Model system	Summary	Simulated or spaceflight	References
Adipose-derived stem cells	Proliferation in Microgravity	Simulated	(Mashiko et al. 2021)
Adipose-derived stem cells	spheroid formation of adipose-derived stem cells in microgravity enhances their stemness properties and increases their therapeutic potential	Simulated	(Zhang et al. 2015)
Adipose-derived stem cells	Functional tissue-engineered microtissue derived from cartilage extracellular matrix for articular cartilage regeneration	Simulated	(Yin et al. 2018)
Bone marrow stromal cells	Sonic hedgehog promotes chondrogenesis of rabbit bone marrow stem cells in a rotary cell culture system	Simulated	(L. Chen et al. 2019)
Cancer stem cells	Simulated microgravity increases polyploid giant cancer cells and nuclear localization of YAP	Simulated	(Arun et al. 2019)
Cancer stem cells	PTEN/FOXO3/AKT pathway regulates cell death and mediates morphogenetic differentiation of Colorectal Cancer Cells under Simulated Microgravity	Simulated	(Arun et al. 2017)
Cardiovascular progenitor cells	Cells cultured under microgravity conditions displayed upregulation of downstream genes involved in the Hippo pathway	Simulated and spaceflight	(Camberos et al. 2019)
Cardiovascular progenitor cells	Spaceflight Activates Protein Kinase C Alpha Signaling and Modifies the Developmental Stage of Human Neonatal Cardiovascular Progenitor Cells	Spaceflight	(Baio et al. 2018a)
Cardiovascular progenitor cells	Simulated Microgravity and 3D Culture Enhance Induction, Viability, Proliferation and Differentiation of Cardiac Progenitors from Human Pluripotent Stem Cells	Simulated	(Jha et al. 2016)
Dental pulp stem cells	Three-dimensional simulated microgravity culture improves the proliferation and odontogenic differentiation of dental pulp stem cell in PLGA scaffolds implanted in mice	Simulated	(Li et al. 2017)
Embryonic stem cells	Space microgravity maintains stemness and long-term survival of embryonic stem cells, promising 3D aggregate formation	Spaceflight	(X. Lei et al. 2018)
Embryonic stem cells	LIF-free embryonic stem cell culture in simulated microgravity	Simulated	(Kawahara et al. 2009)

(Continued)

Table 5.1 (Continued)

Model system	Summary	Simulated or spaceflight	References
Embryonic stem cells	Microgravity simulation activates Cdc42 via Rap1GDS1 to promote vascular branch morphogenesis during vasculogenesis	Simulated	(S. Wang et al. 2017)
Embryonic stem cells	Microgravity reduces the differentiation and regenerative potential of embryonic stem cells	Spaceflight	(Blaber et al. 2015)
Endothelial progenitor cells	Conditioned media from cells in microgravity promote angiogenesis and bone fracture healing	Simulated	(Kong et al. 2021)
Endothelial progenitor cells	Increases in angiogenic factors, including vascular endothelial growth factor, under microgravity	Simulated	(Hagiwara et al. 2018)
Epidermal stem cells	rotary bioreactor enhances proliferation of human epidermal stem cells and supports formation of 3D epidermis-like structure	Simulated	(X. H. Lei et al. 2011)
Hematopoietic stem cells	Spaceflight/microgravity inhibits the proliferation of hematopoietic stem cells by decreasing Kit-Ras/cAMP-CREB pathway networks as evidenced by RNA-Seq assays	Simulated and spaceflight	(P. Wang et al. 2019)
hepatic and biliary tree stem/progenitor cells	Simulated microgravity promotes the formation of tridimensional cultures and stimulates pluripotency and a glycolytic metabolism in human hepatic and biliary tree stem/progenitor cells	Simulated	(Costantini et al. 2019)
hiPSC-Cardiomyocytes	Microgravity caused alterations in hiPSC-CM calcium handling and genes involved in metabolism	Spaceflight	(Wnorowski et al. 2019)
hiPSC-Cardiomyocytes	acute alterations of gravity significantly increase the beating rate	Simulated	(Acharya et al. 2019)
Mesenchymal stem cells	Bio-scaffold can alleviate microgravity-induced osteoblast dysfunction	Simulated	(Avitabile et al. 2020)
Mesenchymal stem cells	MSCs more potent immunosuppressive capacity in microgravity.	Spaceflight	(Huang et al. 2020)
Mesenchymal stem cells	Microgravity-cultured MSCs improve recovery following spinal cord ischemia in rats	Simulated	(Kurose et al. 2019)

Mesenchymal stem cells	Recovery of stem cell proliferation by low-intensity vibration under simulated microgravity requires LINC complex	Simulated	(Touchstone et al. 2019)
Mesenchymal stem cells	In vitro experiments confirmed that the encapsulated MSCs differentiated into osteoblastic cells, formed bone-like tissue under osteogenic microgravity bioreactor conditions	Simulated	(Koc Demir, Elcin, and Elcin 2018)
Mesenchymal stem cells	Simulated microgravity culture enhances the neuroprotective effects of human cranial bone–derived mesenchymal stem cells in traumatic brain injury	Simulated	(Otsuka et al. 2018)
Mesenchymal stem cells	Simulated microgravity significantly enhanced cartilaginous matrix formation in cocultures of primary meniscus cells and adipose-derived MSCs	Simulated	(Weiss et al. 2017)
Mesenchymal stem cells	Under microgravity, limbal fibroblasts showed a higher proportion of MSC characteristics and were easily induced into different linage cells	Simulated	(Pao et al. 2017)
Mesenchymal stem cells	Duration of simulated microgravity affects the differentiation of mesenchymal stem cells	Simulated	(Xue et al. 2017)
Mesenchymal stem cells	Mechanical unloading of bone in microgravity reduces mesenchymal and hematopoietic stem cell–mediated tissue regeneration	Simulated	(Blaber et al. 2014)
Mesenchymal stem cells	Simulated microgravity enhances multipotential differentiation capacity of bone marrow mesenchymal stem cells	Simulated	(N. Wang et al. 2014)
Mesenchymal stem cells	Microgravity potentiates stem cell proliferation while sustaining the capability of differentiation	Simulated	(Yuge et al. 2006)
Muscle stem cells	Microgravity influences maintenance of the human muscle stem/progenitor cell pool	Simulated	(Hosoyama et al. 2017)
Neural crest cells	Simulated microgravity impairs cardiac autonomic neurogenesis from neural crest cells	Simulated	(Hatzistergos et al. 2018)
Planarian worms	Altered gravity elicits cell homeostasis imbalance	Simulated	(Salvetti et al. 2021)

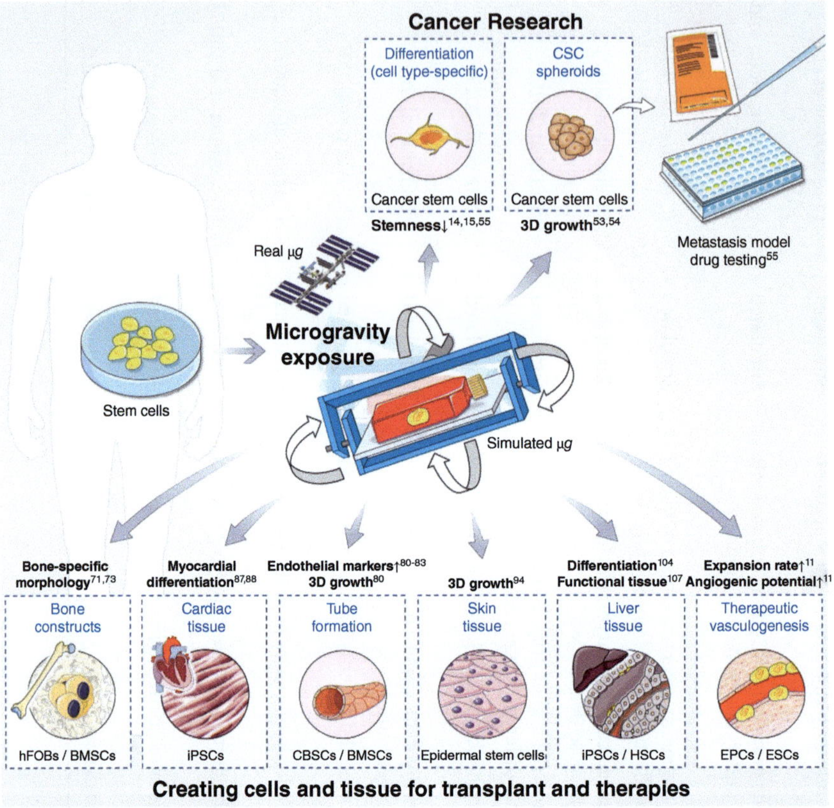

Figure 5.1 The effects of microgravity on differentiation and cell growth in stem cells and cancer stem cells. Effects of real or simulated microgravity on stem cell cultures (lower panel) and cancer stem cells (upper panel) together with the possible applications in medicine. Arrows indicate increases/reductions. BMSC, bone marrow–derived stem cell; CBSC, umbilical cord blood stem cell; CSC, cancer stem cell; EPC, endothelial progenitor cell; ESC, endothelial stem cell; hFOB, human fetal osteoblast; HSC, hepatic stem cell; iPSC, induced pluripotent stem cell; μg, microgravity. Parts of the figure are drawn using pictures from Servier Medical Art (https://smart.servier.com), licensed under a Creative Commons Attribution 3.0 Unported License (https://creativecommons.org/licenses/by/3.0). Source: STEM CELLS Translational Medicine, Volume 9, Issue 8, Pages: 882–894, First published: 30 April 2020, 10.1002/sctm.20-0084.

5.3 Expansion of Select Cell Types

While stem cells and stem cell-derived tissues hold great promise for use in research and as clinical-grade therapeutic agents, safe and efficient expansion of stem cells and their derivatives continues to be a major challenge. Multipotent mesenchymal stem cells (MSCs) are one of the most common stem cell types utilized within the field of regenerative medicine, even though significant concerns remain in regards to their heterogeneity, identity, derivation (Andrzejewska, Lukomska, and Janowski 2019). Recent studies have demonstrated that human MSCs can be cultivated aboard the ISS, and that in LEO, can exhibit improved immunosuppressive capabilities compared to MSCs grown on Earth (Huang et al. 2020). Additionally, MSCs cultured under simulated microgravity conditions on Earth have shown increased therapeutic potential in a pre-clinical model of traumatic brain injury (Otsuka et al. 2018). Investigations into the advantageous and deleterious effects of culturing pluripotent and multipotent stem cells in microgravity could identify novel molecular and cellular mechanisms capable of improving biomanufacturing or clinical therapy on Earth, or perhaps serve as a catalyst for future large-scale, in-orbit biomanufacturing or stem cell-derived products. Other major applications of stem cells for in vitro disease modeling, drug discovery, and cell therapy revolve around common diseases such as cardiovascular disease (Sharma, Wu, and Wu 2013). These diseases include cardiomyopathies, ischemic heart disease, stroke, and chronic obstructive pulmonary disease, which continue to grow as leading causes of mortality worldwide. Microgravity disease modeling experiments have indicated that this environment can lead to the identification of novel targets for enhancing the therapeutic benefit of heart muscle cells (cardiomyocytes) (Hughson, Helm, and Durante 2018), and have also demonstrated the feasibility of long-term microgravity culture of induced pluripotent stem cell–derived cardiomyocytes (Wnorowski et al. 2019) (Figure 5.2). Other cell types also hold significant potential for microgravity-induced expansion, but this remains to be explored. For example, hematopoietic stem cells (HSCs) have been used clinically as stem cell therapies for decades, and perhaps could be considered the greatest success story in the stem cell field. However, their clinical utility is limited by their scalability, availability, and compatibility with potential recipients. Unfortunately, the number of patients needing clinical therapy exceeds the population of appropriately matched donors. Significant work also remains to be done in optimizing *in vitro* expansion of either primary HSCs or stem cell–derived HSCs (Kunisaki and Frenette 2012). A long-investigated, but still sought-after permanent solution for these issues would come from being able to generate an

Figure 5.2 Astronaut Dr. Kate Rubins Culturing Beating iPSC-Derived Cardiomyocytes in Space. Source: NASA.

unlimited number of HSCs from pluripotent stem cells (Demirci, Leonard, and Tisdale 2020). Although limited data exist at this time, future experiments in sustained microgravity may uncover mechanisms to enable the mass expansion of stem cells and their derivatives, thus enabling the use of established clinically viable stem cell populations such as hematopoietic stem cells.

5.4 Altering Differentiation Capacity

Another area of discussion is whether a sustained microgravity environment can elicit a beneficial effect on stem cell differentiation, including differentiation of multipotent and pluripotent stem cells. Although differentiation protocols for a variety of stem cell populations have dramatically improved over the years, methods of improving the yield, purity, maturity, and function of stem cell–derived cells are all highly sought after. There are major questions of particular interest in regard to the impact of microgravity on stem cell differentiation. These include: (i) Could microgravity further enhance the differentiation efficiency and yield of stem cells to multiple somatic cell types? (ii) Could microgravity allow for the generation of cell types not currently possible to make from terrestrially based differentiation of stem cells? (iii) Could differentiation in microgravity alter the maturity or immaturity profiles of stem cell–derived cells? From terrestrial-based, basic research efforts, it is well understood that differences in culturing conditions and cell source can have significant impacts on stem cell differentiation (Zakrzewski et al. 2019). Stem cell differentiation capacity can be influenced by a variety of environmental cues such as cell culture nutrient quality, the stiffness of extracellular matrix on which 2D cultures are grown, cell passage number, freeze/thaw cycles, reagent expiration, and general human technical error during the stem cell manipulation process (Volpato and Webber 2020). Sourcing can also alter the capacity of stem cell function. For example, hiPSC production is notoriously variable with only select clones surviving the reprogramming process, which can yield less than 10% efficiency (Qin et al. 2014). The reprogramming process thus generates an inherent bottleneck into subsequent culture, whereby only genetically and epigenetically "superior" clonal populations are propagated during extended culture. All of these factors contribute to the general difficulty of developing highly reproducible differentiation protocols. With the recent push toward suspension-based and 3D cultures, this variability in differentiation has only been compounded. While there have been many studies elucidating the impacts of sustained microgravity on stem cells (Grimm et al. 2020), they can be difficult to interpret due to the variable nature of the conditions, equipment, and cell sources utilized. To fully evaluate how microgravity influences stem cell differentiation into the three primary germ layers (mesoderm, ectoderm, and endoderm), a comprehensive, parallel analysis of stem cell differentiation in LEO compared to groundside control conditions should be conducted utilizing real-time analysis techniques, -omics level analyses, and current state-of-the-art culture conditions. These can include modern bioreactor-based or 3D suspension culture, as well as high-throughput, real-time analyses of stem cell and differentiated cell integrity and function.

5.5 Altering Stem Cell Potency and Progenitor Cell Function

In similar fashion to the role of the environment on altering stem cell differentiation capacity, environmental stimuli can also alter stem cell potency, including the ability to self-renew and maintain the expression of key stem cell potency proteins, such as the critical pluripotency transcription factor OCT4 (Carey et al. 2011). Potency is closely linked to differentiation capacity, as highly pluripotent hiPSC lines are typically those with the highest capacity for differentiation into a lineage of interest. Some hiPSC lines, however, may be strongly pluripotent but epigenetically or transcriptionally inclined to differentiate into a germ layer of interest, such as mesoderm. Thus, the role of microgravity can be further investigated for its influence on these properties of pluripotency or multipotency. For example, microgravity may induce bone loss by inhibiting the differentiation capacity and cell cycle activity of osteoblasts, the canonical progenitor cells that create bone (Blaber et al. 2013; Louis et al. 2017). This has been documented in both in vitro and in vivo studies. Simulated microgravity can also alter the differentiation capacity, viability, and proliferative potential in cardiovascular progenitor cells derived from pluripotent stem cells (Baio et al. 2018a, b; Fuentes et al. 2015; Jha et al. 2016). The differentiation, morphology, and migration of mesenchymal stem cells are also changed in simulated microgravity, which may further be critical to osteoblast function since osteoblasts may be derived from MSCs (Z. Chen et al. 2016; Luna, Yew, and Hsieh 2015; Xue, Li, and Chen 2017). These microgravity-induced alterations in differentiation capacity could be the result of epigenetic shifts in methylation patterns and chromatin structure, which has been seen with rat skeletal myoblasts under simulated microgravity (Furukawa et al. 2018). Outside of muscle and bone-focused studies, fundamental stem cell culture conditions are reportedly altered in microgravity, such as the reduced requirement of leukemia inhibitory factor for maintaining pluripotency in mouse embryonic stem cells (Kawahara et al. 2009). In separate, confirmatory studies, spaceflight was shown to maintain pluripotency and long-term survival of mouse embryonic stem cells (X. Lei et al. 2018) while simultaneously inhibiting their capacity to differentiate (Blaber et al. 2015). In relation to hematopoietic stem cells, DNA damage repair pathways are inhibited in multipotent human hematopoietic stem cells, which also show impaired immune cell differentiation capacity (Low et al. 2018). Additionally, a handful of stem cell–derived, terminally differentiated tissues have been used to model the effects of microgravity on human physiology (Figure 5.3). For some cell types such as neural or cardiac tissues that cannot be readily obtained or expanded from primary tissue samples, pluripotent stem cell differentiation is the most efficient means by which to mass produce these cells for modeling the effect of microgravity on human cells. For example, primary neural and cardiac tissues are notoriously difficult to obtain and maintain long-term in active

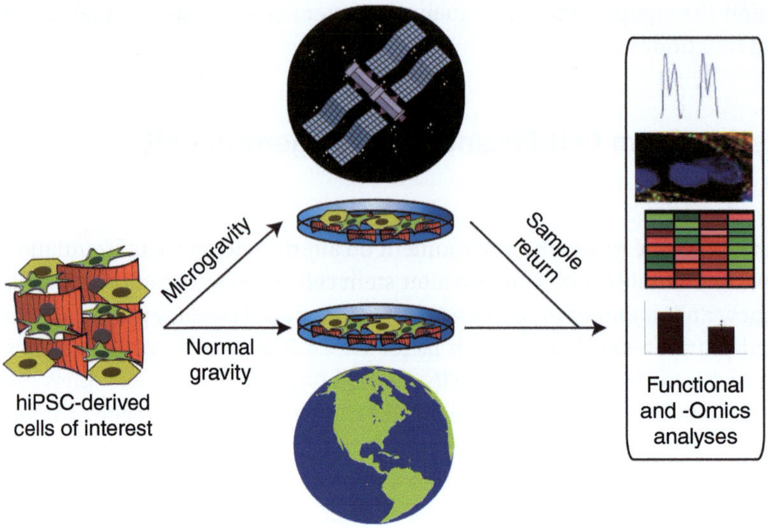

Figure 5.3 Example approach of hiPSC-based studies of cellular function in LEO. Source: Original content by Arun Sharma.

culture, necessitating the use of stem cell–derived neurons and cardiomyocytes to mass produce analogous neural and cardiac cells of interest. Neural organoids have been generated from human embryonic stem cells using a rotary cell culture system simulating microgravity (Mattei et al. 2018), and ongoing spaceflight studies continue to examine the effect of spaceflight on pluripotent stem cell–derived cardiomyocytes (heart muscle cells) (Wnorowski et al. 2019), neurons, and 3D models of neurodegenerative diseases using self-aggregating organoid cultures. For organoid modeling, microgravity may serve as a naturally accelerating environment for 3D organoid production through enhanced cellular aggregation, although more data are required before this hypothesis can be definitively proven.

5.6 Commercial Opportunities for Stem Cells and their Derivatives in LEO

Currently, space travel is expensive and traditionally largely funded by government organizations. However, with an increasing number of private and commercial enterprises working toward developing sub-orbital and orbital vehicles for research, development, and interplanetary exploration, there is a need to understand the influence of microgravity on human health for astronauts and future citizens in space, as well as for discovering new modalities to improve human health on Earth. The commercial spaceflight and LEO research market is poised to expand dramatically in the next decade, with stem cell research being near the forefront of research and development activities conducted in LEO. On Earth, stem cells and stem cell–derived products have two primary customers: the first being those who utilize them as research tools and the second being those who utilize them as therapeutic applications. In the near future, the largest return on investment may be derived from the data that can be translated to improving terrestrial development of stem cell–derived products. However, large-scale production of certain types of stem cells and stem cell–derived products could benefit from being done in a LEO facility, if the environment would confer certain advantages that could not be replicated Earthside. Given the short- and long-term potential of translational stem cell biology, this area of research would benefit in the near term from substantial interactions with government agencies and others enabling this fundamental work. The unique microgravity environment could provide novel insights that would have downstream impacts on the greater stem cell research and development community, eventually leading to translation into novel therapeutic options. A major catalyst here will be specific, successful use cases where large-scale manufacturing in space provides potential for returns on investment, leading to a domino effect in further commercial engagement. To nurture and prepare for these potential opportunities, discussions with biomanufacturing facilities, commercial space station developers, the stem cell and basic research community, and governmental agencies should begin immediately. These interactions should focus on the infrastructure needed to scale to LEO-based, good manufacturing processes (GMP)-grade manufacturing facilities and the requirements needed to enable regulatory approval for these facilities. There are also opportunities for private funding aimed at securing the unique intellectual property and data generated from conducting this research. Again, these possibilities hinge on critical successful "use-cases" where microgravity enhanced stem cell culture, function, differentiation, and other properties. The advent of these successful use-cases is critical to further commercial success in biomanufacturing and translational biomedical research in LEO.

Major gaps identified for potential downstream commercial opportunities center around the current relatively low amount of primary research data from stem cell experiments in LEO, as well as a need for universally accepted standards for stem cell culture, manipulation, and manufacturing. Much can be gleaned from Earthside research efforts where lack of standards, disparate data, and irreproducible results have hindered further research processes in the stem cell field. For example, groups like the International Stem Cell Initiative (ISCI) have worked to establish standards for assessing pluripotency of stem cell populations (Andrews et al. 2005) and the International Society for Stem Cell Research (ISSCR), the largest international organization devoted to stem cell research, has published guidelines to promote best practices in stem cell research and therapy (Soares et al. 2014). A research community engaging in LEO-based stem cell research and development should align with current best practices to ensure results are translatable. Similarly, the importance of cell source and its influence on results should be emphasized, and in general, more than one stem cell donor source should be utilized for the sake of experimental reproducibility. The sources of donor somatic cellular material for iPSC derivation, for example, should be widely available and well characterized using standard genetic analyses. Additionally, population-level genetic backgrounds should ideally be reflected in the experiments being conducted in LEO, to reflect the diverse populations that will ultimately experience spaceflight, and those who can benefit from potential therapeutic opportunities unlocked from LEO stem cell research. As is standard practice on Earth, hiPSC lines from a broad range of ethnic backgrounds should be utilized whenever possible.

The general lack of data on stem cell biology in LEO is an inherent issue due to the challenge of conducting research in a space-based laboratory. However, by adopting standards and applying recent advances in high-throughput experimental assays, miniaturization, microfluidics, robotics, machine learning (ML), and artificial intelligence (AI), the quality and quantity of data return would be amplified. Many of the technological advancements required to enable these space-based platforms would readily translate to terrestrial applications, such as the development of compact, robust, high-throughput systems capable of working autonomously and remotely. Thus, the commercial return on in-space biomedical research will extend beyond the development of new therapeutics and may also encompass the development of new research-facilitating technologies.

5.7 Tissue Engineering

While the emergence of hiPSCs and modern stem cell technologies provides an unprecedented and unique opportunity to derive patient-specific somatic tissues for disease modeling, drug discovery, and therapeutic applications, standard 2D cell monoculture does not always recapitulate a physiological tissue-like environment that harbors, for example, extracellular matrix enabling cellular assembly into complex multicellular structures. Tissue-engineered, multi-lineage 3D models that better recapitulate the architecture, microenvironment, and functionality of living human tissues and organs provide a solution to this research need (Sharma et al. 2020) (Figure 5.4). The International Space Station and future commercial LEO platforms offer an opportunity to study the effects of microgravity at a multiscale tissue level. Scientific and technological advances in stem cell biology, the development of tissue chips, organoids, assembloids, 3D bioprinting, and other tissue-engineered constructs (Sharma et al. 2020) are helping to accelerate drug discovery, advanced tissue modeling, and disease treatment on Earth, while bringing the possibility of bioengineering whole organs for transplantation closer to reality. Developments in each of these topics could be further accelerated using the unique microgravity environment onboard the International Space Station, although more research needs to be conducted to confirm these possibilities.

Recapitulating the biology of complex 3D multicellular structures is essential for a thorough appreciation of *in vivo* organ function and the advancement of biomimetic strategies to develop accurate, prognostic *in vitro* disease models. For example, to study the influence of microgravity on human health, it is critical to simulate the weightlessness and gravitational unloading that occurs during spaceflight. As mentioned previously, several tissues such as muscle, lung, bones and cartilage are altered under the presence of microgravity (Y. Chen et al. 2017; Kang et al. 2013; Stamenkovic et al. 2010; Ulbrich et al. 2014). Since exposing cells and tissues to

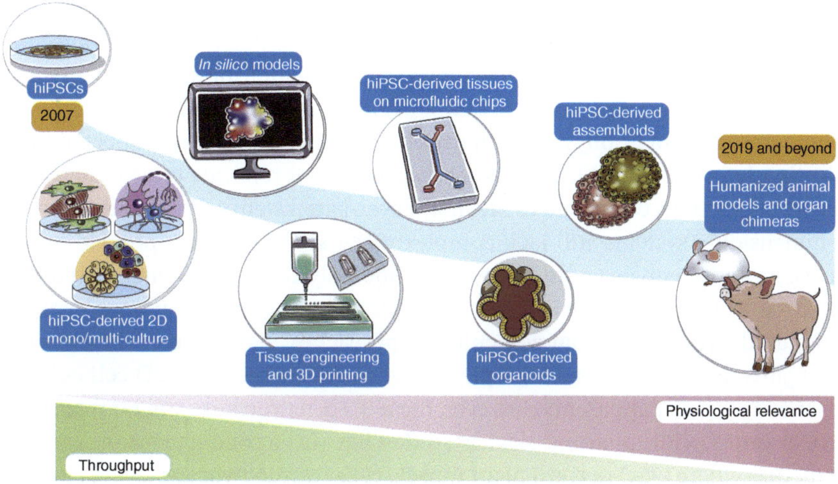

Figure 5.4 Evolution of Multi-lineage hiPSC-Derived Platforms and Comparison of Model Throughput versus Physiological Relevance. Since the discovery of hiPSCs in 2007, hiPSC-derived multi-lineage platforms have grown in complexity from simple, two-dimensional cultures to engineered platforms and complex organoids. These hiPSC-derived multi-lineage platforms strive to maintain a balance between high-throughput and high physiological relevance, with an ideal model exhibiting both characteristics. Future work may incorporate hiPSCs in humanized animal models and organ chimeras to enable human tissue production in large animals. Source: Cell Stem Cell 2020 26309–329. 10.1016/j.stem.2020.02.011.

microgravity in LEO for research purposes is typically costly, time consuming, and risky, most microgravity studies are performed using rotating bioreactors and other microgravity-simulating apparatuses on Earth. These systems mimic microgravity in part by eliminating the hydrostatic fluid stress exerted on the cells and tissues. Hydrostatic force mimics mechanical loading that the cells experience under standard gravity. Loading is critical for tissues like bone and cartilage and should be considered when developing advanced tissue-engineered constructs. Microgravity has been shown to inhibit cartilage development, but hydrostatic pressure and mechanical loading under normal gravity conditions enhance the cartilage formation up to threefold (Mellor et al. 2017). Conversely, human meniscus cells, when co-cultured with MSCs on a 3D collagen-harboring construct, enhanced the induction of chondrogenic differentiation under microgravity (Weiss et al. 2017). The discrepancies between these two studies may be due to several factors including the intercellular and cell–matrix interactions, three-dimensionality, or low-shear conditions in microgravity-mimicking bioreactors with randomized gravitational vectors that restrict proper perfusion of growth factors. Although several approaches have been developed to simulate microgravity on Earth, variability in cellular responses may not replicate the effect of true microgravity in LEO. This could be influenced by the complex effects induced by LEO microgravity, including promoting cell–cell aggregation, altering differentiation (Blaber et al. 2015), or affecting protein turnover (Jha et al. 2016).

In part due to high failure rates in pre-clinical drug development, there is a need for advanced, tissue-engineered, high-content *in vitro* systems that can serve as alternatives for non-human animal models currently employed for drug safety and efficacy testing. There is also an urgent need to develop organ-on-a-chip devices that can adequately represent human physiology, harboring multiple cell types connected by a pseudo-circulatory system. Although such systems better replicate intercellular and cell–ECM interactions compared to 2D monolayer-based stem cell cultures, current three-dimensional systems still lack complex vascular systems that provide nutrients, oxygen, and adequate waste removal. Additionally, these tissue-engineered systems can be technically unwieldy, less scalable, and are typically more expensive than standard two-dimensional cell culture approaches. As has been shown with tissue-engineered cardiac tissue, the physiological advantages over 2D cell culture approaches are readily apparent. However, these 3D constructs are still far from perfect. For example, while pharmacological studies have been conducted using cardiac tissue models where iPSC-derived heart muscle cells (cardiomyocytes) were aligned into confined compartments with realistic, biomimetic dimensions (Mathur et al. 2013, 2015), such systems do not adequately consider the natural 3D environment wherein the cells are enmeshed within the extracellular matrix. Critical components to perfectly replicate 3D architecture, namely physiologically relevant perfusion of nutrients, remain missing in most modern tissue-engineered constructs. Thus, the capacity to keep stem cell-derived tissue functional for extremely long periods of time and to enable longitudinal pharmacological studies remains elusive, especially within a microgravity context that only introduces another layer of technical complexity. Areas of completed, ongoing, and future studies in microgravity related to the bioengineering of human tissues and organs encompass multiple challenges in the field, some of which are also applicable to two-dimensional models. These concerns include issues with (i) stem cell proliferation and differentiation, (ii) biomarker and growth factor regulation, (iii) reliable 3D printing with biocompatible bio-inks and scaffolds, (iv) vascularization during thick tissue formation, (v) cell–matrix interactions and the mechanical loading environment, (vi) bioreactor engineering and design, and (vii) tissue survival and cryopreservation.

5.8 Conclusion

Ultimately, as humans spend more time in space, we must better understand the effects of spaceflight on human physiology and stem cell biology, using such model systems as two-dimensional stem cell derived tissues and three-dimensional tissue-engineered constructs. The long-term cell culture of advanced, highly specialized cells and tissues is now routinely being conducted aboard the ISS, and studies utilizing these technologies can provide informative data. Future studies will employ next-generation technologies, such as three-dimensional organoids, organ or body-on-a-chip systems, and high-throughput screening platforms to more accurately model human physiology in spaceflight. With the recent expansion of commercial spaceflight opportunities, stem cell research and development in LEO and subsequent translation on Earth are becoming more realistic possibilities for private and academic biomedical entities alike.

References

Acharya, A., Brungs, S., Lichterfeld, Y. et al. (2019). Parabolic, flight-induced, acute hypergravity and microgravity effects on the beating rate of human cardiomyocytes. *Cell* 8 (4): https://doi.org/10.3390/cells8040352.

Andrews, P.W., Benvenisty, N., McKay, R. et al., Steering Committee of the International Stem Cell, I.(2005). The International Stem Cell Initiative: toward benchmarks for human embryonic stem cell research. *Nature Biotechnology* 23 (7): 795–797. https://doi.org/10.1038/nbt0705-795.

Andrzejewska, A., Lukomska, B., and Janowski, M. (2019). Concise review: mesenchymal stem cells: from roots to boost. *Stem Cells* 37 (7): 855–864. https://doi.org/10.1002/stem.3016.

Arun, R.P., Sivanesan, D., Patra, B. et al. (2019). Simulated microgravity increases polyploid giant cancer cells and nuclear localization of YAP. *Scientific Reports* 9 (1): 10684. https://doi.org/10.1038/s41598-019-47116-5.

Arun, R.P., Sivanesan, D., Vidyasekar, P., and Verma, R.S. (2017). PTEN/FOXO3/AKT pathway regulates cell death and mediates morphogenetic differentiation of colorectal cancer cells under simulated microgravity. *Scientific Reports* 7 (1): 5952. https://doi.org/10.1038/s41598-017-06416-4.

Avitabile, E., Fusco, L., Minardi, S. et al. (2020). Bioinspired scaffold action under the extreme physiological conditions of simulated space flights: osteogenesis enhancing under microgravity. *Frontiers in Bioengineering and Biotechnology* 8: 722. https://doi.org/10.3389/fbioe.2020.00722.

Baio, J., Martinez, A.F., Bailey, L. et al. (2018a). Spaceflight activates protein kinase C alpha signaling and modifies the developmental stage of human neonatal cardiovascular progenitor cells. *Stem Cells and Development* 27 (12): 805–818. https://doi.org/10.1089/scd.2017.0263.

Baio, J., Martinez, A.F., Silva, I. et al. (2018b). Cardiovascular progenitor cells cultured aboard the International Space Station exhibit altered developmental and functional properties. *NPJ Microgravity* 4: 13. https://doi.org/10.1038/s41526-018-0048-x.

Berthiaume, F., Maguire, T.J., and Yarmush, M.L. (2011). Tissue engineering and regenerative medicine: history, progress, and challenges. *Annual Review of Chemical and Biomolecular Engineering* 2: 403–430. https://doi.org/10.1146/annurev-chembioeng-061010-114257.

Blaber, E.A., Dvorochkin, N., Lee, C. et al. (2013). Microgravity induces pelvic bone loss through osteoclastic activity, osteocytic osteolysis, and osteoblastic cell cycle inhibition by CDKN1a/p21. *PLoS One* 8 (4): e61372. https://doi.org/10.1371/journal.pone.0061372.

Blaber, E.A., Dvorochkin, N., Torres, M.L. et al. (2014). Mechanical unloading of bone in microgravity reduces mesenchymal and hematopoietic stem cell-mediated tissue regeneration. *Stem Cell Research* 13 (2): 181–201. https://doi.org/10.1016/j.scr.2014.05.005.

Blaber, E.A., Finkelstein, H., Dvorochkin, N. et al. (2015). Microgravity reduces the differentiation and regenerative potential of embryonic stem cells. *Stem Cells and Development* 24 (22): 2605–2621. https://doi.org/10.1089/scd.2015.0218.

Camberos, V., Baio, J., Bailey, L. et al. (2019). Effects of spaceflight and simulated microgravity on YAP1 expression in cardiovascular progenitors: implications for cell-based repair. *International Journal of Molecular Sciences* 20 (11): https://doi.org/10.3390/ijms20112742.

Carey, B.W., Markoulaki, S., Hanna, J.H. et al. (2011). Reprogramming factor stoichiometry influences the epigenetic state and biological properties of induced pluripotent stem cells. *Cell Stem Cell* 9 (6): 588–598. https://doi.org/10.1016/j.stem.2011.11.003.

Chen, L., Liu, G., Li, W., and Wu, X. (2019). Sonic hedgehog promotes chondrogenesis of rabbit bone marrow stem cells in a rotary cell culture system. *BMC Developmental Biology* 19 (1): 18. https://doi.org/10.1186/s12861-019-0198-4.

Chen, Y., Xu, C., Wang, P. et al. (2017). Effect of long-term simulated microgravity on immune system and lung tissues in rhesus macaque. *Inflammation* 40 (2): 589–600. https://doi.org/10.1007/s10753-016-0506-0.

Chen, Z., Luo, Q., Lin, C. et al. (2016). Simulated microgravity inhibits osteogenic differentiation of mesenchymal stem cells via depolymerizing F-actin to impede TAZ nuclear translocation. *Scientific Reports* 6: 30322. https://doi.org/10.1038/srep30322.

Costantini, D., Overi, D., Casadei, L. et al. (2019). Simulated microgravity promotes the formation of tridimensional cultures and stimulates pluripotency and a glycolytic metabolism in human hepatic and biliary tree stem/progenitor cells. *Scientific Reports* 9 (1): 5559. https://doi.org/10.1038/s41598-019-41908-5.

Demirci, S., Leonard, A., and Tisdale, J.F. (2020). Hematopoietic stem cells from pluripotent stem cells: clinical potential, challenges, and future perspectives. *Stem Cells Translational Medicine* 9 (12): 1549–1557. https://doi.org/10.1002/sctm.20-0247.

Engel, M., Do-Ha, D., Munoz, S.S., and Ooi, L. (2016). Common pitfalls of stem cell differentiation: a guide to improving protocols for neurodegenerative disease models and research. *Cellular and Molecular Life Sciences* 73 (19): 3693–3709. https://doi.org/10.1007/s00018-016-2265-3.

Fuentes, T.I., Appleby, N., Raya, M. et al. (2015). Simulated microgravity exerts an age-dependent effect on the differentiation of cardiovascular progenitors isolated from the human heart. *PLoS One* 10 (7): e0132378. https://doi.org/10.1371/journal.pone.0132378.

Furukawa, T., Tanimoto, K., Fukazawa, T. et al. (2018). Simulated microgravity attenuates myogenic differentiation via epigenetic regulations. *NPJ Microgravity* 4: 11. https://doi.org/10.1038/s41526-018-0045-0.

Grimm, D., Egli, M., Kruger, M. et al. (2018). Tissue engineering under microgravity conditions-use of stem cells and specialized cells. *Stem Cells and Development* 27 (12): 787–804. https://doi.org/10.1089/scd.2017.0242.

Grimm, D., Wehland, M., Corydon, T.J. et al. (2020). The effects of microgravity on differentiation and cell growth in stem cells and cancer stem cells. *Stem Cells Translational Medicine* 9 (8): 882–894. https://doi.org/10.1002/sctm.20-0084.

Hagiwara, H., Higashibata, A., Ogawa, S. et al. (2018). Effectiveness of endothelial progenitor cell culture under microgravity for improved angiogenic potential. *Scientific Reports* 8 (1): 14239. https://doi.org/10.1038/s41598-018-32073-2.

Hatzistergos, K.E., Jiang, Z., Valasaki, K. et al. (2018). Simulated microgravity impairs cardiac autonomic neurogenesis from neural crest cells. *Stem Cells and Development* 27 (12): 819–830. https://doi.org/10.1089/scd.2017.0265.

Hosoyama, T., Ichida, S., Kanno, M. et al. (2017). Microgravity influences maintenance of the human muscle stem/progenitor cell pool. *Biochemical and Biophysical Research Communications* 493 (2): 998–1003. https://doi.org/10.1016/j.bbrc.2017.09.103.

Huang, P., Russell, A.L., Lefavor, R. et al. (2020). Feasibility, potency, and safety of growing human mesenchymal stem cells in space for clinical application. *NPJ Microgravity* 6: 16. https://doi.org/10.1038/s41526-020-0106-z.

Hughson, R.L., Helm, A., and Durante, M. (2018). Heart in space: effect of the extraterrestrial environment on the cardiovascular system. *Nature Reviews. Cardiology* 15 (3): 167–180. https://doi.org/10.1038/nrcardio.2017.157.

Jha, R., Wu, Q., Singh, M. et al. (2016). Simulated microgravity and 3D culture enhance induction, viability, proliferation and differentiation of cardiac progenitors from human pluripotent stem cells. *Scientific Reports* 6: 30956. https://doi.org/10.1038/srep30956.

Kang, H., Fan, Y., Sun, A. et al. (2013). Simulated microgravity exposure modulates the phenotype of cultured vascular smooth muscle cells. *Cell Biochemistry and Biophysics* 66 (1): 121–130. https://doi.org/10.1007/s12013-012-9460-0.

Kawahara, Y., Manabe, T., Matsumoto, M. et al. (2009). LIF-free embryonic stem cell culture in simulated microgravity. *PLoS One* 4 (7): e6343. https://doi.org/10.1371/journal.pone.0006343.

Koc Demir, A., Elcin, A.E., and Elcin, Y.M. (2018). Osteogenic differentiation of encapsulated rat mesenchymal stem cells inside a rotating microgravity bioreactor: in vitro and in vivo evaluation. *Cytotechnology* 70 (5): 1375–1388. https://doi.org/10.1007/s10616-018-0230-8.

Kong, L., Wang, Y., Wang, H. et al. (2021). Conditioned media from endothelial progenitor cells cultured in simulated microgravity promote angiogenesis and bone fracture healing. *Stem Cell Research & Therapy* 12 (1): 47. https://doi.org/10.1186/s13287-020-02074-y.

Kunisaki, Y. and Frenette, P.S. (2012). The secrets of the bone marrow niche: enigmatic niche brings challenge for HSC expansion. *Nature Medicine* 18 (6): 864–865. https://doi.org/10.1038/nm.2825.

Kurose, T., Takahashi, S., Otsuka, T. et al. (2019). Simulated microgravity-cultured mesenchymal stem cells improve recovery following spinal cord ischemia in rats. *Stem Cell Research* 41: 101601. https://doi.org/10.1016/j.scr.2019.101601.

Lei, X., Cao, Y., Zhang, Y. et al. (2018). Effect of microgravity on proliferation and differentiation of embryonic stem cells in an automated culturing system during the

TZ-1 space mission. *Cell Proliferation* 51 (5): e12466. https://doi.org/10.1111/cpr.12466.

Lei, X.H., Ning, L.N., Cao, Y.J. et al. (2011). NASA-approved rotary bioreactor enhances proliferation of human epidermal stem cells and supports formation of 3D epidermis-like structure. *PLoS One* 6 (11): e26603. https://doi.org/10.1371/journal.pone.0026603.

Li, Y., He, L., Pan, S. et al. (2017). Three-dimensional simulated microgravity culture improves the proliferation and odontogenic differentiation of dental pulp stem cell in PLGA scaffolds implanted in mice. *Molecular Medicine Reports* 15 (2): 873–878. https://doi.org/10.3892/mmr.2016.6042.

Louis, F., Bouleftour, W., Rattner, A. et al. (2017). RhoGTPase stimulation is associated with strontium chloride treatment to counter simulated microgravity-induced changes in multipotent cell commitment. *NPJ Microgravity* 3: 7. https://doi.org/10.1038/s41526-016-0004-6.

Low, E.K., Brudvik, E., Kuhlman, B. et al. (2018). Microgravity impairs DNA damage repair in human hematopoietic stem/progenitor cells and inhibits their differentiation into dendritic cells. *Stem Cells and Development* 27 (18): 1257–1267. https://doi.org/10.1089/scd.2018.0052.

Luna, C., Yew, A.G., and Hsieh, A.H. (2015). Effects of angular frequency during clinorotation on mesenchymal stem cell morphology and migration. *NPJ Microgravity* 1: 15007. https://doi.org/10.1038/npjmgrav.2015.7.

Mashiko, T., Kanayama, K., Saito, N. et al. (2021). Selective proliferation of highly functional adipose-derived stem cells in microgravity culture with stirred microspheres. *Cell* 10 (3): https://doi.org/10.3390/cells10030560.

Mathur, A., Loskill, P., Hong, S. et al. (2013). Human induced pluripotent stem cell-based microphysiological tissue models of myocardium and liver for drug development. *Stem Cell Research & Therapy* 4 (Suppl 1): S14. https://doi.org/10.1186/scrt375.

Mathur, A., Loskill, P., Shao, K. et al. (2015). Human iPSC-based cardiac microphysiological system for drug screening applications. *Scientific Reports* 5: 8883. https://doi.org/10.1038/srep08883.

Mattei, C., Alshawaf, A., D'Abaco, G. et al. (2018). Generation of neural organoids from human embryonic stem cells using the rotary cell culture system: effects of microgravity on neural progenitor cell fate. *Stem Cells and Development* 27 (12): 848–857. https://doi.org/10.1089/scd.2018.0012.

Mellor, L.F., Steward, A.J., Nordberg, R.C. et al. (2017). Comparison of simulated microgravity and hydrostatic pressure for chondrogenesis of hASC. *Aerospace Medicine and Human Performance* 88 (4): 377–384. https://doi.org/10.3357/AMHP.4743.2017.

Otsuka, T., Imura, T., Nakagawa, K. et al. (2018). Simulated microgravity culture enhances the neuroprotective effects of human cranial bone-derived mesenchymal stem cells in traumatic brain injury. *Stem Cells and Development* 27 (18): 1287–1297. https://doi.org/10.1089/scd.2017.0299.

Pao, S.I., Chien, K.H., Lin, H.T. et al. (2017). Effect of microgravity on the mesenchymal stem cell characteristics of limbal fibroblasts. *Journal of the Chinese Medical Association* 80 (9): 595–607. https://doi.org/10.1016/j.jcma.2017.01.008.

Qin, H., Diaz, A., Blouin, L. et al. (2014). Systematic identification of barriers to human iPSC generation. *Cell* 158 (2): 449–461. https://doi.org/10.1016/j.cell.2014.05.040.

Rowe, R.G. and Daley, G.Q. (2019). Induced pluripotent stem cells in disease modelling and drug discovery. *Nature Reviews. Genetics* 20 (7): 377–388. https://doi.org/10.1038/s41576-019-0100-z.

Salvetti, A., Degl'Innocenti, A., Gambino, G. et al. (2021). Artificially altered gravity elicits cell homeostasis imbalance in planarian worms, and cerium oxide nanoparticles counteract this effect. *Journal of Biomedical Materials Research. Part A* https://doi.org/10.1002/jbm.a.37215.

Sharma, A., Sances, S., Workman, M.J., and Svendsen, C.N. (2020). Multi-lineage human iPSC-derived platforms for disease modeling and drug discovery. *Cell Stem Cell* 26 (3): 309–329. https://doi.org/10.1016/j.stem.2020.02.011.

Sharma, A., Wu, J.C., and Wu, S.M. (2013). Induced pluripotent stem cell-derived cardiomyocytes for cardiovascular disease modeling and drug screening. *Stem Cell Research & Therapy* 4 (6): 150. https://doi.org/10.1186/scrt380.

Soares, F.A., Sheldon, M., Rao, M. et al. (2014). International coordination of large-scale human induced pluripotent stem cell initiatives: Wellcome Trust and ISSCR workshops white paper. *Stem Cell Reports* 3 (6): 931–939. https://doi.org/10.1016/j.stemcr.2014.11.006.

Stamenkovic, V., Keller, G., Nesic, D. et al. (2010). Neocartilage formation in 1 g, simulated, and microgravity environments: implications for tissue engineering. *Tissue Engineering. Part A* 16 (5): 1729–1736. https://doi.org/10.1089/ten.tea.2008.0624.

Theunissen, T.W., Friedli, M., He, Y. et al. (2016). Molecular criteria for defining the naive human pluripotent state. *Cell Stem Cell* 19 (4): 502–515. https://doi.org/10.1016/j.stem.2016.06.011.

Touchstone, H., Bryd, R., Loisate, S. et al. (2019). Recovery of stem cell proliferation by low intensity vibration under simulated microgravity requires LINC complex. *NPJ Microgravity* 5: 11. https://doi.org/10.1038/s41526-019-0072-5.

Ulbrich, C., Wehland, M., Pietsch, J. et al. (2014). The impact of simulated and real microgravity on bone cells and mesenchymal stem cells. *BioMed Research International* 2014: 928507. https://doi.org/10.1155/2014/928507.

Volpato, V. and Webber, C. (2020). Addressing variability in iPSC-derived models of human disease: guidelines to promote reproducibility. *Disease Models & Mechanisms* 13 (1): https://doi.org/10.1242/dmm.042317.

Vunjak-Novakovic, G., Searby, N., De Luis, J., and Freed, L.E. (2002). Microgravity studies of cells and tissues. *Annals of the New York Academy of Sciences* 974: 504–517. https://doi.org/10.1111/j.1749-6632.2002.tb05927.x.

Wang, N., Wang, H., Chen, J. et al. (2014). The simulated microgravity enhances multipotential differentiation capacity of bone marrow mesenchymal stem cells. *Cytotechnology* 66 (1): 119–131. https://doi.org/10.1007/s10616-013-9544-8.

Wang, P., Tian, H., Zhang, J. et al. (2019). Spaceflight/microgravity inhibits the proliferation of hematopoietic stem cells by decreasing Kit-Ras/cAMP-CREB pathway networks as evidenced by RNA-Seq assays. *The FASEB Journal* 33 (5): 5903–5913. https://doi.org/10.1096/fj.201802413R.

Wang, S., Yin, Z., Zhao, B. et al. (2017). Microgravity simulation activates Cdc42 via Rap1GDS1 to promote vascular branch morphogenesis during vasculogenesis. *Stem Cell Research* 25: 157–165. https://doi.org/10.1016/j.scr.2017.11.002.

Weiss, W.M., Mulet-Sierra, A., Kunze, M. et al. (2017). Coculture of meniscus cells and mesenchymal stem cells in simulated microgravity. *NPJ Microgravity* 3: 28. https://doi.org/10.1038/s41526-017-0032-x.

Wnorowski, A., Sharma, A., Chen, H. et al. (2019). Effects of spaceflight on human induced pluripotent stem cell-derived cardiomyocyte structure and function. *Stem Cell Reports* 13 (6): 960–969. https://doi.org/10.1016/j.stemcr.2019.10.006.

Xue, L., Li, Y., and Chen, J. (2017). Duration of simulated microgravity affects the differentiation of mesenchymal stem cells. *Molecular Medicine Reports* 15 (5): 3011–3018. https://doi.org/10.3892/mmr.2017.6357.

Yin, H., Wang, Y., Sun, X. et al. (2018). Functional tissue-engineered microtissue derived from cartilage extracellular matrix for articular cartilage regeneration. *Acta Biomaterialia* 77: 127–141. https://doi.org/10.1016/j.actbio.2018.07.031.

Yuge, L., Kajiume, T., Tahara, H. et al. (2006). Microgravity potentiates stem cell proliferation while sustaining the capability of differentiation. *Stem Cells and Development* 15 (6): 921–929. https://doi.org/10.1089/scd.2006.15.921.

Zakrzewski, W., Dobrzynski, M., Szymonowicz, M., and Rybak, Z. (2019). Stem cells: past, present, and future. *Stem Cell Research & Therapy* 10 (1): 68. https://doi.org/10.1186/s13287-019-1165-5.

Zhang, S., Liu, P., Chen, L. et al. (2015). The effects of spheroid formation of adipose-derived stem cells in a microgravity bioreactor on stemness properties and therapeutic potential. *Biomaterials* 41: 15–25. https://doi.org/10.1016/j.biomaterials.2014.11.019.

Part III

Space Chemistry

6

Chemistry-Related Innovations in Space, Benefits of Flow Chemistry

Ferenc Darvas[1,3], *Paul G. Mezey*[2], *György Dormán*[3,4], *Balázs Buchholcz*[3], *Csaba Janáky*[4,5], *Richard V. Jones*[5], *Gellért Sipos*[6], *Tamás Peidl*[3], and *Gergo Mezohegyi*[3]

[1] *ThalesNano, Inc., Zahony utca 7, H-1031 Budapest, Hungary*
[2] *Kyoto University, Yukawa Institute for Theoretical Physics, International Research Unit of Advanced Future Studies, Kitashirakawa-Oiwakecho, Sakyo-Ku, Kyoto 606-8502, Japan*
[3] *InnoStudio, Inc., Fűzfőgyártelep, H-8184 Balatonfűzfő, Hungary*
[4] *University of Szeged, Department of Physical Chemistry and Materials Science, Aradi tér 1, H-6720 Szeged, Hungary*
[5] *ThalesNano Energy Inc, 3031 SE Lexington Lakes Drive, # 204, Stuart, FL 34994, USA*
[6] *ComInnex, Inc., Zahony utca 7, H-1031 Budapest, Hungary*

6.1 Introduction

Space chemistry is an emerging field, i.e., performing chemistry in space as distinguished from astrochemistry or planetary chemistry. First, our chapter discusses the unique properties (microgravity, low temperature, cosmic ray) of space pointing out the extraordinary chemistry opportunities. Furthermore, the radical expansion of space industry in the last decades has been generating the scientific, technological, and also commercial infrastructure for realizing chemical processes in space.

The "drug-on-demand" capacity will supply astronauts on long-duration spaceflights with the necessary pharmaceuticals, and the deep understanding of the cellular and disease mechanisms also play an important role in developing countermeasures to mediate the response to microgravity and/or space radiation. The exploitation of the advantages of microgravity in protein crystallization and formulation is a rapidly developing area in the space pharma sector; various examples demonstrate that drug uniformity stability has been increased.

Electrochemical CO_2 conversion in space allows to recover oxygen from exhaled carbon dioxide, while generating useful carbon-based products. These processes are extremely important for present and future missions in space.

6.2 Challenges of Chemistry Realized in Space

Space chemistry, as a major component for the presently available space technologies, and even more importantly, as a growing field with many surprises and

In-Space Manufacturing and Resources: Earth and Planetary Exploration Applications, First Edition.
Edited by Volker Hessel, Jana Stoudemire, Hideaki Miyamoto, and Ian D. Fisk.

promises for the future, has already provided several results which may affect not only the future development of space chemistry itself, but also our understanding of chemistry in general. In our interpretation of the details and the underlying scientific relations of space chemistry, molecular modeling approaches, with a strong quantum chemistry foundation, will have a major role.

Some of the not fully explained and unusual aspects of chemical and biochemical processes observed in satellites and during space experiments on the International Space Station (ISS) are bringing up some fundamental questions concerning the possibilities that some of the expected features of such processes, based on the accumulated experiences and interpretations on the surface or near to the surface of Earth, are affected by some of our biases.

6.2.1 Unique Features of Space: Lack of Directionality (Cosmic Rays, Microgravity)

The most natural is to focus on the differences in the levels of intensity of cosmic rays as well as on the manifestations of gravity in the two types of environments. Many aspects of radiation chemistry have been well studied for a wide range of radiation types, including cosmic rays, on the other hand, much less emphasis has been placed on a very "natural" condition, usually taken for granted: the role of the gravity of the Earth.

One special aspect of gravity may play a role: the presence of a dominant directionality of gravitational forces near the surface of the Earth. This directionality is far less important in space, at a larger distance from the center of the Earth, than it is within the "usual" environment near the Earth's surface, where most of our experiences and scientific development have taken place during our past history. On the other hand, even the radiation effects from cosmic rays are affected by differences not only in intensity, but also by differences in directionality, if one moves away from the surface of the Earth to space. By being at a significant distance away from the Earth, the dominant "from up to down" directionality of cosmic rays on the surface of the Earth is replaced by a more multidirectional radiation; hence, again, in space there is a much lesser dominance of directionality for cosmic rays, just as a much lesser dominance of directionality for the role of gravity. One aspect of the lesser dominance of directionality is likely to be manifested in a fuller realization of randomness, if the direction of "downward" is having a far lesser significance. Entropy, and the manifestation and role of randomness are not independent of the presence or lack of a dominant direction (and of the actual level of its dominance). In the space, this is likely to have some effect even on chemical reactions, and ultimately, the chemistry contribution to space technologies. There is strong evidence from studies on reaction features affected by hydrodynamic instabilities arising in horizontally propagating vertical chemical fronts that dominant directionality often gives rise to some unexpected features (Pópity-Tóth et al. 2013), also providing additional information in more recent studies on reactive fronts in modulated gravity environment (Horváth et al. 2014, pp. 26279–26287) and specifically in microgravity (Bába et al. 2018; Bába et al. 2019, pp. 406–412). Directionality of injections relative to chemical fronts (Tóth et al. 2020, pp. 10278–10285) has a specific role that is influenced by higher or lesser dominance of directionality effects.

6.2.2 Theoretical Aspects: Quantum Chemistry Approach

How molecules can modify their shapes even from relatively weak forces (Mezey 1993), how their parts interact (Mezey 1999, pp. 169–178), and even the details of chemical reactions can be modeled rather accurately by quantum chemistry methodologies. Concerning space chemistry, various constraints, such as local forces acting on individual atoms in a molecule, additionally to the usual forces of chemical bonds keeping the parts of the molecule together and being responsible for the various vibrational motions can be introduced, to see how a change in the role of the directionality of gravity may influence molecular behavior.

Molecular shapes, their variations during actual molecular vibrations, their conformational changes which may actually play a major role in space chemistry, where even weak effects are assumed to be able to influence molecular behavior in dominant way, and the actual distortions of molecular atomic arrangements during chemical reactions have been studied in great detail by quantum chemistry calculations, providing precise descriptions how individual atomic nuclei move relative to the other atomic nuclei in such processes (Mezey 1993; Mezey 1999, pp. 169–178). What is relevant to our present problem, it is also possible to introduce some artificial forces on individual atomic nuclei, imitating some enhancement of directional forces, or introducing atomic mass dependent counter-forces, in order to gain understanding of the actual effects of the presence of "privileged" directions, such as "downward" direction. The model of assumedly "unbiased" randomness of orientations of molecules in a solution can be tested and possibly refined by such computations.

For example, by introducing artificial "counter-forces" into such quantum chemistry models, and modeling their effects on the speed and outcome of chemical reactions, or, on the speed of development in crystallization and the lesser frequency of local irregularities in crystal grows, can provide new hints how to interpret and exploit better some of the remarkable and unexpected phenomena in space chemistry. Besides learning from actual quantum chemistry modeling approaches, such models may also motivate new experiments. For example, when the effects of one type of change are studied, it is often useful to study the exactly opposite effect. In our case, by imitating the effects of much higher levels of gravity on chemical reactions and crystal growth. Whereas placing chemical experiments within environments of much higher gravity, such as it is present near the planet Jupiter is technically not yet feasible, nevertheless, by using experiments in centrifuges: by placing the chemically reacting molecules as well as the monitoring equipment within a centrifuge, one may imitate high-gravity environments, and when compared to conditions on Earth, a far more dominant effect of forces with dominant directionality (within the rotating system) can be achieved that may help to explain the opposite phenomenon: the lesser dominance of directionality in space chemistry.

In space chemistry, the role of symmetry may also deserve special attention: if a dominant directionality, such as the direction of the force of gravity, is of lesser importance, this fact alone may indicate that possibly, a better approximation to perfect symmetry is manifested there. As follows from quantum chemistry studies of approximate symmetries and symmetry deficiency measures for molecules (Lijie et al. 2006, pp. 145–153; Mezey et al. 1998, pp. 99–105), the distortions of molecular parts have strong relations with the distortions of the entire molecules. For example, locally no perfect mirror plane can exist for any CH_2 group in any molecule, if the complete molecule does not have a mirror plane. Especially, measures for the degree of chirality may also be affected by the reduced role of the directional forces of gravity in space chemistry that may also affect some selectivity aspects of chemical reactions, for example, in polypeptide or protein synthesis reactions. Symmetry and chirality are suggested for special investigations within the space chemistry context.

An alternative approach may involve reaction studies along surfaces where directionalities can be monitored macroscopically, or those catalytic reactions involving zeolites. If such studies are carried out in space, or in a centrifuge, the directionality effects can be studied from a new perspective. For example, if the zeolite framework is given, and the actual reactions occur within the cavities of zeolites, then the directionality constraints within zeolites cavities, with possible enhancement or hindrance of the directionality bias represented by gravity or by centrifugal forces can provide a combination that may reveal some new aspects of space chemistry. Earlier quantum chemistry studies on zeolites have provided some insight for the non-random distribution of the Si and Al atoms within their frameworks (Hass et al. 1981, pp. 389–399; Mezey 1984 pp. 147–156; Derouane et al. pp.145–156), providing a quantum chemistry explanation of the so-called Loewenstein rule on zeolites: in these alumino-silicates the statistically expected Al–O–Al linkages do not occur, as such linkages would be not only energetically unfavorable, but would also generate a major strain within the zeolite framework, adding to the forces already present, that in the space-chemistry context would involve the modified forces of gravity.

The randomly taken few examples here in this section serve just to illustrate the deep world of the theoretical questions which are emerging when chemistry is performed in space. It is expected that theoretical chemists are going to focus more on responding to the emerging number of the challenging and sometimes astonishing observations which are gathered during systematic on orbit chemical experimentations. See, for example, the comment on the difference in bubble formation on orbit and in the Earth, furthermore its practical consequences, see below.

6.3 Space Chemistry – Concept, Historical Summary, and Current Efforts

Space chemistry, for decades, has been equivalent to astrochemistry or planetary chemistry (i.e. the study of the chemical behavior of atoms and molecules in space or on planets) (Yamamoto 2017; Rehder 2011). Today, the term space chemistry is applied for *performing* chemistry in space. The radical expansion of space industry in the last decades has been generating the scientific, technological, and also commercial infrastructure for realizing chemical processes, like syntheses and analyses in space. It has become necessary to develop novel chemical compounds, formulation methodologies and/or reaction routes, for supporting space traveling, drug discovery and manufacturing, developing effective ISRU processes, initiating asteroid and planet mining (Hessel et al. 2021, pp. 3368–3388; Wouters et al. 2021, pp. 118238–118252), all in an off-Earth environment.

6.3.1 Foundation of the Space Chemistry Consortium

A significant step toward bringing chemistry to space was realized in 2014 by the International Flow Chemistry Society (www.flowchemistrysociety.com) that initiated the worldwide academic–industrial Space Chemistry Consortium (https://innostudio.org/space-chemistry-consortium) at an informal meeting associated with the IMRET13 Conference in Budapest (IMRET13 2014). The Consortium targeted to bring together people from the chemical community interested in doing chemistry in space. The founders' common denominator was their experience and proficiency in flow chemistry and microfluidics; furthermore, their shared view was that these two fields can solve elementary process issues (like imperfect mixing) impeding the chemical experimentation in space for decades. The initial targets of the Consortium were formulated as follows (FROST5 2015):

1. Create a standardized and flexible space laboratory with a capability for a wide range of chemistry discoveries coupled with plug and play "pharmacy on demand" chemistry system (transition from individual to cascade reactors) where pharmaceuticals are synthesized, analyzed, purified, and formulated when needed.
2. Create new chemistry in space: Utilize the extreme conditions of space (tunneling effect) on different chemical processes to discover new routes to novel molecules or more efficient chemical applications.
3. Recycle biological waste (especially carbon monoxide to oxygen). Convert human biological waste materials to reusable chemicals in order to sustain human life.

These objectives of the Consortium harmonize well with the global space research goals of the International Space Exploration Coordination Group (ISECG) (www.globalspaceexploration.org). The Consortium since then has grown to a substantial organization of more than 40 internationally renowned members from academia and industry exerting influence to the newly born space chemistry community. One of the most important and early results of the Space Chemistry Consortium was to create a scientific platform in order to share the latest advancements and results of performing chemistry in space. Together with the Flow Chemistry Society, the Consortium approached the American Chemical Society (as the world's largest chemical association) and initiated the Space Chemistry Symposium series (https://innostudio.org/space-chemistry-symposium) that has regularly been held since 2015 as part of the ACS National Meeting and Exposition. The event is dedicated to key topics and research areas such as pharmaceutical production in space, flow chemical reactors and syntheses for space applications, chemical instruments and analytics for microgravity environment, CO_2 capturing and transformation for applications on Mars, and in-space manufacturing of diverse materials and chemicals. In April 2021, the 6th Symposium was held, keeping the ACS Space Chemistry Symposia the world No.1 scientific event for the space chemistry community today.

6.3.2 Recent Activities and Successes in Space Chemistry

Scientific literature within the topic of performing chemistry in space so far has been scarce. The first papers about the overall concept and promises of flow chemistry for microgravity applications were published by the ThalesNano Group together with the University of Szeged (Jones et al. 2017, pp. 1–3; Sipos et al. 2017, pp. 151–156) where the authors proposed a scheme to prioritize chemical experimentations in space (see Figure 6.1).

The history of chemical experimentation in space goes back to a half century. One of the first, chemistry-related "experiments," although unplanned, had taken place in the lunar module of Apollo 13, which saved the lives of the crew. Following explosion of one of the oxygen tanks in the spacecraft, a makeshift adaptor for the lithium hydroxide canisters to remove toxic CO_2 build-up was made out of space parts, and

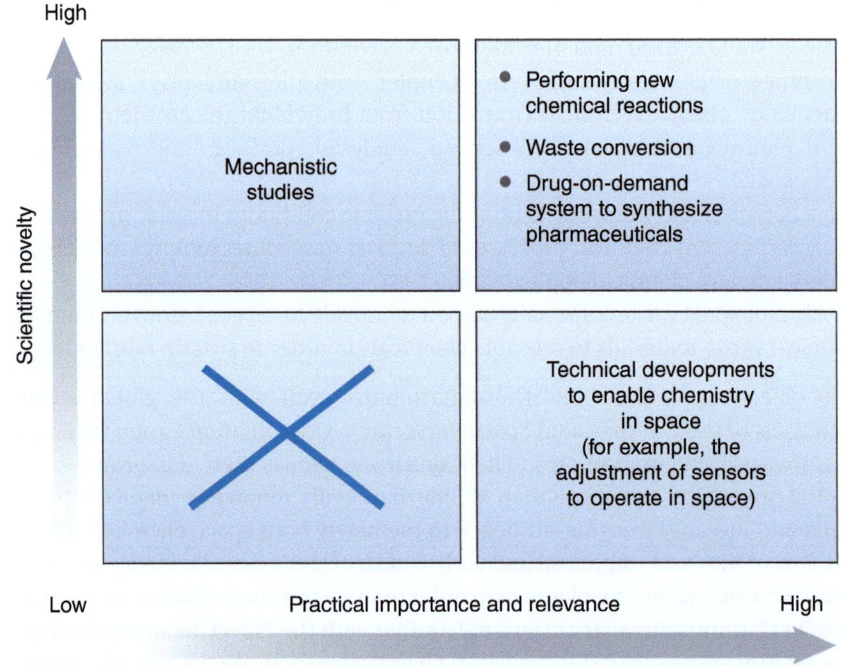

Figure 6.1 Selection criteria for chemical experiments to be performed in space. Source: Republished with permission of Jones et al. 2017, Springer Nature.

residual power from batteries was used to restart the command module and take the crew back to Earth (Li 2018). In the following three decades, sporadic experiments related to chemical research were carried out on Salyut 6, Mir, and the Space Shuttle. One of the earliest chemical reactions investigated on the ISS was about the combustion of fuels (Williams 2017). Mattioda and coworkers first reported photostability results of organic compounds in orbit, including a polycyclic aromatic hydrocarbon, amino acid, metalloporphyrin, and quinone (Mattioda et al. 2012, pp. 841–853).

In the last years, chemistry-related scientific research in space and for space purposes has gained wide interest – that is not surprising, considering the improved access opportunities to space platforms (Joseph and Wood 2021, pp. 123–138). The Israeli–Swiss company SpacePharma (www.space4p.com) developed a miniaturized and remotely controlled laboratory system which can be utilized to conduct diverse research experiments under microgravity either on satellites, spacecraft, or on the ISS. Chemistry-related experiments in SpacePharma's equipment have included enzymatic oxidation, drop coalescence study, and also crystallization (Samburski 2020). One recent experiment on the ISS by researchers from The Pennsylvania State University included mixing an aqueous solution with tricalcium silicate for the first time, in order to study how the chemistry of concrete is changed by microgravity (Neves et al. 2019, pp. 1–12). Understanding how concrete works in space has direct bearing on future plans regarding building on the Moon and Mars.

The Beeler Research Group (www.beelerlab.com) at Boston University, partnering with Space Tango (spacetango.com), investigated three fully automated chemical reactions on the ISS and made the first effort to synthesize organic compounds onboard (Beeler 2020). The University of Adelaide has recently teamed up with Space Tango and Alpha Space (www.alphaspace.com) to conduct a pharmaceutical stability study on ISS (Rosales 2020). The expected results will allow to generate highly innovative data sets that will help direct the future in-orbit and on-demand production of medicines. The first anti-COVID-19 experiments in space were recently carried out in the European module on ISS (ESA 2020). The promising study has recently been led by InnoStudio (www.innostudio.org) and CycloLab (www.cyclolab.hu), in collaboration with Japan Manned Space Systems Corporation (www.jamss.co.jp) and Space Applications Services (www.spaceapplications.com). The researchers, among others, investigated how remdesivir, the active ingredient of the COVID medicine Veklury (www.gilead.com), interacts with its solubilizing cyclodextrin derivative in order to improve the drug's efficiency.

Chemical experimentation is already present on Mars: the MOXIE experiment (see Section 6.6), developed by NASA, is using a device that extracts small amounts of oxygen from the Martian atmosphere via electrolysis (https://mars.nasa.gov/mars2020/spacecraft/instruments/moxie/).

Space agencies today are expecting more science teams to collaborate with industry in order to accelerate their research and development efforts by using commercial space-based platforms on a more intensive level. This trend in parallel defines the urgent need for designing and setting up platforms ensuring chemical experimentation, syntheses and research in space.

6.4 Space Chemistry Results: Life Sciences, Pharma Industry, Agro Industry, Cosmetics, and Others

Space chemistry is presently the single available approach to create a "drug-on-demand" production system in space in which a variety of pharmaceuticals can be synthesized, analyzed, purified, and formulated fresh on orbit when needed. The "drug-on-demand" capacity will supply astronauts on long-duration spaceflights with the necessary pharmaceuticals. It is also expected that flow chemistry in space will allow development of more efficient drugs, syntheses and formulation processes, and new molecules by known chemical reaction or by novel routes (Mezohegyi et al. 2021).

In the area of life sciences, the primary goal is the deep understanding of the cellular and disease mechanisms, biological pathways, and potential protein targets that play a role in mediating the response to microgravity and/or space radiation (Braddock 2020; Ruyters et al. 2017; Ryder and Braddock 2020).

It seems the unique conditions of human spaceflight require substantial biological adaptation (Schmidt et al. 2019). After a year-long spaceflight an extensive study conducted by NASA led to the conclusion that microgravity produces numerous changes in the musculoskeletal, visual, and immune systems, including decreased body mass, bone density, elongation of telomeres, genome instability, transcriptional and metabolic changes, altered drug-drug interactions, and drug resistance (Garrett-Bakelman et al. 2019).

One particular area is to investigate the immune response and drug resistance during space flights. Many reports underline that the immune system in astronauts may be reduced, including suppressive effects on mediators, change in leukocyte distribution, and reduction in T cell function. Determination of the effective dose of antibiotics in microgravity is particularly important while exposing to pathogens. Tissue engineering in space (e.g. three-dimensional aggregates from different cell types) is another potential area of the applied life science developing novel regenerative medicine tools.

6.4.1 Investigating and Exploiting Microgravity Effects

The pharma industry benefited much from the innovative space technologies in many ways; for example, identification of protein targets involved in complications occurring during spaceflights or in enhancing human radio-resistance is particularly important. Rapid alteration in gene expression in osteoblasts and the atrophy tissue metabolism in space help the understanding of the aging process on earth and could lead to the development of novel drugs and therapies (Braddock 2017). Indeed, cardiovascular, musculoskeletal, nervous, gastrointestinal, and immune responses to microgravity are similar to changes associated with aging, but seem to occur about 10 times faster in space than on Earth (Demontis et al. 2017, pp. 547).

The exploitation of the advantages of microgravity in protein crystallization is a very rapidly developing area in the space pharma sector (Reichert et al.

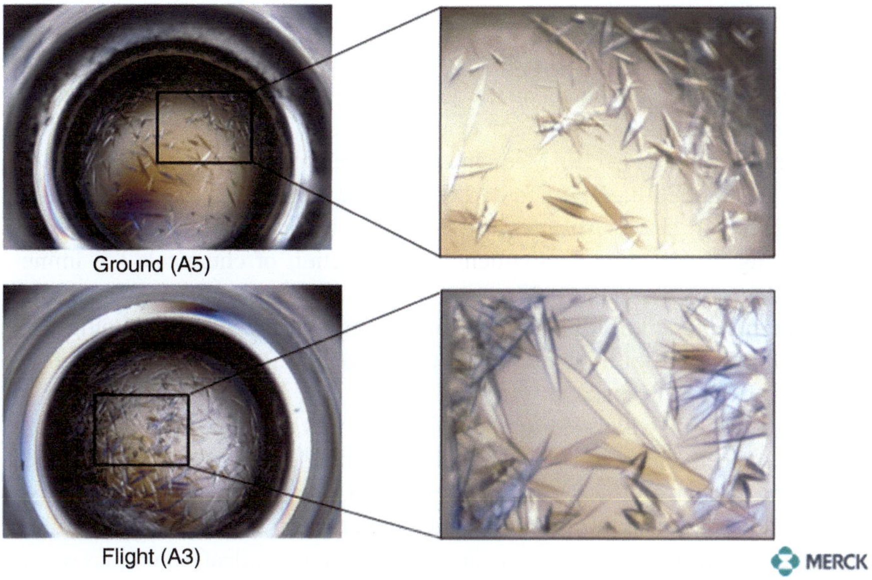

Figure 6.2 Differences in crystal size and morphology of monoclonal antibodies in space and on earth (Merck experience: https://www.nasa.gov/mission_pages/station/research/benefits/mab/). Source: NASA.

2019, pp. 1–8). Structure-based drug discovery relies on the 3D structure of the protein–ligand interaction (Figure 6.2).

Most protein structures are determined by X-ray crystallography, which requires diffraction-quality protein crystals. Convective currents and sedimentation occur under gravity and disturb crystal growth; thus, producing diffraction-quality protein crystals is a particular challenge (Moreno 2017, pp. 51–76). Microgravity has produced crystals that grow larger with more regular morphology and diffract to higher resolution than crystals of the same proteins grown on Earth (McPherson and DeLucas 2015). Furthermore, producing ultra-pure, crystalline biological medicines (e.g. therapeutic antibodies) in space has much potential. The drug substance is often stored as frozen diluted solutions or frozen lyophilized powders. For areas of the world, where refrigeration is limited, stable, high-purity crystalline drug substances are needed (Daugherty and Mrsny 2010, pp. 103–129), and only crystallization of proteins in space could provide such quality (Reichert et al. 2019, pp. 1–8).

In the area of drug formulation producing new polymorphs and co-crystals could enhance druggability and activity of existing or new active chemical ingredients (Yamaguchi et al. 2021). See details in Section 6.5.

Lyophilization is commonly used in drug formulation to increase the chemical and physical stability of pharmaceutical products. Pharma companies are interested in understanding what effects gravity has on the properties and physical state of lyophilized materials (http://www.iss-casis.org/iss360/taking-chemistry-out-of-this-world/).

6.4.2 Drug Stability in Space

Drug stability during space flights is the other critical issue. Astronauts going to long-term space travels or staying on the Moon and Mars are facing the problems of the fast decomposition of active pharmaceutical ingredients (APIs) in space (Mehta and Bhayani 2017, pp. 111–119). Medicines should remain active at least up to their designated Earth shelf-life and should be protected from the degradation caused by the radiation in space. An alternative solution is the on-demand continuous manufacturing of APIs. On-site-on-demand production of chemicals are immediately used after generation (Dallinger et al. 2020, pp. 1330–1341). A robotically controlled experimental flow chemistry platform, combining artificial intelligence, with planning and self-optimization, quality control and formulation (Coley et al. 2019) could be useful in space.

Solar concentrators which led to photomicroreactors seem to be an appropriate tool for pharmaceutical, chemical, and fuel production in space (Noël 2021). Space chemistry furthermore allows the study of chemical reactions that exploit the harsh conditions (such as vacuum, high-intensity UV radiation, extreme temperatures, and microgravity) that are prevalent in outer space. Flow chemical technology seems to be suitable to support *in situ* resources utilization (ISRU) (Schlüter and Cowley 2020; Starr and Muscatello 2020, Advincula 2019) and also in-space propellant production (Kornuta et al. 2019).

When talking to medicinal chemists at conferences, and reading the overly dispersed literature on on-orbit applications of space chemistry, it is easy to draw the conclusion that after the first breakthrough results the interest for chemistry-related space applications will increase enormously.

6.5 Space Chemistry Results: Formulation

Formulation has vital role in several industries, like pharmaceutical (Devalapally et al. 2007, pp. 2547–2565), agricultural (DeRosa et al. 2010, p. 91), food (Chaudhry et al. 2017), and cosmetics (Santos et al. 2019, pp. 313–330), to name a few. Generally, an active ingredient cannot be utilized by itself without formulation because of its chemical stability, low solubility under physiological conditions, biocompatibility, volatility, or other factors. In addition, the release of an active ingredient often needs to be controlled in time and/or needs to be target specific.

Crystallization has been researched for a long time and it was among the first topics in the fields of space chemistry and space chemical technology (see also Section 6.4). The first space crystallization results were obtained by the Apollo experiments more than 50 years ago, where the role of microgravity in this process has been revealed (Walter 2012). At the beginning of space crystallization experiments (Benz and Dold 2002, pp. 1638–1645) three challenges emerged:

(1) How to apply the same experimental conditions on Earth and in space?
(2) Are the optimal conditions for crystallization under microgravity the same as on Earth?

(3) Since in most cases microgravity experiments are singular, what kind of statistical approach at data processing can be applied?

6.5.1 Crystallization in Space

In the meantime, numerous microgravity platforms have been developed for space research, such as the American Skylab, the Spacelab missions, D1 and D2 missions (Germany), the Japanese sounding rocket programs (Benz and Dold 2002, pp. 1638–1645), and a Chinese platform (Chen et al. 2001, pp. 478–479). While the platforms and crystallization equipment underwent vital evolution over time, the research targets have also changed. Nowadays, the use of microgravity is not considered anymore as the Holy Grail for perfect crystal growth, but still it remained a significant opportunity in improving crystal quality (McPherson and DeLucas 2015).

Crystallization in space has soon triggered the development of special instrumentations. An experiment based on the TEXUS hardware was flown on STS-9 in 1983 and grew crystals of lysozyme and β-galactosidase (Littke and John 1986). The Vapor Diffusion Apparatus (VDA), invented by the University of Alabama (Birmingham, USA) first flew in 1985 (DeLucas et al. 1986). The design of this hardware was meant to mimic the hanging drop vapor diffusion experiments most utilized for crystallization on Earth (Karr et al. 2015). The device enabled to maintain 24 experiments with no temperature stabilization and/or control capability. The use of VDA resulted in crystals of significantly higher quality than of the terrestrial ones. The second generation of VDA equipment was able to keep constant temperature at 4 or 22 °C and maintain 3 × 20 vapor diffusion experiments (DeLucas et al. 1986, pp. 681–693).

The STS-42 (International Microgravity Laboratory), flown in 1992, was the first flight dedicated to the maintenance of a microgravity environment. On this flight, both VDA and the German Cryostat (liquid-liquid diffusion) hardware were flown. In the Cryostat, a Satellite Tobacco Mosaic Virus (STMV) crystal was produced that was 30 times larger than any of the STMV crystals ever been grown on Earth (Karr et al. 2015). Nowadays, microgravity crystallization entered into the commercial phase, where affordability is a key factor to encourage industry to explore crystallization in space. The Japan Manned Space Systems Corporation provides a high-quality protein crystal growth service, so-called Kirara. The 1U (https://www.nasa.gov/content/what-are-smallsats-and-cubesats) volumed incubator is able to maintain 96 individual experiments at fix 20 °C (Sato 2020). The "Kirara box" has a dedicated platform on the ISS, in which some of the present authors performed formulation related research successfully, including protein crystallization, co-crystal formation for small molecules, and also optimization of an anti-SARS-CoV-2 drug formulation with remdesivir and certain cyclodextrin derivatives.

Three particular application areas of space crystallization are worth mention: protein crystallization, co-crystal formation, and inorganic crystal formation.

In the past decades X-ray crystallography underwent a vital evolution. The time required for structural analysis decreased from days/weeks to hours which encouraged the researchers to prepare more and more complex structures (Pflugrath 1992, pp. 811–815). Preparation and structural investigation of protein crystals has vital contribution to medicinal chemistry, especially to the target identification (Braddock 2020). Meanwhile, the parallel/high throughput crystal growing methods have also undergone meaningful development with automatization targeting the preparation and analysis (Deller and Rupp 2014, pp. 133–155). Formation of co-crystals has vital importance in the pharmaceutical industry. These structures can be made up of small specimens (molecules, ions), macromolecules (proteins, polymers), and the combination of these compounds. In the field of drug discovery, co-crystals have an important role in target identification, where the protein is crystallized in the presence of nucleic acids, small molecules, or drug candidates (Zheng et al. 2014, pp. 125–137).

Semiconductor materials have wide interest in the industrial field. Enhancing the optical, electrical, and magnetic properties of them has been an evergreen topic up to now. In the early stage of space crystallography experiments, many experiments have been conducted for the improvement of crystalline semiconductors in space (Benz and Dold 2002, pp. 1638–1645). On-board microchip synthesis of nanocrystalline quantum dots have been proposed in order to apply as a spectral decoy to divert rocket attacks. (Nijhuis et al. 2021, pp. 471–485).

In 1993, the German Spacelab mission D-2 GaAs crystal growth experiments have been launched by the German Space Agency. Using floating-zone technique the obtained sample contained significantly lower amount of dislocation networks with relatively larger cell size than its Earth-growth counterpart (Herrmann and Müller 1995, pp. 350–360). In the same mission, Cröll et al. also reported about decreasing dislocations in the case of GaSb using floating-zone method (1998, p. 365). In 2001, a Chinese research group reported on GaAs crystals with considerably better electrical properties after crystallization in space compared to their terrestrial grown reference samples (Chen et al. 2001, pp. 478–479).

6.5.2 Nanoformulation in Space

Among the several formulation techniques, nanotechnology is one of the most promising methods to enhance druggability, increase chemical stability (Dantuma et al. 2015), and enable target specific formulae, especially in the field of cancer treatment. Since the SARS-CoV-2 outbreak, nanotechnology-based medicines have received considerable attention because mRNA-based vaccines use lipid nanoparticles as carrier for increasing stability and bioavailability (Laczkó et al. 2020, pp. 724–732). Nanoformulation can be done by flow chemical instruments which provide precise drug load and size regulation, continuous manufacturing, and easy handling. Organic flow nanoformulation technologies are on the market as well, for example the Canadian Precision Nanosystems NanoAssemblr® platform that is suitable for nano vaccine synthesis using lipid nanoparticles (Walsh et al. 2014, pp. 109–120). The present authors also have a patented technology which is called SpeedyNano® platform (Manek et al. 2019), in which the nanonized organic compound is embedded in a biocompatible polymer matrix. This method is based on the so-called flash nanoprecipitations, where the composite particle is formed *in situ* by the supersaturation of the active ingredient. The selection of appropriate chemical composition, type of polymer, and flow rate depends on plenty of factors. Artificial intelligence is essential to reduce the number of experiments during the optimization phase.

Colonies in space and on Mars would require vast quantities of food; therefore, growing plants in a greenhouse is the adequate solution. Tailored fertilizers and specific types of bacteria and fungi are beneficial for plant cultivation. However, contamination and mutation would lead to microorganism that are harmful to the plantation and require specific protective agents. On-demand synthesis and nanoformulation of agrochemicals would be essential (Manek et al. 2019).

In summary, we believe that the significance of formulations performed in space will continue to grow rapidly and within a few years crystallization, co-crystal formation, clathrate formation, nanoformulation, and others will be part of the routine toolkit of the formulation experts.

6.6 Electrochemical CO_2 Conversion in Space: a Key Step to *In Situ* Resource Utilization

To recover oxygen from exhaled carbon dioxide, while generating useful carbon-based products is of high importance both at the International Space Station (ISS) and possibly for future missions (Jones et al. 2017). NASA has the current CO_2 Reduction Assembly (CRA) fully operational on-board the ISS since 2011 (Nelson et al. 2020) as part of the Environmental Control and Life Support System (ECLSS). The CRA is based on the Sabatier reaction ($CO_2 + 4H_2 \rightarrow CH_4 + 2H_2O$), to convert CO_2 to methane (CH_4) and water. The water produced is fed to the water purification system and then to an electrolyzer system to produce H_2 and O_2. Hydrogen is recycled to the CRA, O_2 is released into the cabin, and CH_4 is vented externally. The CRA recovers less than 50% of O_2 from CO_2 due to H_2 loss in the vented CH_4 (Nelson et al. 2020). However, there is an urgent need for new technologies that can increase the oxygen recovery rate to at least 75% or more. A better recovery rate means less oxygen needs to be stored and would free up precious cargo space on prolonged (Mars) missions. Finally, as an additional strategic motivation, we mention that if mankind ever aims to colonize Mars, the key available feedstock for any chemical reactions will be CO_2, which represents 96% of the Martian atmosphere (Rapp 2008).

Electrochemical processes in space offer a way to form valuable chemicals with no by-product formation, and without using additional chemicals. Different oxidation and reduction powers can be achieved just by changing the cell voltage (Nelson et al. 2020). Beyond the well-established water splitting, CO_2 conversion, C—C coupling reactions, CO addition or even the synthesis of pharmaceutical compounds can be achieved. The abundance of solar radiation and our ability to convert significant fractions of it to electrical power enables the use of electrical power for electrochemical life support and also in-situ electrochemical material processing. Such processes use water and carbon dioxide as reactants and produces oxygen (on the anode) and small organic molecules, such as ethylene or methanol (on the cathode). The gases can be vented into space (or used for other chemical purposes) and the oxygen is used for breathing. In terms of thermodynamics of the electrolysis, the necessary energy input can be provided by a mixture of electrical and thermal energy. Consequently, electrochemical transformation of CO_2 can be carried out in two fundamentally different configurations: high-temperature and low-temperature methods (Endrődi et al. 2017, pp. 133–154).

High-temperature solid oxide electrolysis cells (SOECs) offer high process efficiencies in the direct electrochemical conversion of CO_2 to CO or water to H_2. Due to the high operating temperatures (750–900 °C), both the kinetics and thermodynamics of these processes are very favorable. High temperatures are needed because the electrolyte materials only become obviously ionically conductive at elevated temperatures, while the anode (usually made of metal oxides) also needs high temperatures to reach good electrical conductivity. The SOEC technology is approaching maturity, with larger pilot plants already installed worldwide (Hauch et al. 2020). This technology necessitates the application of heat-tolerant structural materials and high reaction temperatures, which results in high operational and capital investment cost. A common problem that occurs in high-temperature electrolysis is degradation and carbon deposition, resulting in shorter cell lifespan and lower efficiency. Such a system is already functioning on Mars, on the Perseverance rover, launched in 2021 (https://mars.nasa.gov/mars2020/spacecraft/instruments/moxie/; https://photojournal.jpl.nasa.gov/catalog/PIA24203). The Mars Oxygen *In Situ* Resource Utilization Experiment, or MOXIE, is small, about the size of a car battery (3.9 × 23.9 × 30.9 cm). It harvests the excess heat from an onsite radioisotope thermoelectric generator, which also provides the electric power needed for electrolysis (300 W altogether). It can produce up to 10 g O_2 per hour, and the duration will be per experiment, which will be scheduled intermittently over the duration of the mission. The aim of this experiment is to demonstrate the viability of oxygen generation for future Mars missions. This includes the oxygen needs of the astronauts, and liquid oxygen propellant for sending anything back to Earth (Figure 6.3).

While high-temperature electrolysis has certain benefits, it is limited to the generation of CO, and it needs significant heat input. At the same time, low-temperature electrolyzers can generate a wealth of other products (hydrocarbons/alcohols/formic acid, etc.) beyond CO. These chemicals can act as building blocks of functional materials (e.g. polyethylene) (Endrődi et al. 2017, pp. 133–154). Notably, water has to be transported from Earth or harvested from the crust of Mars in order to utilize the benefits of this technology on Mars. There is already a low-temperature, polyelectrolyte membrane (PEM)-based, water electrolyzer system on ISS, which employs a 28-cell stack, at a current density of 216 mA /cm^2 to produce oxygen at ambient pressure with 80% efficiency (Nelson et al. 2020). Among the low-temperature cells, two types have emerged over the years for CO_2 electrolysis: (i) microfluidic cells, in which thin liquid layer(s) flow between the electrodes, (ii) zero-gap designs, in which only the anode is fed with liquid electrolyte, but all the cell constituents are pressed together (Endrődi et al. 2017, pp. 133–154). In both cases, CO_2 is fed as a gas through a gas diffusion electrode. This technology

Figure 6.3 Photograph showing the MOXIE unit (NASA/JPL-Caltech) and different sized zero gap cells (from left to right: $A = 8\,cm^2$, $A = 100\,cm^2$, $A = 1000\,cm^2$) and a tennis ball having a diameter of 6.5 cm. Source: Endrődi et al. 2020, Royal Society of Chemistry.

is still in the R&D phase, but larger and/or multi-layer cell stacks have already been presented to operate at sufficiently high current densities (Endrődi et al. 2020). In such configurations, one stack is made of several electrochemical cells, where the key component is a bipolar plate (acting as cathode in one cell, and anode in the subsequent cell). This approach avoids the need for manufacturing very large electrodes and cell components, and also avoids difficulties related to the management of the electrochemical process itself. Recently we have demonstrated a zero gap CO_2 cell stack, which consists of multiple layers and can operate with a pressurized CO_2 gas feed, without the need for any liquid catholyte (Endrődi et al. 2019). As a very recent development, we have shown that a key factor limiting long-term stability of low-temperature CO_2 electrolyzers is the formation of precipitates in the porous cathode from the alkaline electrolyte and the CO_2 feed. We presented that while precipitate formation is detrimental for the long-term stability, the presence of alkali metal cations at the cathode improves performance. To overcome this contradiction, we developed an operando activation and regeneration process, where the cathode of a zero-gap electrolyzer cell is periodically infused with alkali cation-containing solutions. This enables deionized water-fed electrolyzers to operate at a CO_2 reduction rate matching those using alkaline electrolytes (CO partial current density of $420 \pm 50\,mA\,cm^{-2}$ for over 200 hours) (Endrődi et al. 2021).

Finally, we note that beyond the obvious practical importance of this topic, there is an additional scientific aspect to be studied. It was shown in the example of water splitting that microgravity significantly influences both fluid dynamics and bubble formation, and, therefore, the kinetics of the whole process (Davenport et al. 1991; Sakuraia et al. 2013). As there is no spontaneous mixing in microgravity, it becomes crucially important to tailor the electrolyzer structure, such as the gas and liquid channels, gas diffusion layers, and catalyst layers to induce sufficient gas and liquid flow, and to ensure that the products leave the cell. Overall, there is an accelerated development in the field, and we foresee the appearance of novel technologies to be deployed in space mission soon.

6.7 Synthetic Chemistry, Purification, and Analytics in Space

6.7.1 Synthetic Efforts in Space

Synthetic chemistry experiments performed in microgravity environment have been scarce so far. In this chapter we discuss recent examples of synthetic chemistry in space including purification techniques and analytical aspects for space chemistry. Until recent years, performing controlled chemical syntheses in space has not been possible due to the lack of an appropriate equipment to do that. In 2019–2020, the first flow chemistry platform for synthetic reactions on the ISS was developed by the Beeler Research Group (Boston University) and Space Tango 2019 (https://www.nasa.gov/mission_pages/station/research/experiments/explorer/Investigation.html?#id=8172; Beeler 2020). Three different organic reactions were carried out inside the automated Flow Chemistry Cube Lab: a peptide coupling, a monophasic, and a biphasic colorimetric reaction. Peptides are composed of amino acids connected by peptide (amide) bonds. Peptide couplings, or more generally amide bond constructions are among the most frequently performed chemical reactions in drug discovery (Brown and Boström 2016, pp. 4443–4458). It is not surprising, therefore, that in the first experiment this reaction was investigated (Scheme 6.1). The samples from the experiment were returned to Earth and analyzed by liquid chromatography – mass spectrometry (LC–MS) and nuclear magnetic resonance (NMR) spectroscopy. The coupling reaction in this initial work gave comparable results to those obtained on Earth (Beeler 2020) indicating directional information for futures study using the automated flow-chemistry reactor system in microgravity. Reaction

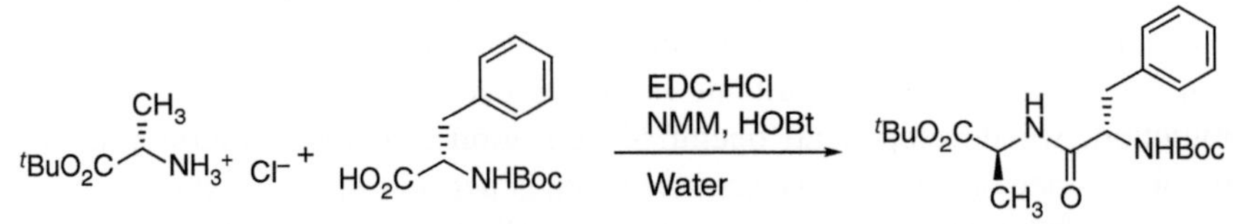

Scheme 6.1 Peptide coupling reaction performed on ISS.

of alanine and phenylalanine in microgravity environment resulted in 85% yield and 95% mass recovery, respectively, at a residence time of three hours.

Phenolphthalein is an organic dye often used as an indicator in acid–base titrations. The compound is pink in basic solutions and turns colorless in acidic solutions. In the second experiment, a pink solution of phenolphthalein (i.e. at basic pH) was reacted with acetic acid in water to give a colorless solution at the end of the reaction (Scheme 6.2). In the third experiment, phenolphthalein was reacted with benzoic acid dissolved in toluene to afford a biphasic reaction mixture. These experiments allowed the researcher team to investigate fundamental aspects of fluid dynamics, and in particular mixing properties in microgravity environment. The outcome of the colorimetric reactions was followed by using the onboard cameras inside the Flow Chemistry Cube Lab.

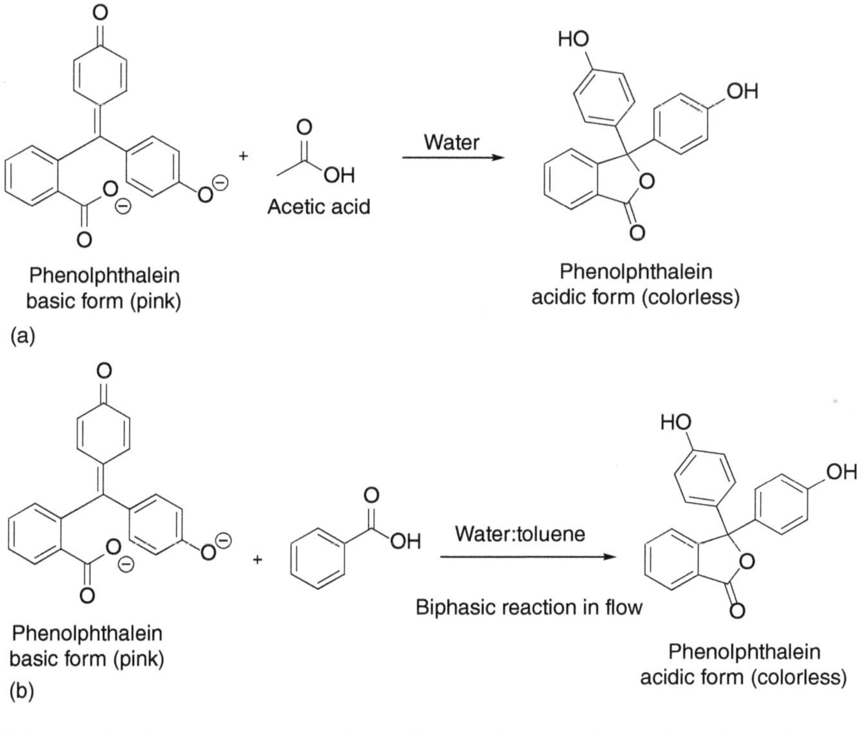

Scheme 6.2 Acid–base colorimetric reactions under monophasic (a) and biphasic (b) reaction conditions performed on ISS.

In 2017, in the payload of DIDO two free orbit satellite launched by SpacePharma (www.space4p.com), the oxido-reductase catalyzed enzymatic oxidation of glucose was studied (Scheme 6.3). Resazurin was employed as an indicator and the reaction was monitored by spectrometry (https://www.space4p.com/missions). Results showed that kinetics was tenfold slower compared to ground reference, and different ratio of secondary oxidation was observed.

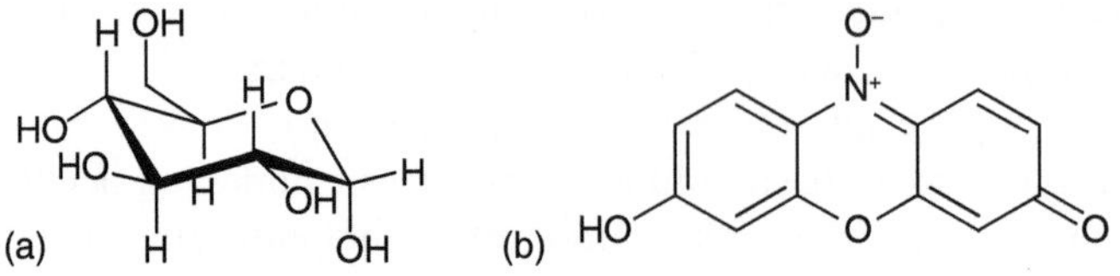

Scheme 6.3 Glucose (a) and resazurin dye (b).

6.7.2 Purification and Analysis in Space

Liquid separation is one of the critical steps for chemical (flow) synthesis. Liquid–liquid extractions allow chemists to separate compounds with different relative solubilities in immiscible liquids. Sufficient phase separation is thus fundamental to achieve successful extractions. Commonly liquid–liquid separations rely on sedimentation. The Zaiput (www.zaiput.com) Separator (see Section 6.8), which relies on separation based on surface tension, has been tested for continuous liquid–liquid separation on the ISS (https://www.nasa.gov/mission_pages/station/research/experiments/explorer/Investigation.html?#id=7696; https://www.zaiput.com/zaiput-in-space).

In most cases, after a reaction has taken place, the product needs to be purified. To do this, purification techniques such as extraction, crystallization, distillation, and chromatography are available for chemical practitioners and should be considered for future syntheses applications in space. Protein crystallization, as a physico-chemical formulation process (discussed in Sections 6.4 and 6.5), has been investigated in detail in microgravity environment in order to access high-quality crystals for single-crystal X-ray diffraction studies (Mohan et al. 2010; McPherson and DeLucas 2015; Strelov et al. 2014, pp. 781–806; McPherson and DeLucas 2015, pp. 1–20). Nevertheless, we are not aware of experiments utilizing crystallization as a standard purification technique in space.

Polymer research is one of the few areas of synthetic chemistry where researchers deliberately have sought microgravity conditions for three reasons: (i) microgravity greatly reduces buoyancy-driven fluid flow; (ii) the pressure head effects are virtually eliminated; and (iii) sedimentation is virtually eliminated. The possibilities of synthesizing more homogeneous polymer materials in space compared to terrestrial specimens have been discussed briefly (Downey and Pojman 2001; Briskman et al. 2001, pp. 169–180).

Thanks to its generality, in small-scale chemical drug discovery labs, column chromatography is the go-to method for compound purification. Automatic flash chromatograph systems are available for this purpose. Their footprint, however, is relatively big, and the method can generate high amount of solvent waste and significant human interaction may still be required. Therefore, it is expected to take some time and development before column chromatography becomes widely available for space applications. Focusing on the method's analytical potential instead of purification, The JET Propulsion Laboratory (JPL) (www.jpl.nasa.gov) researched and suggested microfluidics high-performance liquid chromatographic chip integrated to an ion trap mass spectrometer for answering fundamental questions in astrobiology (Kidd et al. 2015).

On-orbit analysis of chemical reactions is not straightforward, since most earth-based analytical instruments used in organic chemistry have to face a number of challenges when applied in space: besides issues originated from microgravity, the energy consumption, the large weight or size are prohibitive factors. Nevertheless, a number of analytical instruments have either been used in space or considered already for space chemistry application.

Mass spectrometers have a rich history and promising future in planetary exploration (Arevalo et al. 2019, pp. 1–17). It was NASA's Phoenix Mars Lander, equipped with a high-temperature furnace and mass spectrometer (thermal and evolved gas analyzer [TEGA]) that confirmed the existence of water ice on Mars in 2008 (Thompson 2008). The International Space Station uses mass spectrometry to monitor air quality. The so-called Spacecraft Atmosphere Monitor (SAM) is one of the smallest autonomous gas chromatograph-mass spectrometer instruments ever built, which is able to immediately detect trace contaminants that pose potential threats to crewmembers' well-being (https://www.nasa.gov/feature/nasas-spacecraft-atmosphere-monitor-goes-to-work-aboard-the-international-space-station).

NMR spectroscopy is suitable for characterizing ferromagnetic materials, thus a potential tool for analyzing samples of planets or asteroids. This technology can be operated even without containing permanent magnets by exploiting the natural magnetic fields of the mineral phases, resulting in their much lighter weight and making the technology a candidate for space research (https://www.techbriefs.com/component/content/article/tb/techbriefs/physical-sciences/6416). NMR systems and their miniaturization are intensively researched nowadays (Zalesskiy et al. 2014, pp. 5641–5694).

Spectrometers have widely been adapted for diverse space applications and have become indispensable components of space and planet exploration missions. The FTIR-based ANITA system represented a versatile air quality monitor on the ISS, allowing for the first time the detection and monitoring of trace gas dynamics, with high time resolution, in a spacecraft atmosphere (Honne et al. 2011). The CubeSat-based SpectroCube, a miniaturized *in situ* space exposure platform (Elsaesser et al. 2020, pp. 275–288), is an infrared spectrometer interfaced with a sample handling system that was designed to measure photochemical changes of organic molecules. The Mars Curiosity Rover's ChemCam fires a laser and analyzes the elemental composition of vaporized materials by an on-board spectrograph from areas smaller than 1 mm on the surface of Martian rocks and soils (https://mars.nasa.gov/msl/spacecraft/instruments/chemcam). The Perseverance Rover is equipped with Raman and fluorescence spectrometers, which upon modification should also be suitable for analysis of small organic molecules (https://mars.nasa.gov/mars2020/spacecraft/instruments). During ESA's ExoMars rover mission, a Raman laser spectrometer will investigate the mineralogical composition of powders sampled on Mars (Veneranda et al. 2020, pp. 1–14). The Microdevices Laboratory (MDL) at JPL has recently developed the "Chemical Laptop," the first battery-powered, automated, reprogrammable, portable instrument, that houses the microfluidics, electronics, and optics needed to perform highly sensitive microchip electrophoresis coupled to laser-induced fluorescence detection of organic acids and other organic biomarkers (Mora et al. 2020, pp. 12959–12966).

The establishment of flow chemical equipment for space research, in-space processing, and synthetic chemistry will evidently demand the inherent analytical tools soon, giving rapid information about the performance and/or quality of the chemical syntheses and processes realized. Although most analytical developments already validated in space have been targeted for space exploration, several of them will be applicable for tailoring to follow specific chemical processes or routes, or *in situ* analyze chemical syntheses. This means nonetheless a great challenge, considering that chemical reactions and compounds, such as of pharmaceutical production, usually requires individual methods for their reliable analysis.

6.8 Flow Instruments for Space Chemistry

6.8.1 The Flow Chemistry Advantage in Space

Flow chemistry has matured over the past two decades from early demonstrations of simple chemical transformations in microreactors to complex, multistep synthesis which are applied routinely at fine chemistry and pharmaceutical fields (Darvas et al. 2014). The technology has received a remarkable amount of attention in recent years with many reports and publications on what can be done in flow (Darvas et al. 2014; Plutschack et al. 2017, pp. 11796–11893; Luis and Garcia-Verdugo 2019; Noël and Luque, 2020). Flow chemical processes offer numerous advantages to organic synthesis over the traditional “in-a-flask” batch operation, such as assured safety, the environmentally benign green nature, shorter reaction times, excellent yields and selectivity, reduced amount of by-products and improved product quality, use of elevated reaction conditions, easy reaction control, enhanced optimization and faster scale-up opportunities, cost efficiency, and overall production flexibility. Evidently, translation of batch chemistries into continuous-flow processes is not an easy task (Akwi and Watts 2018, pp. 13894–13928).

Chemistry on Earth can thus be performed either in flasks or flow systems https://thalesnano.com/applications. This is not the case under microgravity where the solution in a flask would not mix well and reactions would not be reproducible. The emergence of flow reactors thus provided revolutionary new options to perform chemical reactions in space (Sipos et al. 2017, pp. 151–156; Mezohegyi and Darvas 2019, Hessel, et al. 2020, pp. 115774–115794). The main advantages of flow reactors vs. batch synthesis instruments in space are the following:

- ensuring efficient mixing of fluids and higher reproducibility of chemical experimentation;
- operability via complete automation and by remote control;
- ability to operate in an enhanced chemical reaction space: at high pressure and temperature which are difficult to produce by batch reactors;
- allowing reactions to be performed on very small scale, then producing a fast scale up.

In 2019, the ISS National Lab recognized that flow chemistry is an important application for future discovery and commercial efforts (https://spacetango.com/latest/space-tango-announces-iss-flow-chemistry-collaboration-with-boston-university-beeler-research-group).

6.8.2 Flow Reactors for Space Applications

It is worth to note the wide variety of flow reactor products on the market, which after appropriate modification, seem to be applicable for chemical synthesis and production in space (Figure 6.4). A simple assembly of a continuous-flow reactor includes a pump to transport the reaction reagent(s) throughout the setup and the reactor unit itself in which the chemical transformation takes place. Reactions with two or more reagents in separate streams require additional pumps to be included, as mixers to ensure efficient contact between the reagents. A chemical reaction often requires either cooling or heating in order to dissipate the excess heat or to provide energy to initiate and maintain the reaction. A back-pressure regulator is required in general to maintain a steady flow, which is critical for controlling residence times. A mass-flow controller is usually employed in flow chemistry to control the fluid flow to a set point. Inline analytics are often used to provide information about the progress of reaction performance. A detailed overview about the technology devices, configurations, and reactor solutions was given by Fekete and Glasnov (2014, pp. 95–140).

On the other hand, to design a flow reactor which is destined for operation in space is a challenging task. In practice, the fundamental aspects of constructing a flow reactor should be reconsidered. The main factors of designing a flow reactor on orbit are to define properly the material requirements of the reactor to adapt the reactor to the nature of the reactions to be performed. The pressure and temperature conditions within the reactor must follow strict safety requirements of the in-space applications. The second factor is the size of the reactors, as only limited space is available for chemical instruments brought to the orbit. The third component to be considered is the limited nature of the supplying resources such as electricity and storage volumes.

When designing the reactions in space, we should consider the material changes of the physicochemical and other processes governed by mass transfer due to microgravity (see Section 6.2). The dynamics of fluids in space is also changing relative to a terrestrial laboratory (Monti 2001). An innovative approach for flow chemistry in space is an instrument design with modular assembly. Here modules can be connected to each other in reconfigurable combinations to create a versatile chemical reactor system (Guidi et al. 2020, pp. 8910–8932). Recently, a miniaturized and modular-flow chemical reactor prototype for microgravity applications has been developed by ThalesNano (www.thalesnano.com) together with Innostudio (www.innostudio.org). The 2U-sized equipment scheduled for space qualification enables the performance of all of the important flow chemical synthesis type in a single reactor. In the down-scaled modules of the instrument, liquid–liquid reactions, photochemical/photocatalytic reactions, heterogeneous catalytic (packed-bed) transformations, and also electrochemical syntheses can be performed (Mezohegyi and Darvas 2021). Significant section of the potential on-demand syntheses of

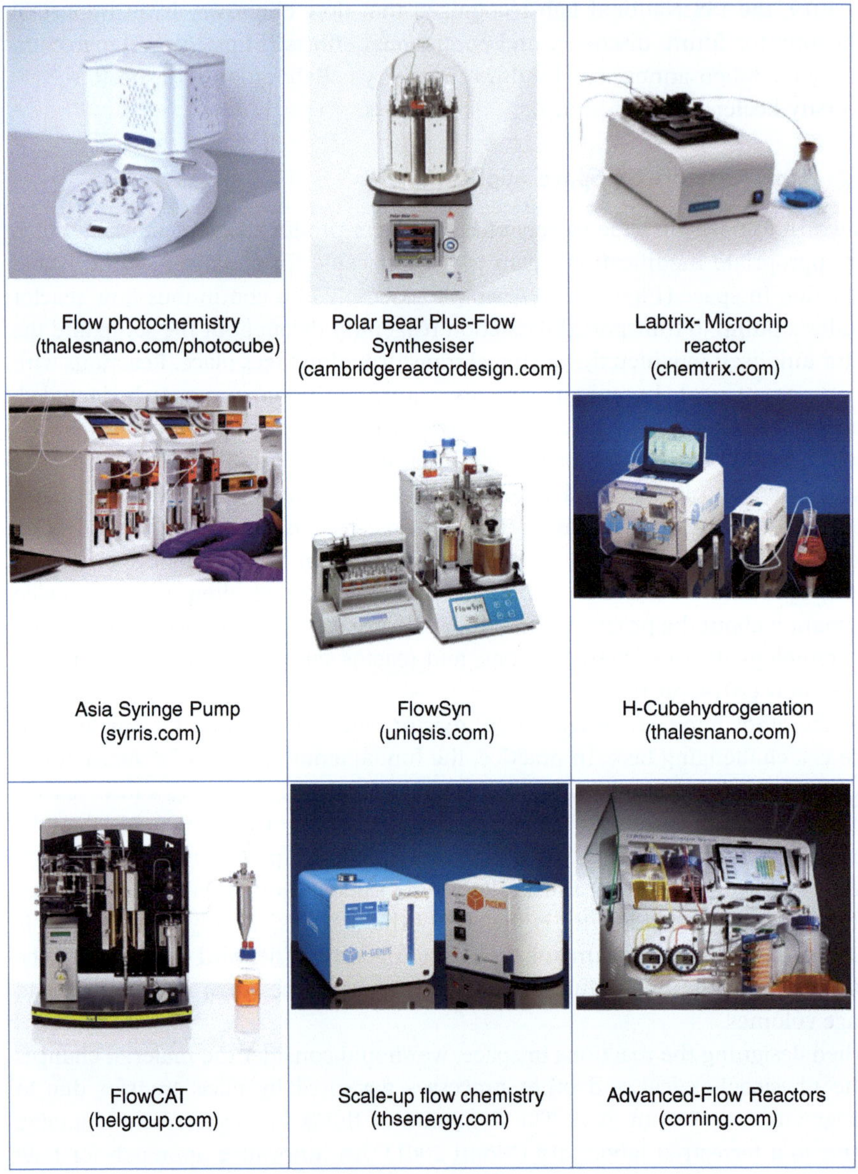

Figure 6.4 Examples of flow chemical reactors available on the market.

organic chemicals in space is expected to be covered by the modular prototype. Its chemical performance under microgravity, including the synthesis of a world top API, and testing of the reactor will be targeted on ESA's Space Rider (https://www.esa.int/Enabling_Support/Space_Transportation/Space_Rider) maiden flight, the researchers stated (Mezohegyi and Darvas 2021).

The first systematic approach to perform systematic organic synthesis experiments in space by using flow reactors was made by the Beeler research group (https://www

.beelerlab.com/research.html) in 2020. Partnering Space Tango (www.spacetango .com), the team investigated three fully automated chemical reactions on the ISS and made the first effort to synthesize organic compounds onboard (Beeler 2020). The sealed 9U tool was equipped with cameras, pumps, flow meters, pressure sensors, and sophisticated valving to ensure safe and efficient operation. The flow system was capable of ensuring a reaction temperature between 4 and 100 °C, a flow rate between 0.25 and 2.0 ml /min, and also end-product capture (Stoudemire 2020). The reactor, developed by the Beeler group, at its size of 8.0 cm × 5.5 cm, had a useful volume of 1 ml. The reactions included peptide coupling and a water-phase colorimetric reaction (synthesis of phenolphthalein) (see details in Section 6.7).

Today, the scientific and commercial opportunities via flow chemical equipment and applications in space research and space industry are getting more and more clear. Among others, the ICE Cubes Service (www.icecubesservice.com), available in the Columbus module of ISS, is a capable experiment platform offering to host a wide range of experiments such as reaction for drug discovery, organ-on-a-chip, cell biology, protein crystallization, 3D tissue engineering, additive manufacturing, nanoparticles etc. An extension of the platform by a multi-user chemistry platform in the near future is being considered, which will include also a flow chemistry module, among others (Stenuit et al. 2021).

6.9 Conclusions

Space chemistry is a fast-emerging novel field, already with some important applications in the pharmaceutical industry. It is expected that versatile flow reactors operating in space will open new avenues to establish novel synthetic routes to existing drugs and new reaction classes leading to unusual chemical structures.

On the other hand, as our chapter demonstrates systematic studies of indigenous synthetic opportunities offered exclusively by space have yet to be started. The reasons are multiple: theoretical chemistry should more precisely pinpoint the unique synthetic opportunities which will be available exclusively in space; on instrumental side, synthesis, purification, and analysis should be provided by complex and compact instruments. A louder voice expressing the need for breakthrough results to the multiple important unsolved industrial problems will definitely stimulate further efforts of space chemistry on orbit.

Nevertheless, the space chemistry community believes that the versatile flow reactors under development will open new avenues to many fields in chemistry to the practical benefit of humankind on Earth.

References

Advincula, R.C. (2019). Polymers and nanomaterials for space manufacturing: A flow chemistry demonstration. *Oral Presentation at the 4th Space Chemistry Symposium in Orlando, USA (1 April 2019).*

Akwi, F.M. and Watts, P. (2018). Continuous flow chemistry: where are we now? Recent applications, challenges and limitations. *Chemical Communications* 54: 13894–13928.

Arevalo, R. Jr., Ni, Z., and Danell, R.M. (2019). Mass spectrometry and planetary exploration: a brief review and future projection. *Journal of Mass Spectrometry* 55 (4454): 1–17.

Bába, P., Rongy, L., De Wit, A. et al. (2018). Interaction of pure marangoni convection with a propagating reactive interface under microgravity. *Physical Review Letters* 121: 024501-1–024501-6.

Bába, P., Tóth, Á., and Horváth, D. (2019). Surface-tension-driven dynamic contact line in microgravity. *Langmuir* 35: 406–412.

Beeler, A. (2020). Flow chemistry on the ISS-NL. *Oral Presentation at the (virtual) International Space Station Research and Development Conference (22 September 2020).*

Benz, K. and Dold, P. (2002). Crystal growth under microgravity: present results and future prospects towards the International Space Station. *Journal of Crystal Growth* 237: 1638–1645.

Braddock, M. (2017). Ergonomic challenges for astronauts during space travel and the need for space medicine. *Journal of Ergonomics* 7 (221): 2.

Braddock, M. (2020). From target identification to drug development in space: using the microgravity assist. *Current Drug Discovery Technologies* 17 (1): 45–56. https://doi.org/10.2174/1570163816666190112150014.

Briskman, V.A., Yudina, T.M., and Kostarev, K.G. (2001). Polymerization in microgravity as a new process in space technology. *Acta Astronautica* 48 (2–3): 169–180.

Brown, D.G. and Boström, J. (2016). Analysis of past and present synthetic methodologies on medicinal chemistry: Where have all the new reactions gone? *Journal of Medicinal Chemistry* 59: 4443–4458.

Chaudhry, Q., Castle, L., and Watkins, R. (2017). *Nanotechnologies in Food*, 2nde. Royal Society of Chemistry.

Chen, N.F., Zhong, X., Lin, L. et al. (2001). Comparison of field effect transistor characteristics between space-grown and earth-grown gallium arsenide single crystal substrates. *Applied Physics Letters* 78: 478–479.

Coley, C.W., Thomas, D.A., Lummiss, J.A. et al. (2019). A robotic platform for flow synthesis of organic compounds informed by AI planning. *Science* 365 (6453), eaax1566.

Cröll, A., Kaiser, T., Schweizer, M. et al. (1998). Floating-zone and floating-solution-zone growth of GaSb under microgravity. *Journal of Crystal Growth* 191: 365.

Dallinger, D., Gutmann, B., and Kappe, C.O. (2020). The concept of chemical generators: on-site on-demand production of hazardous reagents in continuous flow. *Accounts of Chemical Research* 53 (7): 1330–1341. https://doi.org/10.1021/acs.accounts.0c00199.

Dantuma, D.M., Elmaddawi, R., Pathck, Y. et al. (2015). Impact of simulated microgravity on nanoemulsion stability? A preliminary research. *American Journal of Medical and Biological Research* 3 (49): 102–106. https://doi.org/10.12691/ajmbr-3-4-4.

Darvas, F., Dormán, G., and Hessel, V. (ed.) (2014). *Flow chemistry – Volume 1: Fundamentals*. Berlin: De Gruyter.

Daugherty, A.L. and Mrsny, R.J. (2010). Formulation and delivery issues for monoclonal antibody therapeutics. In: *Current Trends in Monoclonal Antibody Development and Manufacturing*, 103–129, New York. Springer.

Davenport, R.J., Schubert, F.H., and Grigger, D.J. (1991). Space water electrolysis: space station through advanced missions. *Journal of Power Sources* 36: 235–250. https://doi .org/10.1016/0378-7753(91)87004-U.

Deller, M.C. and Rupp, B. (2014). Approaches to automated protein crystal harvesting. *Acta Crystallographica Section F: Structural Biology and Crystallization Communications* 70: 133–155.

DeLucas, L.J., Suddath, F., Snyder, R. et al. (1986). Preliminary investigations of protein crystal growth using the space shuttle. *Journal of Crystal Growth* 76: 681–693.

Demontis, G.C., Germani, M.M., Caiani, E.G. et al. (2017). Human pathophysiological adaptations to the space environment. *Frontiers in Physiology* 8: 547. https://doi.org/ 10.3389/fphys.2017.00547.

DeRosa, M.C., Monreal, C., Schnitzer, M. et al. (2010). Nanotechnology in fertilizers. *Nature Nanotechnology* 5: 91.

Devalapally, H., Chakilam, A., and Amiji, M.M. (2007). Role of nanotechnology in pharmaceutical product development. *Journal of Pharmaceutical Sciences* 96 (10): 2547–2565.

Downey, J.P. and Pojman, J.A. (2001). *Polymer Research in Microgravity: Polymerization and Processing*, ACS Symposium Series No. 793. Washington, D.C., USA: American Chemical Society.

Elsaesser, A., Merenda, F., Lindner, R. et al. (2020). SpectroCube: a European 6U nanosatellite spectroscopy platform for astrobiology and astrochemistry. *Acta Astronautica* 170: 275–288.

Endrődi, B., Bencsik, G., Darvas, F. et al. (2017). Continuous-flow electroreduction of carbondioxide. *Progress in Energy and Combustion Science* 62: 133–154.

Endrődi, B., Kecsenovity, E., Samu, A. et al. (2019). Multilayer electrolyzer stack converts carbon dioxide to gas products at high pressure with high efficiency. *ACS Energy Letters* 4: 1770–1777. https://doi.org/10.1021/acsenergylett.9b01142.

Endrődi, B., Kecsenovity, E., Samu, A. et al. (2020). High carbonate ion conductance of a robust PiperION membrane allows industrial current density and conversion in a zero-gap carbon dioxide electrolyzer cell. *Energy & Environmental Science* 13: 4098–4105. https://doi.org/10.1039/D0EE02589E.

Endrődi, B., Samu, A., Kecsenovity, E. et al. (2021). Operando cathode activation with alkali metal cations for high current density operation of water-fed zero-gap carbon dioxide electrolysers. *Nature Energy* 6: 439–448. https://doi.org/10.1038/s41560-021-00813-w.

ESA (2020). COVID-19 drug research and biomining launching to Space Station. https://www.esa.int/Science_Exploration/Human_and_Robotic_Exploration/ COVID-19_drug_research_and_bio-mining_launching_to_Space_Station (accessed 23 June 2021).

Fekete, M. and Glasnov, T. (2014). Technology overview/overview of the devices. In: *Flow Chemistry Volume 1: Fundamentals* (ed. F. Darvas, G.Y. Dormán and V. Hessel), 95–140. Berlin: De Gruyter.

FROST5 (2015). The Space Chemistry Consortium was founded during the FROST5 Conference on 21 October 2015, in Budapest, Hungary, originally called SpaceFlow Consortium. Founding members are Ferenc Darvas (Flow Chemistry Society, Switzerland; ThalesNano, Inc., Hungary), C. Oliver Kappe (University of Graz, Austria), Volker Hessel (University of Adelaide, Australia), Rigoberto Advincula (Case Western Reserve University, USA), Csaba Janáky (University of Szeged, Hungary), Paul Watts (Nelson Mandela University, South Africa), Holger Löwe (Johannes Gutenberg University, Mainz, Germany), Richard Jones (ThalesNano, Inc., Hungary), Mimi Hii (Imperial College, London, UK), András Guttman (Northeastern University, USA), Aaron Beeler (Boston University, USA), Daryl Sauer (University of Wisconsin-Parkside, USA).

Garrett-Bakelman, F.E., Darshi, M., Green, S.J. et al. (2019). The NASA twins study: a multidimensional analysis of a year-long human spaceflight. *Science* 364 (6436), eaau8650. https://doi.org/10.1126/science.aau8650.

Guidi, M., Seeberger, P.H., and Gilmore, K. (2020). How to approach flow chemistry. *Chemical Society Reviews* 49: 8910–8932.

Hass, E.C., Mezey, P.G., and Plath, P.J. (1981). A non-empirical molecular orbital study on Loewenstein's rule and zeolite composition. *Journal of Molecular Structure: THEOCHEM* 76: 389–399.

Hauch, A., Küngas, R., Blennow, P. et al. (2020). Recent advances in solid oxide cell technology for electrolysis. *Science* 370: 6513. https://doi.org/10.1126/science.aba6118.

Herrmann, F.M. and Müller, G. (1995). Growth of 20 mm diameter GaAs crystals by the floating-zone technique with controlled As-vapour pressure under microgravity. *Journal of Crystal Growth* 156: 350–360.

Hessel, V., Sarafraz, M.M., and Tran, N.N. (2020). The resource gateway: microfluidics and requirements engineering for sustainable space systems. *Chemical Engineering Science* 225: 115774–115794.

Hessel, V., Tran, N.N., Orandi, S. et al. (2021). Continuous-flow extraction of adjacent metals—a disruptive economic window for in situ resource utilization of asteroids? *Angewandte Chemie, International Edition* 60 (7): 3368–3388.

Honne, A., Schumann-Olsen, H., Kaspersen, K. et al. (2011). ANITA – An FTIR-based continuous air quality monitoring system on the ISS (International Space Station). In: *Oral presentation at the Applied Industrial Optics: Spectroscopy, Imaging and Metrology, 10-14 July 2011, Toronto, Canada.*

Horváth, D., Budroni, M.A., Bába, P. et al. (2014). Convective dynamics of traveling autocatalytic fronts in a modulated gravity field. *Physical Chemistry Chemical Physics* 16: 26279–26287.

IMRET 13 (2014). qqal Conference on MicroREaction Technology (IMRET 13) in Budapest, Hungary, 23-25 June, 2014, organized by the Flow Chemistry Society in collaboration with Akadémiai Kiadó.

Jones, R., Darvas, F., and Janáky, C.S. (2017). New space for chemical discoveries. *Nature Reviews Chemistry* 1 (0055): 1–3.

Joseph, C. and Wood, D. (2021). Analysis of the microgravity research ecosystem and market drivers of accessibility. *New Space* 9 (2): 123–138.

Karr, L.J., Miller, T.Y. and Donovan, D.N. (2015). Macromolecular crystal growth. Report NP-2015-08-027-JSC. https://www.nasa.gov/sites/default/files/atoms/files/np-2015-08-027-jsc_macromolecular_crystals-iss-mini-book-102715-508c.pdf (accessed 30 June 2021).

Kidd, R.D., Scianmarello, N., Wiley, J. et al. (2015). Miniature liquid chromatographic systems for human and robotic missions. In: *Poster presentation at the 10th Harsh-Environment Mass Spectrometry Workshop, 14 September 2015, Baltimore, USA.*

Kornuta, D., Abbud-Madrid, A., Atkinson, J. et al. (2019). Commercial lunar propellant architecture: A collaborative study of lunar propellant production. *REACH* 13: 100026.

Laczkó, D., Hogan, M.J., Toulmin, S.A. et al. (2020). A single immunization with nucleoside-modified mRNA vaccines elicits strong cellular and humoral immune responses against SARS-CoV-2 in mice. *Immunity* 53: 724–732.

Li, G. (2018). How NASA has contributed to chemistry. https://www.chemistryworld.com/news/how-nasa-has-contributed-to-chemistry/3009574.article (accessed 23 June 2021).

Lijie, W.L., Liqin, W.L., Arimoto, S., and Mezey, P.G. (2006). Large-scale chirality measures and general symmetry deficiency measures for functional group polyhedra of proteins. *Journal of Mathematical Chemistry* 40: 145–153.

Littke, W. and John, C. (1986). Protein single crystal growth under microgravity. *Journal of Crystal Growth* 76 (3): 663–672.

Luis, S.V. and Garcia-Verdugo, E. (ed.) (2019). *Flow Chemistry: Integrated Approaches for Practical Applications.* Cambridge: RSC Publishing.

Manek, E., Jones, R., Fail, J., Darvas, F. (2019). Novel high throughput nanoformulation platform technology to enhance water solubility and leaf/cuticle penetration of nutrients in extra-terrestrial hydroponic plantations. *ACS Spring 2019 National Meeting, Division of Agricultural and Food Chemistry, Session of Food for Space Travel & Extreme Environments (March 31 – April 4, 2019), Orlando, FL, USA.*

Mattioda, A., Cook, A., Ehrenfreund, P. et al. (2012). The O/OREOS mission: first science data from the space environment viability of organics (SEVO) payload. *Astrobiology* 12 (9): 841–853.

McPherson, A. and DeLucas, L. (2015). Microgravity protein crystallization. *NPJ Microgravity* 1: 15010. https://doi.org/10.1038/npjmgrav.2015.10.

Mehta, P. and Bhayani, D. (2017). Impact of space environment on stability of medicines: challenges and prospects. *Journal of Pharmaceutical and Biomedical Analysis* 136: 111–119.

Mezey, P.G. (1984). The future and impact of quantum mechanical calculations in the description, and characterization of zeolites. In: *Catalytic Materials: Relationship Between Structure and Reactivity*, ACS Symposium Series, vol. 248 (9) (ed. E. Derouane et al.), 145–156.

Mezey, P.G. (1993). *Shape in Chemistry: An Introduction to Molecular Shape and Topology*. New York: Wiley-VCH.

Mezey, P.G. (1999). The holographic electron density theorem and quantum similarity measures. *Journal of Molecular Physics.* 96: 169–178.

Mezey, P.G., Fukui, K., Arimoto, S., and Taylor, K. (1998). Polyhedral shapes of functional group distributions in biomolecules and related similarity measures. *International Journal of Quantum Chemistry* 66: 99–105.

Mezohegyi, G. and Darvas, F. (2019). Trends in space chemistry: emerging markets and potential applications. *Oral presentation at the 17th Reinventing Space Conference in Belfast, Northern Ireland (13 November 2019).*

Mezohegyi, G. and Darvas, F. (2021). Triggering economy growth via space chemistry, an emerging sector within the space industry. *Oral Presentation at the 18th Reinventing Space Conference in London, UK (30 June 2021).*

Mezohegyi, G., Buchholcz, B., Sohajda, T. et al. (2021). Space chemistry and global health: Drug development against COVID-19 in space. *Technical Presentation at the (virtual) 58th Scientific and Technical Subcommittee of the United Nations Office for Outer Space Affairs (23 April 2021).*

Mohan, D.A., Ashok, K.B., Ravindra, B.L. et al. (2010). *Handbook of Crystal Growth* (ed. G. Dhanaraj, K. Byrappa, V. Prasad and M. Dudley). Berlin, Heidelberg: Springer-Verlag.

Monti, R. (ed.) (2001). *Physics of Fluids in Microgravity*. New York: Taylor and Francis.

Mora, M.F., Kehl, F., da Costa, E.T. et al. (2020). Fully automated microchip electrophoresis analyzer for potential life detection missions. *Analytical Chemistry* 92 (19): 12959–12966.

Moreno, A. (2017). Advanced methods of protein crystallization. *Protein Crystallography* 1607: 51–76. https://doi.org/10.1007/978-1-4939-7000-1_3.

Nelson, G.J., Vijapur, S.H., Hall, T.D. et al. (2020). Electrochemistry for space life support. *Electrochemical Society Interface* 29 (1): 47–52. https://doi.org/10.1149/2.F06201IF.

Neves, J.M., Collins, P.J., Wilkerson, R.P. et al. (2019). Microgravity effect on microstructural development of tri-calcium silicate (C_3S) paste. *Frontiers in Materials* 6 (83): 1–12.

Nijhuis, J., Tran, Q.D., Tran, N.N. et al. (2021). Toward on-board microchip synthesis of CdSe vs. PbSe nanocrystalline quantum dots as a spectral decoy for protecting space assets. *Reaction Chemistry & Engineering* 6 (3): 471–485.

Noël, T. (2021). Harvesting solar energy for pharmaceutical production in outer space using flow chemistry. *Oral Presentation at the (Virtual) 6th Space Chemistry Symposium (6 April 2021).*

Noël, T. and Luque, R. (ed.) (2020). *Accounts on Sustainable Flow Chemistry*. Switzerland: Springer.

Pflugrath, J.W. (1992). Developments in X-ray detectors. *Current Opinion in Structural Biology* 2: 811–815.

Plutschack, M.B., Pieber, B., Gilmore, K., and Seeberger, P.H. (2017). The hitchhiker's guide to flow chemistry. *Chemical Reviews* 117: 11796–11893.

Pópity-Tóth, É., Pimienta, V., Horváth, D., and Tóth, Á. (2013). Hydrodynamic instability in the open system of the iodate–arsenous acid reaction. *The Journal of Chemical Physics* 139: 164707-1–164707-6.

Rapp, D. (2008). *Human Missions to Mars: Enabling Technologies for Exploring the Red Planet*. Heidelberg, New York: Springer-Praxis Books in Astronautical Engineering.

Rehder, D. (2011). *Chemistry in Space: From Interstellar Matter to the Origin of Life*. Weinheim: Wiley-VCH.

Reichert, P., Prosise, W., Fischmann, T.O. et al. (2019). Pembrolizumab microgravity crystallization experimentation. *NPJ Microgravity* 5 (1): 1–8. https://www.nasa.gov/mission_pages/station/research/benefits/mab/.

Rosales, D. (2020). Space Tango and Alpha Space partner with University of Adelaide for pharmaceutical stability study on the ISS. https://www.issnationallab.org/iss360/space-tango-alpha-space-university-adelaide-pharmaceutical-stability-ng14 (accessed 23 June 2021).

Ruyters, G., Betzel, C., and Grimm, D. (2017). *Biotechnology in Space*. Cham: Springer ISBN: 978-3-319-64053-2.

Ryder, P. and Braddock, M. (2020). Harnessing the space environment for the discovery and development of new medicines. In: *Handbook of Space Pharmaceuticals* (ed. Y. Pathak, M. Araújo dos Santos and L. Zea). Cham: Springer 10.1007/978-3-319-50909-9_32-1.

Sakuraia, M., Sonea, Y., Nishida, T. et al. (2013). Fundamental study of water electrolysis for life support system in space. *Electrochimica Acta* 100: 350–357. https://doi.org/10.1016/j.electacta.2012.11.112.

Samburski, G. (2020). Studying biochemistry in space. *Oral Presentation at the (virtual) 5th Space Chemistry Symposium at ACS Autumn 2020 National Meeting (19 August 2020).*

Santos, A.C., Morais, F., Simões, A. et al. (2019). Nanotechnology for the development of new cosmetic formulations. *Expert Opinion on Drug Delivery* 16: 313–330.

Sato, N. (2020). JAMSS pilot mission of "Kirara" in-space manufacturing using ISS. *Oral Presentation at the (virtual) 5th Space Chemistry Symposium (19 August 2020).*

Schlüter, L. and Cowley, A. (2020). Review of techniques for in-situ oxygen extraction on the Moon. *Planetary and Space Science* 181: 104753.

Schmidt, M.A., Schmidt, C.M., and Goodwin, T.J. (2019). *Pharmacogenomics in Spaceflight. Handbook of Space Pharmaceuticals* (ed. Y. Pathak, M. Araújo dos Santos and L. Zea), 1–39. Springer https://doi.org/10.1007/978-3-319-50909-9_26-1.

Sipos, G., Bihari, T., Milánkovich, D., and Darvas, F. (2017). Flow chemistry in space – a unique opportunity to perform extraterrestrial research. *Journal of Flow Chemistry* 7 (3-4): 151–156.

Space Tango (2019). Space Tango announces ISS flow chemistry collaboration with Boston University Beeler Research Group. https://spacetango.com/latest/space-tango-announces-iss-flow-chemistry-collaboration-with-boston-university-beeler-research-group (accessed 28 June 2021).

Starr, S.O. and Muscatello, A.C. (2020). Mars in situ resource utilization: a review. *Planetary and Space Science* 182: 104824.

Stenuit, H., Sapera, A., and Mulligan, M. (2021). Collaboration on a multi-user chemistry platform in space – Contribution by ICE Cubes. *Oral Presentation at the (virtual) 6th Space Chemistry Symposium (6 April 2021).*

Stoudemire, J. (2020). Flow chemistry on the International Space Station – Novel approaches to chemical synthesis and fully automated microreactor systems. *Oral Presentation at the (virtual) 5th Space Chemistry Symposium (19 August 2020).*

Strelov, V.I., Kuranova, I.P., Zakharov, B.G., and Voloshin, A.E. (2014). *Crystallography Reports* 59: 781–806.

Thompson, A. (2008). Water ice on Mars confirmed. https://www.space.com/5686-water-ice-mars-confirmed.html#sthash.EQONjTIf.dpuf (accessed 05 July 2021).

Tóth, Á., Schuszter, G., Das, N.P. et al. (2020). Effects of radial injection and solution thickness on the dynamics of confined A + B – C chemical fronts. *Physical Chemistry Chemical Physics* 22: 10278–10285.

Veneranda, M., Lopez-Reyes, G., Manrique-Martinez, J.A. et al. (2020). ExoMars Raman Laser Spectrometer (RLS): development of chemometric tools to classify ultramafic igneous rocks on Mars. *Nature Scientific Reports* 10 (16954): 1–14.

Walsh, C., Ou, K., Belliveau, N.M. et al. (2014). Microfluidic-based manufacture of siRNA-lipid nanoparticles for therapeutic applications. In: *Drug Delivery System*, pp. 109–120, Springer.

Walter, H.U. (2012). *Fluid Sciences and Materials Science in Space: A European Perspective*. Springer Science & Business Media.

Williams, F.A. (2017). Cool flames in microgravity. *Oral presentation at the 1st International Workshop on Near-Limit Flames in Boston, MA, USA, 29 July 2017.*

Wouters, M., Rahman, S., Myamoto, H. et al. (2021). Continuous microfluidic solvent extraction of cobalt from mimicked and real asteroid leaching solutions. *Separation and Purification Technology* 260: 118238–118252.

Yamaguchi, S., Sunagawa, N., Matsuyama, K. et al. (2021). Preparation of large-volume crystal of cellulase under microgravity to investigate the mechanism of thermal stabilization. *International Journal of Microstructure and Materials Properties* 38: 380103.

Yamamoto, S. (2017). *Introduction to Astrochemistry – Chemical Evolution from Interstellar Clouds to Star and Planet Formation*. Tokyo: Springer.

Zalesskiy, S.S., Danieli, E., Blümich, B., and Ananikov, V.P. (2014). Miniaturization of NMR systems: Desktop spectrometers, microcoil spectroscopy, and "NMR on a chip" for chemistry, biochemistry, and industry. *Chemical Reviews* 114 (11): 5641–5694.

Zheng, H., Hou, J., Zimmerman, M.D. et al. (2014). The future of crystallography in drug discovery. *Expert Opinion on Drug Discovery* 9: 125–137.

7

Catalysis in Space Environments

Shaumica Saravanabavan, Manpreet Kaur, Claire T. Coulthard, and Katharina Brinkert

University of Warwick, Department of Chemistry, Gibbet Hill Road, Coventry CV4 7AL, UK

7.1 Introduction

It is estimated that 90% of all commercially available products on Earth – from fertilizers to furniture and fuels – involve catalysis in their production cycle [1]. For the *International Space Station* (ISS) and future, long-term space missions and the establishment of Moon and Mars habitats, this represents a significant challenge: resupplies from Earth are limited and essential parts of the life support system have to be synthesized, isolated, purified, and recycled with the highest possible reliability and efficiency, obeying principles of 100% atom economy and sustainability. Catalysis is therefore a key player in the establishment of habitants in low-Earth orbit, but even more so in the realization of long-term space missions. This chapter will provide an overview of key challenges associated with catalysis in reduced gravitational environments such as the absence of buoyancy – an upward force exerted by a fluid that opposes the weight of a partially or fully immersed object – and how they are (at least partially) overcome. Moreover, it will explore the current employment of catalysis for applications in space environments such as its role in the *Environmental and Life Support System* (ECLSS) as part of the oxygen (O_2) production, water recycling and carbon dioxide (CO_2) reduction processes, as well as ongoing research activities for future applications, e.g. in *in situ resource utilization* (ISRU). Furthermore, it will discuss opportunities for (photo-)electrocatalysis and enzyme-based biocatalysis for future space missions and build a bridge to the sustainable application of catalysis on Earth, emphasizing the importance of the governing principle of space habitants: to close the loop.

7.2 Physical Fundamentals of Catalysis in Space Environments

Catalysis is a process facilitating and accelerating reaction kinetics while enhancing the efficiency of desired reaction paths in a chemical reaction. It uses a so-called

In-Space Manufacturing and Resources: Earth and Planetary Exploration Applications, First Edition.
Edited by Volker Hessel, Jana Stoudemire, Hideaki Miyamoto, and Ian D. Fisk.

catalyst, a compound which is not consumed or changed by the reaction. In heterogenous catalysis, the catalyst acts in a different phase than the reactants. Most heterogenous catalysts are substrates in a liquid or gaseous reaction mixture. Heterogenous catalysis involves at least five steps: (i) mass transfer of the reactant to the catalyst via diffusion (a mass transfer phenomenon that causes the distribution of a chemical species to become more uniform in space with time) or convection (the bulk motion of a fluid); (ii) adsorption of the reactant onto the catalyst surface; (iii) the activation of the adsorbed reactant; (iv) the actual reaction of the adsorbed reactant; (v) the desorption of the product from the catalyst; and (vi) the product mass transfer from the catalyst. In homogenous catalysis, catalyst and reactants are in the same phase, i.e. dissolved in a solvent together. During the catalytic process, they form an intermediate species which then decomposes or reacts with another reactant in one or more steps to regenerate the catalyst and form the product. The key difference a catalytic reaction experiences in space is reduced gravitation: the acceleration due to the Earth's gravitational field ($\sim$9.81 m/s^2 or 1g, where g is the acceleration due to gravity) is reduced to 1.62 m/s^2 on the moon [2] and to 3.72 m/s^2 on Mars [3]. On the ISS, chemical reactions experience microgravity (10^{-2} to 10^{-4} g), which fluctuates slightly due to internal perturbations. Another difference is that chemical reactions in space are exposed to the entire electromagnetic spectrum outside the Earth's atmosphere (AM 0, whereas AM is the air mass coefficient, defining the direct optical path length of sunlight through the Earth's atmosphere), approximated by the 5800 K black body radiation. Whereas the latter is likely to influence photocatalytic reactions and the stability of molecules in general due to a higher spectral irradiance in the UV part of the solar spectrum, reduced gravitation or microgravity significantly influences all catalytic processes: the absence of natural convection and therefore, buoyancy-induced convection, results, e.g. in hindered phase separation in space due to the absence of density-driven convection. This is of particular importance to reaction processes within the life-support system, where gaseous reactants such as CO_2 and products such as O_2 are the major reaction participants in the air revitalization system.

Studies of the gravitational effect on water electrolysis have been carried out since the 1960s, especially within the frame of developing a reliable spacecraft environmental control system for oxygen production [4]. Hitherto investigations primarily in drop tower experiments ($\sim 10^{-6}$ g) report the formation of stable H_2 and O_2 gas bubble froth layers on the electrode surface due to the absence of buoyancy with an increasing bubble diameter over time. Bubbles also adhere to the membrane separating the two half-cells [5, 6]. As a result, the ohmic resistance in proximity to the electrode surface increases linearly with froth layer thickness in both acidic and alkaline electrolytes, although different pH levels reveal strong differences. Moreover, larger bubbles are formed in reduced gravitational environments, whereas alkaline electrolytes tend to bubble foam and bubble coalescence remains a dominant process in acid electrolytes [5]. It is further known that bubble-induced flow at large bubbles significantly impacts the mass transfer of substrate and product to and from the electrode surface, which represents a fundamental difference to gas bubbles evolving in boiling processes [7]. Bubble coalescence is also known to vary considerably in different electrolytes [8], influencing the overall gas bubble dynamics. Capillary flow and electric forces are other important factors, which predominantly affect gas bubble dynamics in reduced gravitation [9]. One microconvectional process, which has been investigated in more detail over the past years, is the Marangoni effect, which is thought to mainly cause gas bubble convection dynamics in reduced gravitation. The Marangoni effect results from the formation of a gas concentration gradient upon dissolving gas in proximity to the electrode surface which causes a local change in surface tension. This surface tension change can also form due to temperature gradients generated by ohmic heating of the electrolyte. The resulting shear stress imposes a local convection in the electrolyte and the formed gas bubbles which also influences mass transfer [10].

The absence of macroconvectional processes in reduced gravitational environments leads to a diffusion control of chemical reactions along concentration gradients. Diffusion constants of reactants and products as well as the surface tension of the reaction solution or electrolyte become significantly more important for a chemical reaction than under terrestrial conditions. The absence of buoyancy remains a significant challenge for chemical reactions. Despite that fundamental processes of catalysis such as electron and proton transfer, reactant binding and product release are gravitation-independent processes, mass transfer limitations, and hindered gas bubble desorption can drastically impact the catalytic efficiency of reactions, particularly, in heterogenous catalysis. Whereas some gravity-independent physical processes such as filtration, distillation, ad-, and absorption as well as ion exchange are in use on the ISS, rotational devices remain of key importance to induce phase separation in some chemical reactions such as oxygen generation, where hydrogen gas is separated from water during water electrolysis through the creation of a centrifugal force field in a so-called Rotary Separator/Accumulator (RSA) [11].

Reduced gravitation can on the other hand also provide an ideal environment to study governing interfacial processes, e.g. in (photo-)electrocatalysis, where gas bubble desorption phenomena and the impact on the overall device efficiency are not well understood. Moreover, it opens up a great possibility to study reactions involving gaseous reactants since the absence of buoyancy prevents gases to separate from the reaction solution and they could be "trapped," e.g. in proximity to an electrode surface. This is particularly interesting for catalytic reactions involving the reduction of CO_2 and N_2, where the low gas solubility usually requires the use of pressurized reaction vessel.

In the following, we will show that despite the afore discussed challenges, catalysis already has profound applications in space environments, focusing on its application within life-support technologies on the ISS and IRSU. We will furthermore provide an outlook for potential future applications of catalysis and biocatalysis for the synthesis of chemicals and chemical precursors during long-duration space missions and provide suggestions to improve them in reduced gravitation.

7.3 The Application of Catalysis in Atmosphere Revitalization – Catalytic Oxygen Generation and Carbon Dioxide Reduction for Life Support

The current exploitation and mission architectures of the *Environmental Control and Life Support System* (ECLSS) consist of the *Air Revitalization System* (ARS) and the *Oxygen Generator System* (OGS) (Figure 7.1). The *Water Recovery System* (WRS) is also a compartment within the ECLSS and is discussed in Section 7.4.1. The ARS and OGS systems have been operational on the ISS for several years and are responsible for cleaning and circulating cabin air as well as producing oxygen and hydrogen. As continuous resupply from Earth is limited, it is most crucial to recycle oxygen and water, key pillars of space life-support system [12].

Around 5 kg of O_2 gas are produced by the OGS per day, whereas each crew member consumes about 0.9 kg. The OGS can produce between 2.3 and 9.2 kg/day of O_2 at a selectable rate in continuous operation in a day/night orbital cyclic operation [13].

The (OGS) was launched in 2006 and consists of two major parts – the *Oxygen Generation Assembly* (OGA) and the *Carbon Dioxide Removal Assembly* (CDRA) (Figure 7.1). The OGA is a water electrolyzer system, producing O_2 and H_2. *Elektron* and *Vika* are two additional back-up electrolyzer systems, which can supply O_2. H_2 is recycled back into the *CO_2 Reduction Assembly* (CRA) [14, 15]. The current state-of-the-art method recovers approximately 50% of the O_2 from metabolic CO_2. For long-duration space missions, however, an absolute minimum of 75% recovery is required, with the desired target of 90% for long-term space missions [16]. The CDRA uses a Sabatier reactor in which compressed CO_2 reacts with oxygen from the OGA that will be discussed later in this section [17].

At present, the ISS relies on a two-stage process to produce oxygen, involving external solar cells generating an electric current to electrolyze water in a polymer

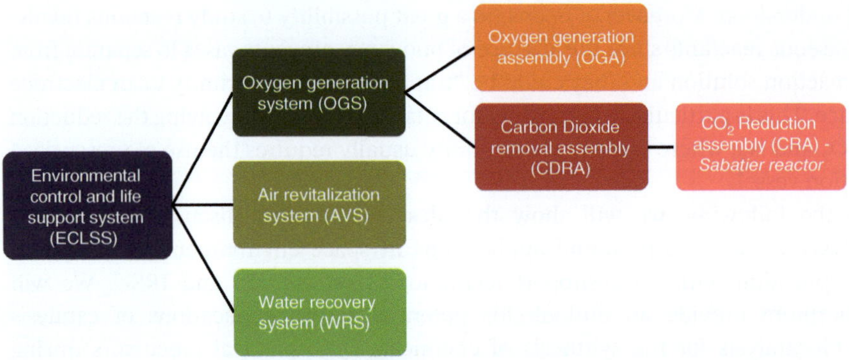

Figure 7.1 Flowchart illustrating selected compartments and their connection in the Environmental Control and Life Support System (ECLSS) employed on the International Space Station (ISS). Source: Manpreet Kaur, Shaumica Saravanabavan, Katharina Brinkert.

electrolyte membrane (PEM) electrolyzer cell as described in Eqs. (7.1)–(7.3).

$$\text{Cathode: } 4H^+(aq) + 4e^- \rightarrow 2H_2(g) \quad E^o = 0.00\,V \tag{7.1}$$

$$\text{Anode: } 2H_2O(l) \rightarrow O_2(g) + 4H^+(aq) + 4e^- \quad E^o = -1.23\,V \tag{7.2}$$

$$\text{Overall cell reaction: } 2H_2O(l) \rightarrow 2H_2(g) + O_2(g) \quad {E^o}_{cell} = -1.23\,V \tag{7.3}$$

Typically, noble metal electrocatalysts based on Pt or Pd are employed for the hydrogen evolution reaction (HER) at the cathode and RuO_2/IrO_2 electrocatalysts are used as anode materials for the oxygen evolution reaction (OER). Water is pumped to the anode where it is oxidized to oxygen, protons, and electrons. The protons diffuse through a thin (20–300 μm) proton-conducting membrane (conductivity for protons 0.1 ± 0.02 S/cm) such as Nafion to the cathode side, where they are reduced to hydrogen (Figure 7.2).

The advantages of a PEM electrolyzer for space applications are the high operating current density (about 2 A cm^{-2} at 1.65–2.5 V), the operation at low temperatures (20–80 °C), and the output of high-purity gas (typically greater than 99.99% H_2) [18, 19]. The low gas crossover makes it furthermore safe to operate. The drawback of the PEM water electrolyzer cell is the corrosive acidic cell operation environment, which requires electrocatalysts and materials that are stable at low pH conditions and sustain the high voltage and oxygen concentration at the anode as well as the required high-purity level of water (~15 MΩ).

Approximately 1.5 kW out of the 4.6 kW used by the entire ECLSS is consumed by the OGA [20]. The high-energy demand results from the electrochemical potential to split water into O_2 and H_2 (+1.23 V) and the present ohmic, concentration, and activation overpotentials of the system which increase this potential further. Ohmic overpotentials are caused by membrane and interfacial resistances in the electrolyzer, the concentration overpotential is due to transport limits of substrates and products to and from the electrode surface and the activation overpotential is directly related to the electrocatalyst exchange current density

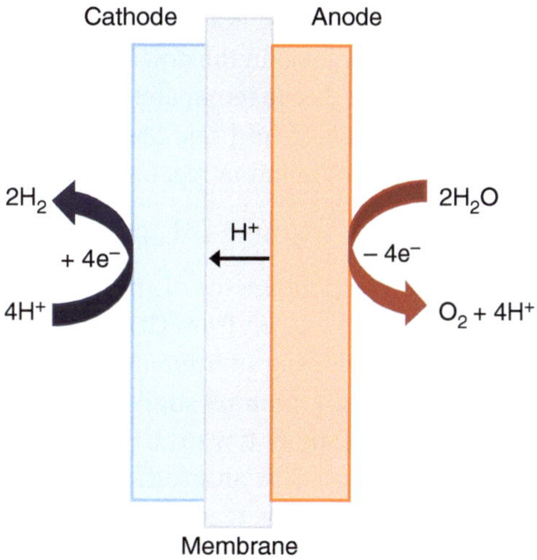

Figure 7.2 Scheme of a PEM water electrolyzer cell. Source: Katharina Brinkert.

which is a measure of the electron transfer rates at anode and cathode. The high energy consumption is also caused by the fact that the PEM electrolyzer system on the ISS operates in contrary to most terrestrial systems at cabin temperature (21–27 °C), whereas the device efficiency significantly increases upon increased temperatures due to the temperature dependency of the electrochemical potential according to the Nernst equation. Chow et al. [20] also emphasized the point that an increase of the electrocatalyst exchange current density (a measure of an electrode's "readiness" to proceed with an electrochemical reaction – the higher the exchange current density, the lower the energy barrier that the charge must overcome in moving from electrolyte to the catalyst surface and vice versa) from 1×10^{-9} to 1×10^{-7} A cm^{-2} at the anode and from 1×10^{-3} to 1×10^{-2} A cm^{-2} at the cathode could already improve the ISS PEM electrolyzer significantly. New, highly efficient, and stable OER electrocatalysts are particularly sought after as well for terrestrial (photo-)electrochemical water splitting, illustrating the similarities of challenges faced in ongoing research efforts for green energy applications and space exploration.

Recently, Evonik Industries and Airbus designed a customized, fixed-bed hydrogenation catalyst that can provide breathable oxygen for manned space missions by direct reduction of CO_2 from cabin air, circumventing the need for water resupply for the electrolyzer compartment from Earth. The so-called Advanced Closed-Loop System (ACLS), which is currently tested on the ISS, traps CO_2 as it passes through small beads of a solid amine resin (*Astrine*) [21]. Steam is furthermore used to extract the CO_2 and process it in a two-step Sabatier reactor to produce methane and water, which in turn is fed again into the electrolyzer. In the current system set-up, methane is vented overboard. The developed catalyst provides high reliability, performance, and mechanical stress resistance over a significant period of time. The system generates 40% of water directly from CO_2 reduction, saving up to 450 kg of water upload for a crew of three per year [21].

Currently, metabolically generated CO_2 is removed from the cabin air by the CDRA using a four-bed molecular sieve. The first two beds are packed with silica gel and zeolite 13X and are used to remove water, which would otherwise interfere with the CO_2 removal in the downstream zeolite beds. This technique has proven to be very effective due to its capability of removing 100% of metabolic CO_2 generation from six crewmembers [22]. The CO_2 then reacts with H_2 from the OGA in a Sabatier CO_2 methanation reactor to produce methane and water:

$$CO_2(g) + 4H_2(g) \rightarrow CH_4(g) + 2H_2O(l) \qquad \Delta H^0 = -165 \text{ kJ/mol} \tag{7.4}$$

Due to the stoichiometry of the Sabatier reaction, the reactor can only recycle about half of the oxygen from CO_2. Therefore, the goal of 75% oxygen recovery can only be achieved using additional technologies [21]. The currently employed reactor on the ISS uses an alumina supported ruthenium catalyst [23], which has a higher activity and selectivity toward CH_4 production than the traditional nickel catalyst. The Sabatier process is an exothermic reaction which is limited by the thermodynamic equilibrium. The reactor currently employed on the ISS is designed to operate at ~515 °C and at slightly sub-ambient pressure to protect against hydrogen leaks.

The CO_2 enters the reactor at ~50 °C and operating pressure to produce the desired products. Lower operating temperatures around 250–400 °C are usually desirable for higher CO_2 conversion rates as well as CH_4 selectivity [24].

A recent development from Precision Combustion, Inc (PCI) with support from NASA includes a Sabatier reactor that is based on a patented Microlith® technology as an alternative to the traditional packed bed and microchannel reactor designs (https://ntrs.nasa.gov/citations/20140012511). This novel approach was made possible by the use of high-heat-transfer and high-surface-area Microlith catalytic substrates, allowing catalyst operation at conditions favorable for thermodynamic equilibrium of CO_2 methanation. This reactor has been demonstrated to produce high CO_2 conversion and near 100% CH_4 selectivity (i.e. ≥90% of the thermodynamic equilibrium values) at high space velocities and low operating temperatures [24].

Historically, the reduction of CO_2 via the so-called Bosch reaction has also been investigated for space applications. Here, CO_2 reacts with H_2 at ~650 °C over an iron catalyst, resulting in the formation of elementary carbon and water:

$$CO_2(g) + H_2(g) \rightarrow CO(g) + H_2O(l) \tag{7.5}$$

$$CO(g) + H_2(g) \rightarrow C(s) + H_2O(l) \tag{7.6}$$

The first part involves the fast reverse water gas shift reaction (7.5), where the reaction of carbon monoxide with hydrogen (step 2) is rate determining (7.6). Apart from the significantly higher reactor temperature compared to the Sabatier reactor, the reaction is largely impracticable as carbon deposits on the catalyst surface can occur, reducing the catalyst's performance and activity [25, 26].

7.3.1 Current Research and Developments within the ECLS System

The ECLS system is one of the most critical and complex systems of a human space mission. To develop the next state-of-the-art ECLS system for a long-duration mission to Mars or a lunar base, more resources have to be allocated to the development and integrated testing of ECLSS hardware. The long-duration, high-reliability, and high-efficiency requirements for future missions beyond low-Earth orbit will drive the system selection and especially the next CO_2 removal system. ISRU employing natural resources available on moon and Mars is hereby key for a successful realization of these missions. Moreover, the development of technological alternatives to the currently existing oxygen production, water recycling, and carbon dioxide conversion hardware is required to close the loop onboard space vehicles and in habitats with less energy input, increased life time and reliability.

7.3.2 Oxygen Generation from Lunar Regolith

Oxygen is the most abundant element in lunar regolith by mass (40–45%) [27]. It is however bound in the regolith as oxides in the form of minerals and amorphous glass and, therefore, it cannot be used directly. The Fray, Farthing, Chen (FFC) – Cambridge process involves the solid-state electrochemical reduction of

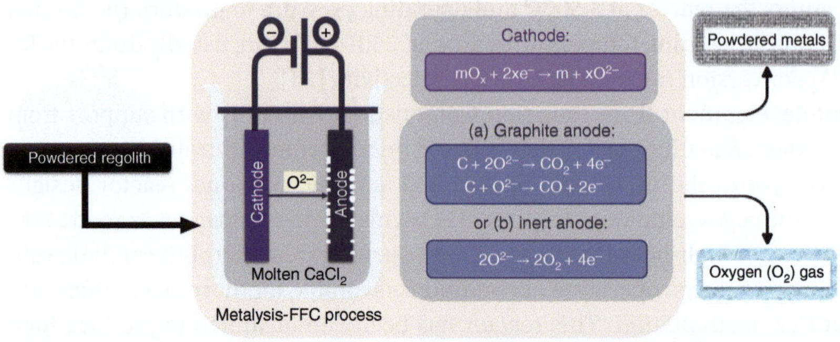

Figure 7.3 Scheme showing the basic principles of the Metalysis-FCC process for O_2 generation from lunar regolith. mO_x is a metal oxide, which is reduced in this process to elementary metal with the co-generation of CO/CO_2 at a graphite anode or O_2 at an inert anode. Source: modified Lomax et al. [28] by Manpreet Kaur and Katharina Brinkert.

metal oxides in molten salts ($CaCl_2$) at temperatures around 900 °C and provides the possibility of reducing all lunar minerals and oxides while storing energy in the molten salt electrolyte during lunar night. The process is already routinely used by the UK-based company Metalysis, which produces pure metals and alloys that are required by commercial specifications to have an oxygen content of <0.5 wt%, exceeding the ISRU requirements (Figure 7.3) [28]. The method is currently scaled up for a wide range of metals such as tantalum and zirconium. Lomax et al. [28] have recently demonstrated that the process can be used to extract O_2 from the lunar regolith simulant JSC-2A to 96%, marking an important milestone in the development of ISRU technology for O_2 generation [28]. The application of this methodology could furthermore co-generate metal alloys that could be used for construction purposes on the lunar surface.

7.3.3 Oxygen Generation via MOXIE on the Mars 2020 Perseverance Rover

The *Mars Oxygen In Situ Resource Utilization Experiment* (MOXIE) is an oxygen-producing device located at the front, right side of the 2020 *Perseverance* Mars Rover [29]. It collects Martian atmosphere, containing about 95.5% CO_2. Ar and N_2 comprise most of the remaining gases [30]. The gas is drawn in through a-High-Efficiency Particulate Air (HEPA) filter to a compressor inlet (Figure 7.4a). It is then pre-heated in a heat exchanger before flowing into a stack of 10 solid oxide electrolysis (SOXE) cells, converting 30–50% of the hot CO_2 to CO and O_2, depending on operational conditions [31]. The scandia-stabilized zirconia (ScSZ) electrolyte [32] has unique properties: zirconia (ZrO_2) is a ceramic, which exists in several crystal arrangements. The unstable cubic crystal structure can be stabilized by replacing some of the Zr^{4+} ions with larger ions such as Sc^{3+}. The larger ions form the equivalent of Sc_2O_3 in the lattice, leaving oxygen vacancies within the lattice at Zr^{4+} sites. When an electric field is applied across a ScSZ sheet, the current is carried by O^{2-} ions, which are transferred from the cathode through the ScSZ

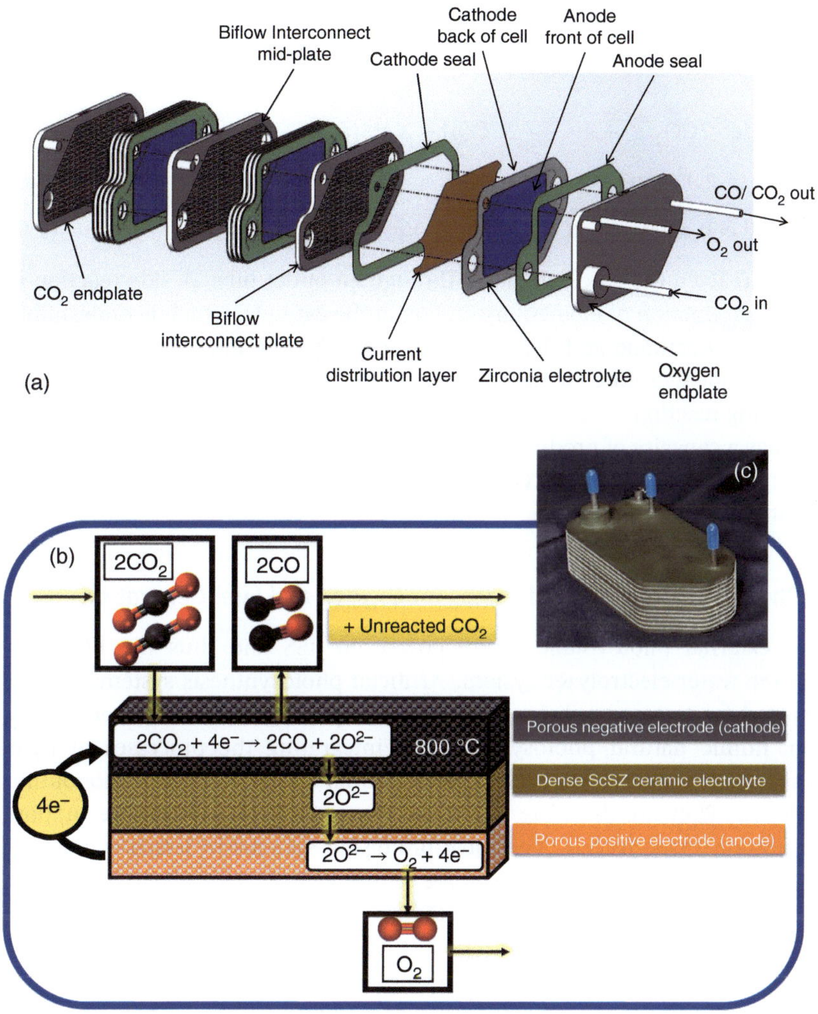

Figure 7.4 (a)View of the top level SOXE assembly of MOXIE. (b) Cross section of a single SOXE cell layer, including reaction mechanisms. (c) Stack identical to the MOXIE stack, where the ports are caped to protect the interior prior to the installation in a test fixture and a development stack that was instrumented to characterize each of the 10-cell in the stack. Source: Courtesy of S. Elangovan, OxEon Energy, Utah/USA.

lattice by "jumping" from vacancy to vacancy. This allows the application of the material in fuel cells and electrolysis cells [33]. A doped lanthanum cobalt ferrite is used for the OER at the anode and a nickel–ceria cermet cathode reducing CO_2. The cathode material choice is crucial; early onwards in the development process, a traditional nickel cermet electrode was used (Ni-YSZ, yttria-stabilized zirconia), as well as Pt and Pt-YSZ in dry CO_2 electrolysis studies. The Pt-based catalyst material exhibited however a large cathodic overpotential, resulting in a high overall cell resistance. Pt coarsening and dewetting of the YSZ was a significant degradation mechanism. When heated CO_2 flows over the Ni-catalyzed cathode surface under an applied potential, CO_2 is reduced to CO and O^{2-}, with the O^{2-} ions being transferred through the electrolyte to the anode, producing gaseous O_2 (Figure 7.4b):

$$\text{Cathode: } 2CO_2(g) + 4e^- \rightarrow 2CO(g) + 2O^{2-}(aq) \tag{7.7}$$

$$\text{Anode: } 2O^{2-}(aq) \rightarrow O_2(g) + 4e^- \tag{7.8}$$

$$\text{Overall cell reaction: } 2CO_2(g) \rightarrow 2CO(g) + O_2(g) \tag{7.9}$$

CO and unreacted CO_2 are exhausted through an outlet tube. A side reaction is however the production of elementary carbon at the cathode, which is undesirable as it coats the Ni cathode and damages it structurally. The long-term operational challenges include therefore the maintenance of the voltage and gas supply favoring the CO forming reaction.

MOXIE has a capacity of producing 99.6% O_2 at a rate of 10 g/h. For future space missions, it is estimated that MOXIE requires a hundredfold scale-up from its current car battery size [29] .

7.3.4 (Photo-)electrocatalytic Devices for Oxygen and Fuel Generation

Currently, external photovoltaic arrays power the ISS and thus the ECLSS – including the water-electrolyzer system. Artificial photosynthesis systems, which are developed for terrestrial, light-assisted water splitting and simultaneous CO_2 reduction, mimic natural photosynthesis by integrating the processes of light absorption, charge separation, and transfer as well as catalysis in a monolithic (single) device. They are therefore particularly interesting as well for life support in long-duration space travels. The most promising systems in terms of reliability, stability, and long-term efficiency are technologically advanced, integrated semiconductor–electrocatalyst systems (Figure 7.5). The semiconductor device generates electrons and holes upon photoexcitation, which are transferred to the integrated (metallic) catalyst, catalyzing oxidation–reduction reactions to produce oxygen (photoanode) and hydrogen or carbon-based fuels (photocathode, Figure 7.5a). The catalytic rate of the electrode reaction is strongly influenced by the molecular (photo-)electrocatalytic processes at the electrified solid/liquid interface, i.e. the strong interactions between reactants and electrode, ions and solvents as well as gas bubble nucleation, growth, and detachment.

7.3.5 (Photo-)electrocatalytic Hydrogen Production in Microgravity Environment

Brinkert et al. [34] have recently demonstrated efficient, light-assisted hydrogen production on *p*-indium phosphide photoelectrodes coated with a rhodium electrocatalyst in microgravity environment realized at the Bremen Drop Tower (10^{-6} g) [34]. Hydrogen gas bubble detachment was hereby observed to occur from the electrode surface due to a nanostructured Rh electrocatalyst layer fabricated by shadow nanosphere lithography, a lithographic technique using the self-assembly

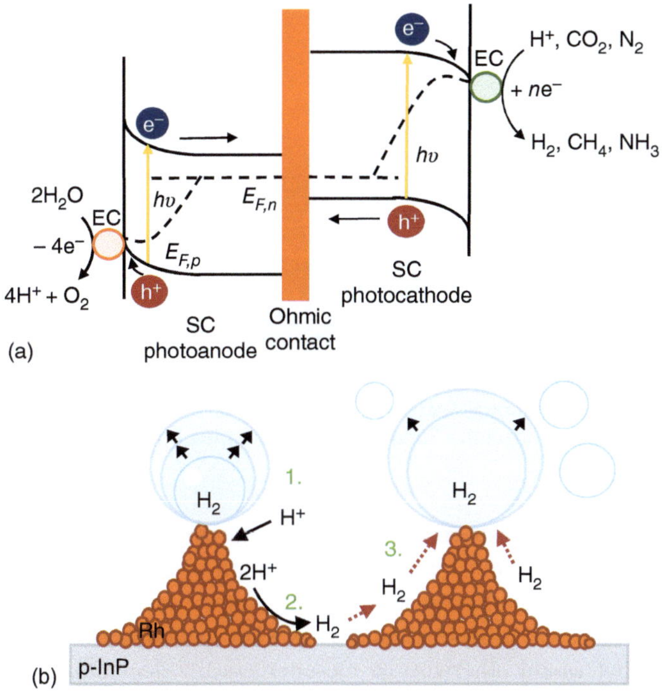

Figure 7.5 (a) Energetic scheme of a solar water-splitting device for hydrogen production/carbon dioxide and dinitrogen reduction. SC is the semiconductor material, EC is the integrated electrocatalyst, and $E_{F,n}/E_{F,P}$ corresponds to the Fermi level of the respective semiconductor. (b) Hypothetical scheme of hydrogen gas bubble evolution on nanostructured *p*-InP-Rh photoelectrodes, illustrating that the formation of catalytic hot-spots might support enhanced gas bubble detachment in reduced gravitational environments. Source: Katharina Brinkert.

of a hexagonally closed-packed (hcp) monolayer of latex spheres to create the metal nanotopography. This pattern presumably introduced catalytic hot-spots on the electrode surface (Figure 7.5b), preventing gas bubble coalescence during the 9.2 seconds of free fall. The nanostructured electrodes showed nearly the same photocurrent-voltage behavior and half-cell efficiency as their terrestrial control samples. These experiments open up the possibility of utilizing artificial photosynthesis devices as well in space environments and of optimizing gas-producing electrodes for operation in reduced gravitation in general.

7.4 Catalytic Waste Recycling on the International Space Station

7.4.1 Wastewater Treatment – the Water Recovery System (WRS)

Air, food, and water accounts for a resupply equipment mass of 15 kg/d on the ISS. Recycling food and water could reduce this however to 1 kg/d. There is strong

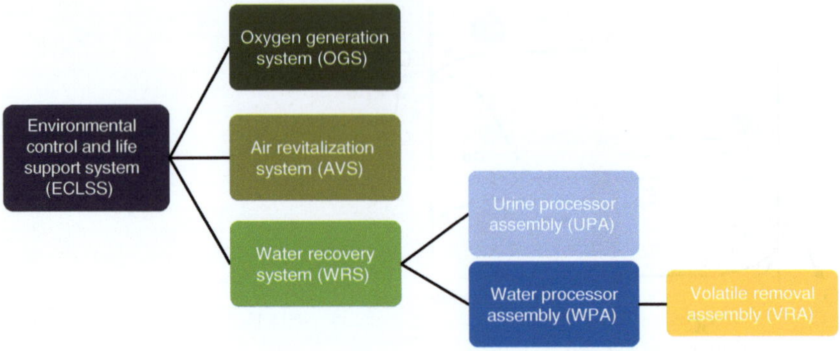

Figure 7.6 Flow chart illustrating the different compartments within the Water Recovery System (WRS) employed on the International Space Station (ISS) for wastewater purification and recovery. Source: Manpreet Kaur, Shaumica Saravanabavan, Katharina Brinkert.

incentive for this since shipment costs remain over \$10, 000 per kg transported into LEO [35]. Water accounts for 92% by mass consumption of all life-support consumables in space. Therefore, effective and reliable water purification and recovery is essential [36]. For these processes, catalysis is a key technology to drive the chemical reactions that would otherwise be too energetically demanding or simply unfeasible in space environments.

Currently, the water recycling technology used on the ISS is called the WRS. WRS consists of two units: the *Urine Processor Assembly* (UPA) that processes crew urine distillate, and the *Water Processor Assembly* (WPA) that processes cabin air humidity condensate, waste hygiene distillate, and product water from the Sabatier reactor discussed in Section 7.3 (Figure 7.6) [37]. The technology makes use of processes including particulate filtration, adsorption, ion exchange and heterogenous catalytic oxidation for wastewater processing [12]. The catalytic reactor within the WPA is responsible for the removal of volatile contaminants, also called the *Volatile Removal Assembly* (VRA) [38]. The VRA employs a noble metal catalyst that oxidizes trace organic contaminants such as low-molecular-weight alcohols and acetic acid in water in the presence of gaseous oxygen at elevated temperatures of roughly 135 °C [12]. A regenerative heat exchanger is coupled with the catalytic reactor to recycle waste heat and feed it back into the reactor [39]. The total water recovery with the WPA is >90% [40].

The WRS on the ISS is made up of several processors, adding complexity to the water recycling system. Tasks such as regular replacement of multi-filtration filters and discharge of brine filtration tanks impose a maintenance penalty on the crew members [41]. Besides the complexity, the WRA is also an energetically demanding system, requiring high power to function. Catalysts in the WRA currently have insufficient activity for oxidation at ambient pressures and temperatures lower than the boiling point of water. Therefore, efforts have been directed toward developing innovative catalyst supports that improve their activity and efficiency. A catalyst support with sub-mesopores and macropores has been recommended to improve the surface area as well as the diffusion rate [42].

The *Vapor-Phase Catalytic Ammonia Removal* (VAPCAR) technology is a highly integrated processor configuration currently under development to produce potable water from wastewater that employs two catalytic reactor beds. This is a single-step water processor that accepts a combined stream of humidity condensate, hygiene water and urine to produce potable water with ~97% efficiency [43]. On the first catalytic bed, the oxidative reactor converts volatile hydrocarbons to carbon dioxide and water, while ammonia is oxidized to nitrogen and nitrous oxide. This catalyst contains 0.5% platinum on alumina oxidation catalyst pellets and operates at ~250 °C. The second catalyst bed consists of 0.5% ruthenium on alumina catalyst pellets and operates at a higher temperature of 450 °C. This segment decomposes the pollutant nitrous oxides to nitrogen and oxygen, which are both used to replenish crew cabin air [44]. In this way, the system is designed to eliminate the need for expendable chemicals. The VAPCAR yields the highest quality water in terms of ammonia content, conductivity, and total organic carbon content in the recovered water. This is particularly beneficial because non-regenerable reactor beds would not be required for ammonia removal and savings on cost and mass can be made. The VAPCAR system is however in need of further developments, such as a lower specific energy demand for water production [44].

Commercially available catalysts have been designed with 1–3% Pt loadings, and studies have been conducted to find the optimal catalyst for the VAPCAR technology. Instead of separate catalytic beds, current efforts focus on improving the selectivity and activity of the catalyst. A single catalyst that selectively oxidizes ammonia to nitrogen over NO_x species would be beneficial. A more active catalyst would require a lower operating temperature and may also have a lower bed volume thus leading to energy and mass savings. Other investigations focus on catalyst support, be it spheres, pellets, or a monolith structure. Among the commercially available catalysts, 1% Pt on alumina spheres at 200 °C were found to be the most optimal catalytic composition for selective ammonia oxidation to dinitrogen. Pelleted catalysts have also been developed with improved activity and selectivity indicating that there is room for improvement and optimization of the VAPCAR catalyst [45]. Main factors for optimization of the catalyst include the T_{100} (temperature at which 100% of ammonia is oxidized), percentage of catalyst loading, operating temperature, oxygen pressure, and catalyst support medium and structure [45].

7.4.2 Removal of Trace Contaminants

Aside from potable water, clean air to breathe is another necessity that requires meticulous care and sophisticated infrastructure to maintain. In the confined space of a spacecraft cabin, the quantities of any contaminants can be damaging to health. It is therefore necessary to maintain low levels of gaseous impurities. The spacecraft maximum allowable concentration (SMAC) defines the exposure limit for maintaining a healthy cabin environment, and there are extensive data available on chemical load for trace gas contaminant control in spacecrafts [46]. Among the total load on

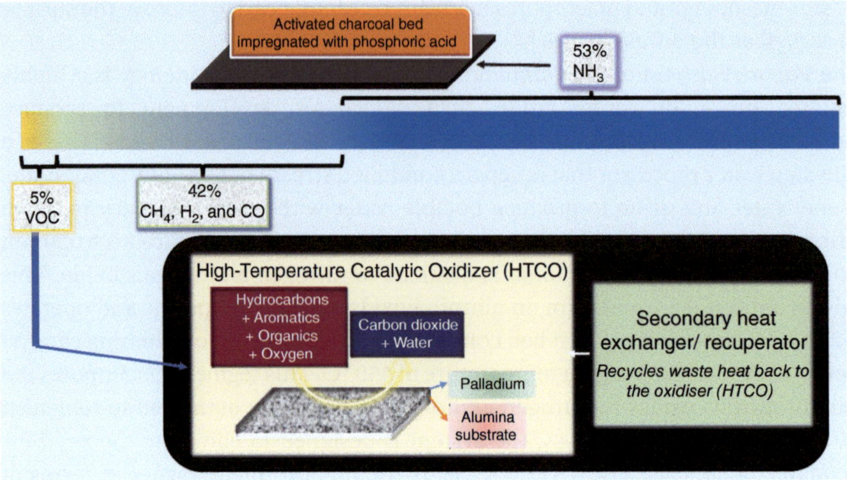

Figure 7.7 Relative composition of the contaminant gases in the spacecraft cabin air and their removal methods. VOC refers to volatile organic compounds such as short-chain hydrocarbons, ketones, esters, alcohols, and aromatics. Source: Manpreet Kaur, Katharina Brinkert.

the ISS almost 53% of the metabolic load comes from ammonia gas (Figure 7.7). The sources of ammonia are crew members, the external thermal control system and payloads or scientific equipment. A further 42% of the load is accounted for by methane, hydrogen, and carbon monoxide, and the final five per cent comprises short-chain saturated hydrocarbons, ketones, esters, alcohols, and aromatics [47]. In general, the flow rate of cabin air at the catalytic oxidation stage is dependent on ammonia concentration. The four compounds – methane, hydrogen, carbon monoxide, and ammonia – account for 94% of the human metabolic waste whereas ammonia has a high generation rate and a low SMAC [48]. In addition, ammonia build-up poses great health risk to the crew, including irritations of eyes, mucous membranes, and the respiratory tract (100–200 ppm), with higher concentrations being potentially lethal in long exposure [47]. The *Trace Gas Contaminant Control* (TCCS) system on the ISS comprises several charcoal and LiOH beds, each of which is designed to handle certain contaminants and eliminate odors [47]. Physical and chemical adsorption on activated carbon and zeolite is not effective at removing volatile compounds such as CO, light hydrocarbons (e.g. methane), or alcohols. For the treatment of these contaminants, a high-temperature catalytic oxidizer (HTCO) is employed [46]. The HTCO contains an oxidation catalyst such as palladium on alumina and is operated at 438 °C. During thermal catalytic oxidation, the process stream is heated in the presence of the catalyst and the contaminants are oxidized to carbon dioxide and water.

The post-sorbent LiOH bed is used to remove acid gases that may have formed by oxidation [44].

Methane, carbon monoxide, and formaldehyde are secondary compounds that inform the flow rate at the catalytic oxidation stage, as well as the temperature of the HTCO. A catalytic oxidizer operating at ambient temperature is also available which is lighter, simpler, and uses negligible power. It is mainly to control carbon monoxide levels and is preferable for use in cabins with air volumes exceeding 111 m^3 that can therefore maintain safer levels of hydrogen and methane. For smaller cabins, HTCO is preferable to control the levels of hydrogen and methane gas, likely because any trace contaminants would pose a greater risk at a lower volume due to higher concentration [49].

For the HTCO to reduce its energy requirements and therefore be viable for space missions, a secondary heat exchanger/ recuperator must be coupled with the system to recycle waste heat back to the oxidizer. Currently, the catalytic oxidizer and the recuperator are two separate units that are coupled together in the TCCS [46]. The company PCI patented a Microlith technology with promising results to combine the HTCO and the heat recuperator. This technology consists of a series of discrete, ultra-short-chain-length channels with very short diameters. These channels make up a fine metal mesh where they are catalytically coated, therefore, providing a larger geometric surface area than monolith support as well as high heat and mass transfer coefficients. This means that the reactor is able to rapidly respond to gas temperatures and provides an improved rate of contact between reactant and catalyst [50]. The same technology has also been used to develop the CDRA systems as previously discussed in Section 7.3 [26]. The two systems could also be combined in future devices to achieve a more compact and lightweight reactor design [51].

7.5 Enzyme Catalysis for Space Applications

In order to meet stringent safety and consistency requirements, the ECLSS on board the ISS relies exclusively on physical and chemical technologies, even at the cost of significant energy inputs. In contrast, biological-based technology mimicking parts of natural catalytic processes offers new possibilities for sustainable, closed-loop systems capable of operating at ambient conditions. Despite issues with high maintenance and durability, bioreactors have been well researched, with developments leading to reliable alternatives for industrial processes [52]. For over 30 years, efforts have been made to develop hybrid biological–physiochemical devices for efficient waste, water, food, and atmospheric recycling and to this end, we shall examine the role of enzymes as sustainable biological catalysts for synthetic applications in space.

Developing bioregenerative technology for resource recovery is a key motif in projects such as Micro-Ecological Life Support System Architecture (MELISSA), an international foundation that aims at developing an efficient, closed-loop life-support system by exploiting the natural metabolic processes of microorganisms and higher plants. For instance, nitrogen-converting bacteria such as *Nitrosomonas Europaea* have been incorporated as part of the nitrogen recovery architecture, which oxidize ammonia to nitrate using ammonia monooxygenase and hydroxylamine oxidoreductase enzymes [53]. Microorganisms also dominate research in biologically assisted waste recovery, yet the desired metabolic processes of these organisms depend on the enzymatic reactions occurring within the cell. Thus, it stands to reason that in utilizing enzyme technology, it is possible to circumvent the need for complex organisms while selectively generating high-quality products with minimal energy input.

Enzymes are complex proteins that are crucial for the cellular metabolism and are thus directly responsible for controlling biochemical reactions in living organisms. Enzyme-catalyzed reactions can be accelerated by 100 million to 10 billion times that of the uncatalyzed reaction and due to distinct molecule–molecule interactions within an active site, enzymes possess uniquely high substrates specificities. Terrestrially, enzymes are an essential tool for the bioprocessing of many products, for instance, over 500 industrial products, ranging from food to biofuels require biological catalysts [54]. The qualities that are attractive for industry, i.e. high specificity, rapid turnover, and biodegradability, may also prove beneficial for application in space technology. Immobilized biocatalysts are frequently used in both industry and in environmental protection efforts, namely in pollutant removal and the conversion of biomass. Immobilized urease bioreactors have been used for urea removal in water sources and were tested for potential use on a space station; more research and starting materials not free of chemicals (chromium) used to neutralize station waste products - this is a significant issue limiting biocatalysis and full utilization of starting materials on ISS that would be helpful to note since it may help to change thinking for future systems; however, more research is required before this type of technology can be fully integrated into the water reclamation architecture of a spacecraft [55]. Efficient fermentation of municipal food waste into useful chemicals with common enzymes, such as glucoamylase or carbohydrase for ethanol production, may also be useful in closing the loop on resource recycling [56]. With the onset of new molecular engineering tools, it may be possible to tailor enzymes to overcome challenges associated with biocatalytic devices (e.g. durability, product specificity, high maintenance) and generate advanced chemicals for wider applications [57]. Particularly, procedures such as directed enzyme evolution allow the optimization of enzyme activity, substrate binding affinity, and product specificity [58]. This process modifies well-characterized and studied enzymes in a way that they can catalyze reactions, which are known to synthetic chemists, but which do not occur naturally, thus allowing the production of a wider range of chemicals with high efficiencies by organisms, which are easily cultivatable.

The investigation of enzyme kinetics in reduced gravitation may therefore offer invaluable insight into the physical parameters, which would otherwise be significantly modified or masked under terrestrial conditions. Indeed, it has been well documented that biological functions can be altered in space, particularly, the kinetics of enzymatic reactions. Thus, an understanding of the mechanisms, which might be affected by gravitation, will be necessary for the design of high-performance devices. Optimizing biocatalytic devices for space applications will therefore require a more comprehensive investigation into the effects of space on enzyme behavior; however, research in this regard is not only limited, but conflicting.

7.5.1 Enzyme Kinetics in Reduced Gravitation

The mechanism of action for enzyme catalysis involves reactant(s) binding within an active site to form a transition state complex in a highly ordered environment, eventually leading to covalent bond breaking and formation to synthesize product(s). Regarding the forces that govern enzyme–substrate binding, the contribution of gravitation at the molecular scale is negligible even compared to the weak interactions between enzyme and substrate (i.e. ionic bonding, hydrogen bonding, and hydrophobic interactions), and it is therefore unlikely to substantially affect the rate of reaction. Reduced gravitation does also not alter rates of electron transfer nor any parameter that involves chemical bonding/interactions apart from macro-level processes such as crystal formation. As discussed earlier in Section 7.2, diffusion effects in microgravity have long been studied, and the absence of buoyancy causes mass transport retardation and poor phase separation due to the absence of density-driven convection. The information available on enzyme kinetics and the changes observed in reduced gravitation predominantly involve proteins that are expressed by cells on the ISS/sounding rockets and not of isolated enzyme–substrate reactions. It is most likely that this is a consequence of the cells compensating for the drastic environmental change as opposed to a direct effect of gravity on enzyme kinetics. Very few investigations have been made to specifically identify the effect of microgravity on the function of the enzyme itself, with observations made in the early 2000s showing conflicting results. Isocitrate lyase (a protein in the glyoxylate cycle) maintained the same forward rate of conversion (isocitrate to glyoxylate and succinate) and backward rate of condensation in both terrestrial and microgravity environment [59]. This contradicts results by Maccarrone et al. [60], whereby the activity of soybean lipoxygenase-1 (an enzyme involved in plant development and mammalian metabolism) was found to increase fourfold in microgravity [60]. One would expect to see poorer rates of a reaction if substrate diffusion was affected; however, this is not the case. There could be unanticipated counteractive effects that alter the solution conditions to the point where the reaction rate is maintained or greater than expected. For instance, Marangoni convection is shown to be the dominant mechanism for fluid motion in microgravity, which can significantly impact mass transport rates of substrates/products [61]. This effect is greatest where there is a large difference in surface tensions [62]; density differences between substrates and products, coupled with minute changes in temperature arising from thermodynamic effects, could potentially induce local Marangoni convection. Given that there are only two known kinetic experiments of this type, it is however difficult to make predictions. The initial results seem to suggest that the response to microgravity is dependent on the specific enzyme–substrate system. The large increase in activity for lipoxygenase-1 in reduced gravitation is however not insignificant; it shows that the effect of reduced gravitation on enzymatic activity has to be examined thoroughly, which could allow the optimization of these biocatalytic reactions for space applications at improved productivities and potential scaling-up procedures. Given the variety of reactions enzymes are involved in, there is certainly a tremendous possibility for their application in spacecrafts and space habitats, e.g. in contaminant and pollutant removal systems, the conversion of municipal wastes into high-value chemicals such as ethanol or hydrogen or the conversion of lignocellulosic matter from higher plants into value added products. Advanced enzyme engineering [57, 58] combined with improved system designs could furthermore significantly improve enzyme compatibility for long-duration missions as well.

7.6 Conclusions

Catalysis has always played a major role in space exploration and it will be even more important for the realization of long-term space travel. Apart from catalyst optimization and the development of additional technologies for oxygen production, CO_2 reduction, and water production and recycling, a key role could be the involvement in the loop closure of waste recycling, including wastewater, municipal waste, and materials such as food containers or batteries. Current waste management methods involve prolonged storage and later waste-return to Earth or incineration during atmospheric re-entry, which will not be feasible for long-duration space missions. Efforts to convert waste and waste gases into high-value products such as life support oxygen and water, fuels, or energy sources are paramount. Catalysis also offers an efficient and reliable means for the synthesis of chemicals and chemical precursors in LEO such as pharmaceuticals, building materials, and fertilizers. A major issue faced by space plant growth is for instance that the required mass of fertilizer and soil media is too heavy to be flown to extraterrestrial locations beyond low-Earth orbit (estimated to be about 90 kg per crew member per year) [63]. The *in situ* production of nitrogen-based fertilizers using catalysis and resources available such as ammonia could provide a solution. Also, polymer synthesis, e.g. via electropolymerization, could be a viable way to synthesize a variety of new, polymer-based materials.

The challenges we face on Earth, e.g. with respect to creating a sustainable, green energy economy, are closely related to the challenges faced in long-term space missions and the construction of habitats on Moon and Mars. Sustainability, reliability, and long-term efficiency are among the key words summarizing the requirements for systems in both terrestrial and space environments. Building a strong connection between the space and terrestrial research communities is therefore fundamentally important to closing loops.

References

1 Armor, J.N. (2011). A history of industrial catalysis. *Catalysis Today* 163 (1): 3–9.

2 di Prampero, P.E. (2000). Cycling on earth, in space, on the moon. *European Journal of Applied Physiology* 82: 345–360.

3 Hirt, C., Claessens, S.J., Kuhn, M. et al. (2012). Kilometer-resolution gravity field of Mars: MGM2011. *Planetary and Space Science* 67 (1): 147–154.

4 Clifford, J. E., Gates, J. T., McGallum, J., et al. (1962). Research on the electrolysis of water under weightless conditions. Wright-Patterson Air Force Base, AMRL-TDR-62-44, May 1962. http://contrails.iit.edu/reports/7451 (accessed 7 March 2022).

5 Matsushima, H., Fukunaka, H., and Kuribayashi, K. (2006). Water electrolysis under microgravity: Part II. Description of gas bubble evolution phenomena. *Electrochimica. Acta* 51 (20): 4190–4198.

6 Kiuchi, D., Matsushima, H., Fukunaka, Y., and Kuribayashi, K. (2006). Ohmic resistance measurement of bubble froth layer in water electrolysis under microgravity. *Journal of the Electrochemical Society* 153 (8): 138–143.

7 Matsushima, H., Kiuchi, D., Fukunaka, Y., and Kuribayashi, K. (2009). Single bubble growth during water electrolysis under microgravity. *Electrochemistry Communications* 11: 1721–1723.

8 Sides, P.J. and Tobias, C.W. (1985). A close view of gas evolution from the back side of a transparent electrode. *Journal of the Electrochemical Society* 132: 583.

9 Bashkatov, A., Hossain, S.S., Yang, X. et al. (2019). Oscillating hydrogen bubbles at Pt microelectrodes. *Physical Review Letters* 123: 214503.

10 Yang, X., Baczyzmalski, D., Cierpka, C. et al. (2018). Marangoni convection at electrogenerated hydrogen bubbles. *Physical Chemistry Chemical Physics* 20: 11542–11548.

11 Samplatsky, D.J. and Clark Dean, W. (2002). Development of a rotary separator accumulator for use on the international space station. In: *32nd International Conference on Environmental Systems, 15-18 July 2002*. San Antonio, Texas, USA: SAE Technical Paper.

12 Carrasquillo, R. (2005). ISS ECLSS technology evolution for exploration. In: *43rd AIAA Aerospace Sciences Meeting and Exhibit*. Nevada: ARC-AIAA https://doi.org/10.2514/6.2005-337.

13 Bowman, M.E., Bowman, N.D., Cramblit, L.E. et al. (2017). Analysis of chemical and microbial components adsorbed. In: *47th International Conference on Environmental Systems*. AIAA-SES.

14 Brown, B.R. (2019). Electrochemistry's Role in Current and Future Environmental Control and Life Support Systems. ECS Meeting Abstracts. https://iopscience.iop.org/article/10.1149/MA2019-02/57/2469 (accessed 18 May 2021).

15 Takeda, C.K., Ghariani, E.A., and Keuren, V.S. (2015). Advancing the oxygen generation assembly design to increase reliability and reduce costs for a future long duration mission. In: *45th International Conference on Environmental Systems*.

16 Karakaya, C., Jackson, G.S., Kee, R.J. (2019). Opportunities for Electrochemical and Catalytic Processes That Suppport Human Activity on Mars. ECS Meeting Abstracts. https://iopscience.iop.org/article/10.1149/MA2019-02/57/2456 (accessed 18 May 2021).

17 Noyes, G. (1988). *Carbon dioxide reduction processes for spacecraft ECLSS: a comprehensive review. SAE Transactions* 97: 374–381. http://www.jstor.org/stable/44470335 (accessed 18 May 2021).

18 Miller, H., Bouzek, K., Hnat, J. et al. (2020). Green hydrogen from anion exchange membrane water electrolysis: a review of recent developments in critical materials and operating conditions. *Sustainable Energy & Fuels* 4: 2114–2133.

19 Buttler, A. and Spliethoff, H. (2018). Current status of water electrolysis for energy storage, grid balancing and sector coupling via power-to-gas and power-to-liquids: a review. *Renewable and Sustainable Energy Reviews* 82 (3): 2440–2454.

20 Chow, R., Nelson, G.J., and Perry, J.L. (2018). Electrolyzer exergy analysis for an environmental control and life support system. Proceedings of the ASME 2018. In: *International Mechanical Engineering Congress and Exposition*. Pittsburgh, PA, USA: ASME.

21 Bockstahler, K., Hartwich, R., Matthias, C. et al. (2017). Status of the advanced closed loop system ACLS for accommodation on the ISS. In: *47th International Conference on Environmental Systems*. Charleston, South Carolina, USA.

22 Mattox, E.M., Knox, J.C., and Bardot, D.M. (2013). Carbon dioxide removal system for closed loop atmosphere revitalization, candidate sorbents screening and test results. *Acta Astronautica* 86: 39–46.

23 Greenwood, Z.W., Abney, M.B., Brown, B.R. et al. (2018). State of NASA oxygen recovery. In: *48th International Conference on Environmental Systems*. Albuquerque, New Mexico, USA.

24 Junaedi, C., Hawley, K., Vilekar, S.A. et al. (2016). Evaluation of CO_2 adsorber Sabatier reactor, and solid oxide stack for consumable, propellant, and power production - potential in ISRU architecture. In: *46th International Conference on Environmental Systems*. Vienna, Austria.

25 Kenn, F. (2011). New concepts for the avoidance or utilisation of methane in life support systems. *Advances in Space Research* 48: 457–464.

26 Junaedi, C., Hawley, K., Walsh, D. et al. (2011). Compact and lightweight Sabatier reactor for carbon dioxide reduction. In: *41st International Conference on Environmental Systems*. Portland, OR: ARC-AIAA https://doi.org/10.2514/6.2011-5033.

27 McKay, D.S., Heiken, G., Basu, A. et al. (1991). The lunar regolith. In: *Lunar Sourcebook* (ed. G.H. Heiken, D.T. Vaniman and B.M. French), 285–356. Cambridge: University Press.

28 Mellor, I. and Doughty, G. (2016). Novel and emerging routes for titanium powder production an overview. *Key Engineering Materials* 704: 271–281.

29 Hecht, M. 2020, *MOXIE*. NASA. https://mars.nasa.gov/mars2020/spacecraft/instruments/moxie/ (accessed 17 May 2021).

30 Potter, S. 2021. *NASA's Perseverance Mars Rover Extracts First Oxygen from Red Planet*. NASA, https://www.nasa.gov/press-release/nasa-s-perseverance-mars-rover-extracts-first-oxygen-from-red-planet (accessed 17 May 2021).

31 Hecht, M., Hoffman, J., Rapp, D. et al. (2021). Mars oxygen ISRU experiment (MOXIE). *Space Science Reviews* 217: 9.

32 Elangovan, S., Hartvigsen, J.J., Larsen, D., et al. (2019). Electrochemical Production of Oxygen and Fuel on Mars. IOPScience. ECS Meeting Abstracts. https://iopscience.iop.org/article/10.1149/MA2019-02/57/2440 (accessed 18 May 2021).

33 Hecht, M.H. and Hoffman, J.A. (2021). The Mars oxygen ISRU experiment (MOXIE) on the Mars 2020 rover. In: *46th Lunar and Planetary Science Conference*. NASA https://www.hou.usra.edu/meetings/lpsc2015/pdf/2774.pdf.

34 Brinkert, K., Richter, M.H., Akay, Ö. et al. (2018). Efficient solar hydrogen generation in microgravity. *Nature Communications* 9: 2527.

35 Volpin, F., Badeti, U., Wang, C. et al. (2020). Urine treatment on the International Space Station: current practice and novel approaches. *Membranes* 10 (11): 327.

36 Tamponnet, C., Savage, C.J., Amblard, P. et al. (1999). Water recovery in space. *ESA Bulletin* 97: 56–60. http://www.esa.int/About_Us/ESA_Publications/ESA_i_Bulletin_i_97_March_1999 (accessed 21 May 2021).

37 Carter, L., Pruitt, J., Brown, C.A. et al. (ed.) (2015). Status of ISS waste management and recovery. In: *45th International Conference on Environmental Systems.* Washington: ARC-AIAA https://doi.org/10.2514/6.2011-5223.

38 Nalette, T., Snowdon, D., Pickering, K. et al. (2007). The ISS water processor catalytic reactor as a post processor for advanced water reclamation systems. In: *37th International Conference on Environmental Systems.* Illinois: SAE International https://doi.org/10.4271/2007-01-3038.

39 Carter, L. (ed.) (2012). Status of the regenerative ECLS water recovery system. In: *40th International Conference on Environmental Systems.* Spain: ARC-AIAA https://doi.org/10.2514/6.2010-6216.

40 Lindeboom, R.E.F., Paepe, J.D., Vanoppen, M. et al. (2020). A five-stage treatment train for water recovery from urine and shower water for long-term human Space missions. *Desalination* 495: 114634.

41 Flynn, M.T. and Borchers, B.A. (1996). An evaluation of the VAPCAR process for use in space flight. *SAE International* 105: 871–878. https://doi.org/10.4271/961545.

42 Yu, P., Nalette, T., and Kayatin, M. (2016). Development of advanced ISS-WPA catalysts for organic oxidation at reduced pressure/temperature. In: *46th International Conference on Environmental Systems.* Vienna, Australia: http://hdl.handle.net/2346/67603 (accessed 21 May 2021).

43 Jones, H.W. and Kliss, M.K. (2010). Exploration life support technology challenges for the crew exploration vehicle and future human missions. *Advances in Space Research* 45 (7): 917–928.

44 Wydeven, T. (1988). A survey of some regenerative physico-chemical life support technology. *Ames Research Center, National Aeronautics and Space Administration, NASA Technical Memorandum* 101004.

45 Thibaud-Erkey, C. and Nalette, T. (ed.) (2003). Testing and development of new catalysts for vapor phase ammonia oxidation. In: *33rd International Conference on Environmental Systems.* Canada: SAE International https://doi.org/10.4271/2003-01-2502.

46 Vilekar, S.A., Hawley, K., Junaedi, C. et al. (ed.) (2015). Microlith®–based catalytic reactor for air quality and trace contaminant control applications. In: *45th International Conference on Environmental Systems.* Washington: http://hdl.handle.net/2346/64520 (accessed 21 May 2021).

47 Macatangay, A.V., Prokhorov, K.S., and Sweterlitsch, J.J. (2007). Strategies to mitigate ammonia release on the international space station. *SAI International* 116: 344–355. https://doi.org/10.4271/2007-01-3186.

48 Perry, J.L. and Kayatin, M.J. (ed.) (2015). Trace contaminant control design considerations for enabling exploration missions. In: *45th International Conference on Environmental Systems.* Washington: http://hdl.handle.net/2346/64385 (accessed 21 May 2021).

49 Stapleton, T., Heldmann, M., Torres, M. et al. (ed.) (2017). Environmental control and life support system developed for deep space travel. In: *47th International*

Conference on Environmental Systems. South Carolina: http://hdl.handle.net/2346/72885.

50 Vilekar, S.A., Hawley, K., Junaedi, C. et al. (ed.) (2017). Energy efficient Microlith®-based catalytic reactor and recuperator for air quality control applications. In: *47th International Conference on Environmental Systems*. South Carolina: http://hdl.handle.net/2346/73015 (accessed 21 May 2021).

51 Roychoudhury, S., Walsh, D., and Perry, J. (2005). Resistively-heated Microlith-based adsorber for carbon dioxide and trace contaminant removal. *SAE Transactions* 114 (1): 197–204. https://doi.org/10.4271/2005-01-2866.

52 Le-Clech, P., Chen, V., and Fane, T. (2006). Fouling in membrane bioreactors used in wastewater treatment. *Journal of Membrane Science* 284 (1-2): 17–53.

53 Stein, L. and Arp, D. (1998). Loss of Ammonia monooxygenase activity in the *Nitrosomonas Europea* upon exposure to nitrite. *Applied and Environmental Microbiology* 64 (10): 4098–4102.

54 Liu, X. and Kokare, C. (2017). *Microbial Enzymes of Use in Industry* (ed. G. Brahmachari), 267–298. Biotechnology of Microbial Enzymes.

55 Jolly, C. and Bagdigian, R. (1989). Application of biocatalysts to space station ECLSS and PMMS water reclamation. *Journal of Aerospace* 98: 413–417.

56 Matsakas, L., Qiuju, G., Jansson, S. et al. (2017). Green conversion of municipal solid wastes into fuels and chemicals. *Electronic Journal of Biotechnology* 26: 69–83.

57 Kuchner, O. and Arnold, F. (1997). Directed evolution of enzyme catalysts. *Trends in Biotechnology* 15 (12): 523–530.

58 Arnold, F. (2018). Directed evolution: bringing new chemistry to life. *Angewandte Chemie International Edition* 57: 4143–4148.

59 Ranaldi, F., Vanni, P., and Giachetti, E. (2003). Enzyme catalysis in microgravity: steady-state kinetic analysis of the isocitrate lyase reaction. *Biophysical Chemistry* 103 (2): 169–177.

60 Maccarrone, M., Bari, M., Battista, N., and Finazzi-Agro, A. (2001). The catalytic efficiency of soybean lipoxygenase-1 is enhanced at low gravity. *Biophysical Chemistry* 90 (1): 97–101.

61 Watanabe, T., Melnikov, D., Matsugase, T. et al. (2014). The stability of a thermocapillary-buoyant flow in a liquid bridge with heat transfer through the interface. *Microgravity Science and Technology* 26: 17–28.

62 Bain, C. (2008). The overflowing cylinder sixty years on. *Advances in Colloid and Interface Science* 144 (1-2): 4–12.

63 Lunn, G., Stutte, G., Spencer, L. et al. (2017). Recovery of nutrients from inedible biomass of tomato and pepper to recycle fertilizer. In: *Proceedings of the 47th International Conference on Environmental Systems*.

64 Lomax, A.B., Conti, M., Khan, N. et al. (2019). Providing the viability of an electrochemical process for the simultaneous extraction of oxygen and production of metal alloys from lunar regolith. *Planetary and Space Science* 180: 104748–104758.

Part IV

Space Mining

8

Mining and Microbiology for the Solar System Silicate and Basalt Economy

Charles S. Cockell and Rosa Santomartino

University of Edinburgh, UK Centre for Astrobiology, School of Physics and Astronomy, Edinburgh, EH9 3JZ, UK

8.1 Introduction

If humanity is to achieve a self-sustaining presence beyond the Earth, which is the aim of many space agencies and private corporations, then we will need to mine and process elements and compounds of societal use from indigenous raw materials, wherever we happen to be. Although there is an emerging and rich literature of "space mining," what is often not highlighted is that most of the materials we will be confronted with in most locations in the Solar System will be silicate rocks that on Earth would be regarded as low grade with respect to economic use.

Much of Mars and the Moon is composed of basaltic rock with a mineralogy and elemental composition grossly similar to crustal basaltic and other volcanic rocks on Earth that are not considered ore grade (Ruzicka et al. 2001; McSween et al. 2009). By "ore" we mean material that has sufficiently high concentrations of given elements of interest and sufficient extractability to make it economically viable from a conventional mining point of view. Of course, the definition of whether a material counts as an "ore" will depend on the economic circumstances at any given time. An unprofitable asteroid may become profitable if the price of one of its constituents rises sufficiently.

Other volcanic rocks of the Moon (such as anorthosite) and Mars (such as various ultra-mafic rocks) are similarly not ore grade. As on Earth, the exact content of any given rock type varies depending on its formation pathway and processing. For example, regions of the Moon, such as the Oceanus Procellarum, contain KREEP rocks, which are enriched in potassium (K), rare-earth elements (REEs), and phosphorus (P) (Wieczorek and Phillips 2000). The REEs, which are of considerable interest in a wide range of high-technology electronic and magnetic devices, are enriched ten times more in these rocks than in "typical" basalt. Yet their concentration is still a hundred to a thousand times lower than high-quality REE ores on Earth.

In-Space Manufacturing and Resources: Earth and Planetary Exploration Applications, First Edition.
Edited by Volker Hessel, Jana Stoudemire, Hideaki Miyamoto, and Ian D. Fisk.

One reason for the apparent paucity of high-grade ores beyond Earth is because the crust of most planetary bodies has not been subjected to a hydrological cycle assisted by tectonism, which circulates and concentrates metal-rich fluids on Earth, resulting in the deposition of ores. An exception might be Mars, where orbital and ground-based evidence for a large number of craters, volcanoes, and their associated hydrothermal systems (e.g. Osinski et al. 2013) could imply subsurface or near surface deposits of industrial interest. However, these have not yet been prospected or mapped. We do not know how magmatic processes, which also result in the deposition of ores such as in sulfide units, might have influenced ore formation on an early volcanically active Mars compared to Earth. As with hydrothermal deposition, the lack of long-term magmatic processes on objects such as asteroids is a further reason why many of them will not be rich in ore-quality materials.

In summary, on the Moon, asteroids, and comets, hydrothermal and magmatic concentration of significant ores is not expected to have occurred and Mars remains little understood from this perspective. Thus although we cannot discount places in the Solar System that have economically useful enrichments of certain elements, the fact remains that in the majority of places where humans might construct settlements or attempt to mine, the material they are presented with is low-grade silicate rock. Two of the most abundant of these rock types are basaltic rocks (and other volcanic rocks with different silicate contents ranging from ultramafic to felsic) (Ruzicka et al. 2001; McSween et al. 2009) and other silicate materials, including chondritic materials which comprise most of the asteroids (Reddy et al. 2015).

There is an entire field, that is still in its early stages, that seeks to quantify the economics of extraterrestrial manufacturing processes (Barker 2020). This has not yet been employed to investigate in detail the economics of mining different types of silicate rocks and we will not consider economics in detail here.

In this chapter there are three objectives: (i) To discuss the Solar System silicate economy and what useful elements those materials contain, (ii) To discuss what we know about the mining and processing of such low-grade crustal rocks, (iii) To highlight the use of biology to concentrate and process rocks in the circular silicate economy.

8.2 The Silicate and Basalt Economies

8.2.1 Rationale for Opening the Silicate Economy

We could refer to the extraction of elements from "average" silicate rocks, which in almost any economic scenario on Earth would be of low economic value, as the "silicate economy." One subdivision of this economy is the "basalt economy" which deals with processing of basaltic rocks specifically (for example, on the Earth, Moon, and Mars). In addition to basalt, the silicate economy covers all other rocks, including the great majority of asteroidal materials (chondrites and achondrites) and any other silicate materials to be found in the Solar System (e.g. ultramafic and felsic rocks).

Why would mining silicate rocks that on Earth would be considered economically unviable interest us? There are two points to consider:

First, as discussed in the introduction, plate tectonics, long-term active volcanism, and other geological processes have generated a vast diversity of rocks with concentrated elements on Earth that feed our industries, yet these diversities of ores are not thought to exist on many other planetary bodies where we might wish to establish a human presence, for example, the Moon, many of the asteroids, and the moons of the Outer Solar System. The presence of ores on Mars remains an open question given that the planet did experience impact hydrothermal systems and protracted volcanism in its past.

Furthermore, we might note than even on Earth, despite the huge diversity of ore bodies, it is estimated that about 99.99% of the metal ions in the crust are not in ores (Skinner 1986), illustrating the vast repository of resources bound up in non-ore-quality crustal rocks even on Earth, a situation that is even more extreme on other planetary bodies.

An important point is that even in the case of extraterrestrial objects that do have attractively high elemental concentrations, the abundance of these objects is not high. Elvis (2013) discusses the case of ore-bearing asteroids. Canonically, we regard the asteroid belt as almost limitless sources of resources and in a qualitative way, it is frequently cited as an example of the vast resources that await us beyond Earth. Elvis considers the platinum group elements (PGEs), containing platinum [Pt], rhodium [Rh], osmium [Os], iridium [Ir], palladium [Pd], and rhenium [Re]. These elements tend to associate with iron (they are "siderophile" elements), and hence they tended to have been fractionated into the core of the Earth during global differentiation when the Earth formed, making their concentrations in the terrestrial crust low, and consequently making their economic value high. In iron–nickel asteroids, PGEs concentrations are on the order of tens of parts per million and may be economically viable. However, Elvis notes that only 1% of near-Earth objects (NEOs) are iron–nickel meteorites. Taking into account the number of objects with a diameter greater than 100 m, a size that might make them economically viable (to Earth-based industries), he estimates there to be only 10 NEOs of potential economic interest. Of course, Elvis's calculations are with respect to Earth use. If we were physically located within the asteroid belt, objects less than 100 m in diameter may be of more interest and amenable to collection and processing. Although these smaller objects are of little value to terrestrial industries, processing a small object to top up efficiently recycled planetary stations with certain elements might make them much more attractive than larger more unwieldy objects, so Elvis's conclusions for Earth may be less applicable to a Solar System-wide human presence (discussed further in Chapter 20).

Nevertheless, the general point is that although we can identify certain elements that are enriched in extraterrestrial materials, such as PGEs in certain asteroids and REEs in certain rocks on the Moon, the Solar System is not strewn with an abundant diversity of ores variously containing enrichments of all the industrially useful elements, and those places where we do find certain enrichments are themselves not evenly and abundantly distributed in the Solar System. This situation suggests that the capacity to mine traditionally non-ore-quality silicate rocks for elements would be a powerful industrial capacity for humans beyond Earth.

Second, the vast spatial scales of the Solar System are likely to make it useful to extract elements required by human settlement at the local level, a concept often referred to as *in situ* resource utilization (ISRU). For example, shipping REEs from Earth to the outer Solar System requires removing them from the Earth's gravity well and transporting them over long distances (i.e. long time spans of ~months/years) to the outer Solar System. This incurs high mass, energy, and time costs. If efficient recycling is achieved in planetary stations, the absolute industrial requirement for new elemental provisions may be low, reducing the attractiveness of the effort needed to ship resources at great cost from very far away. It may therefore be useful to devise ways to extract and purify elements from silicate rocks at the local level.

8.2.2 Elements of the Silicate Economy

Despite the low attractiveness of typical crustal and asteroidal silicate rocks for processing, they contain a remarkable number of elements that, with suitable processes, can feed a wide diversity of industries.

For example, Figure 8.1 shows the major elements associated with a typical basaltic rock and some examples of their industrial uses. The astonishing store house of elemental potential in a simple lump of basalt is illustrated. Basalt is a fine-grained extrusive igneous rock formed from the rapid cooling of low-viscosity

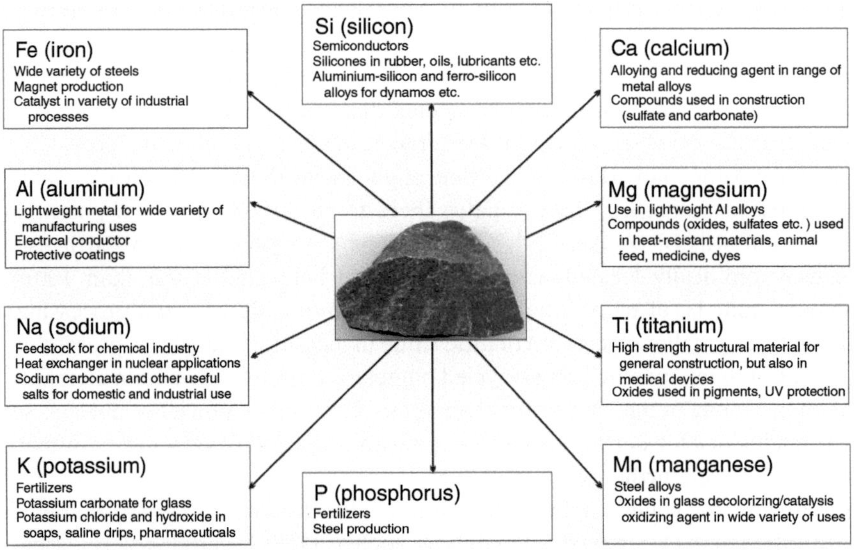

Figure 8.1 Major elements of the silicate economy with examples of industrial uses. Source: Charles S Cockell / ESA image.

Table 8.1 Typical concentrations of a basalt rock showing major elements of the silicate economy. This example is an olivine basalt rock collected near Gufunes, Reykjavik in Iceland (64°08′22.18″N, 21°47′21.27″W). Elements are shown as oxide equivalent as obtained by X-ray Fluorescence (XRF).

Mineral	Element	%
Na_2O	Sodium (Na)	1.92
MgO	Magnesium (Mg)	10.00
Al_2O_3	Aluminum (Al)	15.35
SiO_2	Silicon (Si)	47.48
P_2O_5	Phosphorus (P)	0.128
K_2O	Potassium (K)	0.162
CaO	Calcium (Ca)	11.69
TiO_2	Titanium (Ti)	1.344
MnO	Manganese (Mn)	0.186
Fe_2O_3	Iron (Fe)	12.12

Source: Based on Cockell et al. 2020.

lava rich in magnesium and iron when it is exposed and cooled at or near the surface of a rocky planet or a moon. More than 90% of all volcanic rock on Earth is basalt and it primarily makes up the oceanic crust, the continental crust being primarily composed of granite (formed from magma that has higher concentrations of silica and alkali metal oxides that slowly cool and solidify underground). Basalt is a highly abundant component of the Moon (constituting the mare regions) and makes up most of the Martian crust (Ruzicka et al. 2001; McSween et al. 2009).

The concentrations of elements within basalt vary depending on the specific rock, but Table 8.1 shows concentrations in basalt from Gufunes, Reykjavik in Iceland, the material used to demonstrate biological mining of REEs and vanadium from basalt on the International Space Station (Cockell et al. 2020, 2021). These concentrations are representative of a variety of basalts. Silicon is the most abundant element, followed by aluminum, iron, calcium, and magnesium, all of which have a wide diversity of industrial uses with iron and aluminum having particularly wide use in a variety of metallic structures.

At lower concentrations, yet hugely industrially important, are a variety of trace elements including metal and non-metal elements that range from tens to thousands of parts per million (Table 8.2). At less than 10 ppm are REEs (Table 8.3). All of these elements are subject to considerable variation depending on the originating magma and rock formation conditions. Basaltic rocks can often be found that contain enriched concentrations of these elements compared to mean basalt concentrations. An example is the lunar REE-enriched basaltic KREEP rocks (Wieczorek and Phillips 2000).

Table 8.2 Trace elements of the silicate economy and some examples of their industrial uses. This example is an olivine basalt rock collected near Gufunes, Reykjavik in Iceland (64°08′22.18″N, 21°47′21.27″W). Determined by XRF.

Element	ppm	Examples of industrial uses
Barium (Ba)	85.7 ± 6.8	Drilling fluids, glass, and paint production
Cadmium (Cd)	8.1 ± 0.9	Toxic, although can be used in electroplating
Cobalt (Co)	88.8 ± 10.6	Alloys, magnets, electroplating, pigments
Chromium (Cr)	436.5 ± 81.0	Steel alloy production, catalysis, pigments
Copper (Cu)	166.4 ± 10.6	Alloys, industrial machinery, electrical uses
Nickel (Ni)	222.7 ± 54.8	Use in alloys in many industrial contexts, catalysis
Sulfur (S)	1343.0 ± 52.3	Production of sulfuric acid for use in chemical industry
Strontium (Sr)	246.7 ± 7.6	Magnets, zinc refining, "glow in the dark" applications
Zinc (Zn)	142.2 ± 9.5	Galvanized steel, die-castings, alloys
Zirconium (Zr)	129.2 ± 12.9	Glass and ceramics, superconductive materials

Table 8.3 Rare-earth elements of the silicate economy. This example is an olivine basalt rock collected near Gufunes, Reykjavik in Iceland (64°08′22.18″N, 21°47′21.27″W) and used to demonstrate rare-earth elements mining in microgravity and simulated Martian gravity on the International Space Station (Cockell et al. 2020). Determined by XRF.

Rare-Earth element	Concentration in basalt (ppm)
Lanthanum (La)	6.8 ± 0.5
Cerium (Ce)	13.5 ± 1.7
Praseodymium (Pr)	2.3 ± 0.2
Neodymium (Nd)	11.6 ± 0.8
Samarium (Sm)	3.0 ± 0.2
Europium (Eu)	1.1 ± 0.1
Gadolinium (Gd)	3.7 ± 0.1
Terbium (Tb)	0.6 ± 0.6
Dysprosium (Dy)	3.9 ± 0.2
Holmium (Ho)	0.8 ± 0.0
Erbium (Er)	2.4 ± 0.1
Thulium (Tm)	0.3 ± 0.0
Ytterbium (Yb)	2.1 ± 0.2
Lutetium (Lu)	0.3 ± 0.0

Source: Data from Cockell et al. 2020.

8.3 Processing in the Silicate Economy

Although little attention has been given to mining rocks such as basalt because of their low economic value, the contribution by Steen and Borg (2002) might be regarded as seminal from this point of view as it provides an insight into the economics of the Solar System silicate economy.

Steen and Berg do not address extraterrestrial mining. They sought to calculate the costs of extracting metals from crustal rocks on Earth as a way to illustrate the value we should place on ores, both economic and in terms of caution about the rate of depletion of valuable ores. However, their calculations serendipitously provide a baseline for understanding the Solar System silicate economy.

Steen and Berg do not discuss the use of biology to extract and concentrate elements, an approach discussed in Section 8.4. They focus on chemical methods and specifically the use of hydrochloric acid (HCl) to dissolve the silicate matrix of rocks to access more valuable elements. Their proposed scheme is not necessarily ideal for the extraterrestrial case, but it provides an example.

They imagine mining crustal rocks by blasting and then crushing and grinding into rock fragments of ~80 μm diameter or less. The fines produced in this way are leached using HCl, which they estimate at 54 kg of acid required per ton of rock. This leachate contains many of the metal ions found in the basalt, for example leaching for 24 hours in 30% HCl can liberate 72 kg of Al, 94 kg of Fe, and 38 g of Co. The authors then suggest sulfide precipitation, which causes metal sulfide precipitation of many of the required elements. The sulfide precipitation phase is followed by sodium hydroxide precipitation of oxides and hydroxides such as Mn and Cr. The waste water from this entire process may itself be processed to recover metals.

Although Steen and Berg's process was not designed for the extraterrestrial case, it is intriguing to note that both NaCl (processed to produce the required HCl) and sulfate salts (reduced with charcoal to produce the required sulfides) have been identified on Mars (e.g. Bibring et al. 2006; Osterloo et al. 2010), making their process potentially applicable there. If planetary protection considerations obviate the possibility of large scale biomining on that planet and mining must be achieved chemically, then an adaptation of their process may be viable. It would be optimal to locate a Martian basalt mine in an area with chloride and sulfate salts co-located or close by each other, for example in the southern cratered highlands where chloride deposits are localized.

On the Moon and asteroids, HCl is not easily acquired and would likely have to be imported, although sulfides exist in the form of minerals such as the iron sulfide, troilite and could be processed for H_2S precipitation steps.

Considering all stages of this process, Steen and Berg manage to calculate the projected costs for production of certain metals from basalt. For example, cobalt can be produced at 646 Euro/kg and zinc at 81.4 Euro/kg. The absolute prices are not particularly interesting here, but the comparison to ore-quality material is instructive. For all the metals they examined, the cost of extracting it from basalt is about one (Cr) to five (W) orders of magnitude higher than using ores.

Perhaps of more use is to identify where the cost increases are incurred and where the energy use occurs in the process. The great majority of energy is used in electrolysis of NaCl (1100–2070 MJ/ton) to produce HCl and in grinding (140–400 MJ/ton), which accounts for about half of the associated cost. In the extraterrestrial case, the energetic costs in the chemical steps would depend on the process used, but in any chemical leaching and precipitation process, initial grinding of the material is optimal. The energy budget required to do this (>100 MJ/ton) illustrates that provision of large quantities of energy is a crucial step in enabling access to the silicate economy. As biological extraction of elements in biomining is also optimally performed on high surface areas of material, then the crushing and grinding steps are the most energy intensive in any mode of material production.

Steen and Berg's analysis illustrates that relatively simple leaching and precipitation approaches can be used to extract elements from basaltic and potentially other silicate rocks. Although their paper was used to illustrate the economic barriers to mining basalt and other crustal rocks on Earth, if our goal is merely to access elements, not necessarily for an economic market, but to supply the local needs of an extraterrestrial outpost or settlement, then they adequately demonstrate feasibility. It should be realized, however, that although their cost analysis clearly places the processes they describe out of terrestrial economic viability, the cost of extraction, processing, and launch from Earth could bring them within economic viability in some locations in space. In space, the crucial practical matters to be overcome are the provision of suitable energy, which might be accomplished using nuclear energy and, in the case of chemical leaching, the provision of the required leaching chemicals.

8.4 The Use of Biology in the Silicate Economy

8.4.1 General Observations on Silicate Biomining

One way to overcome the requirements for large concentrations of chemicals such as HCl and H_2S is to use microorganisms to extract elements in the process of biomining. Biomining was devised to overcome these chemical needs and to accomplish mining with less environmental destruction. Although biological mining can also overcome some of the energy requirements (for example, the considerable NaCl electrolysis energy requirements where leaching is carried out with HCl and the HCl is obtained from NaCl), it cannot overcome the energetic needs to crush the rock, which will always be a fundamental baseline requirement of mining where the material is not already highly porous.

Biomining would be a useful way to process silicate rocks in any location where feedstocks for chemical leaching, such as HCl or HF, are either scarce or non-existent, which is most locations in the Solar System. Although microorganisms require nutrients to grow, reproduce and thus carry out biomining, the silicate rock being mined can provide the essential and trace elements required for microbial growth. Microbial growth on meteoritic material, for example, has been demonstrated (e.g. Mautner 1997).

An anomalous, but important case is Mars. The planet is subject to planetary protection regulations which at the current time would disallow the deliberate introduction of microorganisms onto that planet in a biomining facility (although it is an open matter about whether, once humans are on Mars, these regulations can be as restrictive as they are today, given the inevitable introduction of the human microbiome. A biomining facility could be fully sealed, for example). Yet it is an interesting point that the presence of chlorides and sulfates noted earlier that could be used as feedstocks for chemical leaching is the result of ancient liquid water on that planet that has led to its protection under planetary protection regulations in the first place. So in this case, the existence of liquid water, a requirement for habitability and life, has resulted in the formation of the evaporitic minerals that provide the feedstock for chemically based mining which provides a way to carry out mining on an otherwise biologically regulated world.

The case of Mars also illustrates that there is no single approach to mining in the Solar System. Different bodies likely require different processing steps and approaches, illustrating that to access the silicate economy we require a diversity of methods that can be applied to the same rock types in different locations. On Mars, basalt processing might favor chemical processes, on the Moon regulations might allow for biomining of basalt.

8.4.2 Using Biomining in Space

Biomining involves the use of microorganisms to assist in the release of economically important elements from rocks (Rawlings and Silver 1995; Bosecker 1997). On Earth, both copper and gold, for example, are extracted from ore using acidophilic iron-oxidizing bacteria in "biomining" operations (Das et al. 1999; Norris et al. 2000; Stott et al. 2003). The range of conditions in which biomining is accomplished on a commercial scale is still quite specific. Generally, metals within pyritic ores are used because the iron within them can be oxidized by iron-oxidizing bacteria, such as *Acidithiobacillus* species, that can grow within the acidic conditions produced from sulfide oxidation to sulfuric acid.

The traditional principle of biomining, although involving diverse chemical reactions, is quite simple. In sulfide ores (pyritic ores), iron-oxidizing bacteria provide a source of Fe^{3+} which causes oxidation of the sulfidic minerals (Clark and Norris 1996; McGuire et al. 2001), producing sulfuric acid that sustains further rock dissolution, breaking apart the rock matrix to facilitate the release and concentration of useful elements (e.g. Pronk and Johnson 1992; Schroeter and Sand 1993; Solisio et al. 2002; Rawlings 2005) such as nickel, copper, and gold, among other metals. Sulfur-oxidizing microorganisms can also contribute to rock dissolution. The biomining principle is based on the ability of microorganisms to accelerate rock weathering. In the case of pyrite, microorganisms can accelerate rock weathering by up to six orders of magnitude.

Pyrites are rather specific rock types and not abundant in the rest of the Solar System. However, the microbial production of acids, metal-chelating compounds (such as siderophiles and polysaccharides), and local changes in pH, among other mechanisms, allows bioleaching to be achieved at neutral or alkaline pH conditions (Rezza et al. 2001; Willscher and Bosecker 2003). A wide range of elements are now potentially extractable using biomining approaches, an example being the REEs (Fathollahzadeh et al. 2019; Rasoulnia et al. 2021), the subject of the next section of this paper. In principle, any microorganism that is capable of accelerating the weathering and dissolution of a rock that contains a commercially important element would be useful in biomining, even if the rate of weathering was increased only by a few-fold.

The advantages of biomining over other methods are: (i) reduced need for chemicals to enhance rock dissolution or elemental release (this might be offset by the requirements to provide microbial growth nutrients, but if the rock contains bioessential elements, these additional requirements can be minimized), (ii) reduction or removal of the use of poisonous compounds (such as HF and cyanides, traditionally used to leach rocks), (iii) reduction in energy requirements to achieve rock dissolution. In space, because biomining reduces energy and chemical needs, it may be a useful approach in distant locations to extract elements which would make it non-viable on Earth.

Biomining is one application of microbe-mineral interactions in space (Cockell 2010). Other applications include the breakdown of rocks and regolith into agriculturally useful soil, the use of rocks and regolith as nutrient feedstock in life support systems, the use of microbes and "soil crusts" to control dust release from surface materials in enclosed spaces, and others besides.

An example of a space application is provided by the extraction of iron on the Moon. Volger et al. (2020) carried out theoretical calculations on iron extraction from lunar and Martian regolith using the iron-reducing bacterium *Shewanella oneidensis* and demonstrated the potential for this process to release magnetically extractable iron. In laboratory experiments (Castelein et al. 2021), this bacterium was used to reduce three different types of lunar and Martian regolith simulants, allowing for the magnetic extraction of iron-rich materials. Bacterial treatment and magnetic extraction was used to achieve a 5.8-times higher quantity of iron and 43.6% higher iron concentration compared to magnetic extraction alone. The resulting materials were 3D printed into cylinders for which the mechanical properties demonstrated a 400% improvement in compressive strength in the bacterially treated samples compared to non-bacterial samples.

8.4.3 Demonstrating Biological Access to the Silicate Economy on the International Space Station

In an experiment to investigate bioleaching of elements and biofilm growth in microgravity and simulated Martian gravity ($0.38 \times g$), Cockell et al. (2020, 2021) demonstrated the biologically enhanced extraction of REEs and vanadium from basaltic rock (Tables 8.1–8.3). The REEs include the lanthanides, scandium, and yttrium. Because of their physical properties, including ferromagnetism and luminescence, REEs are used in electronic devices such as cell phones and computer

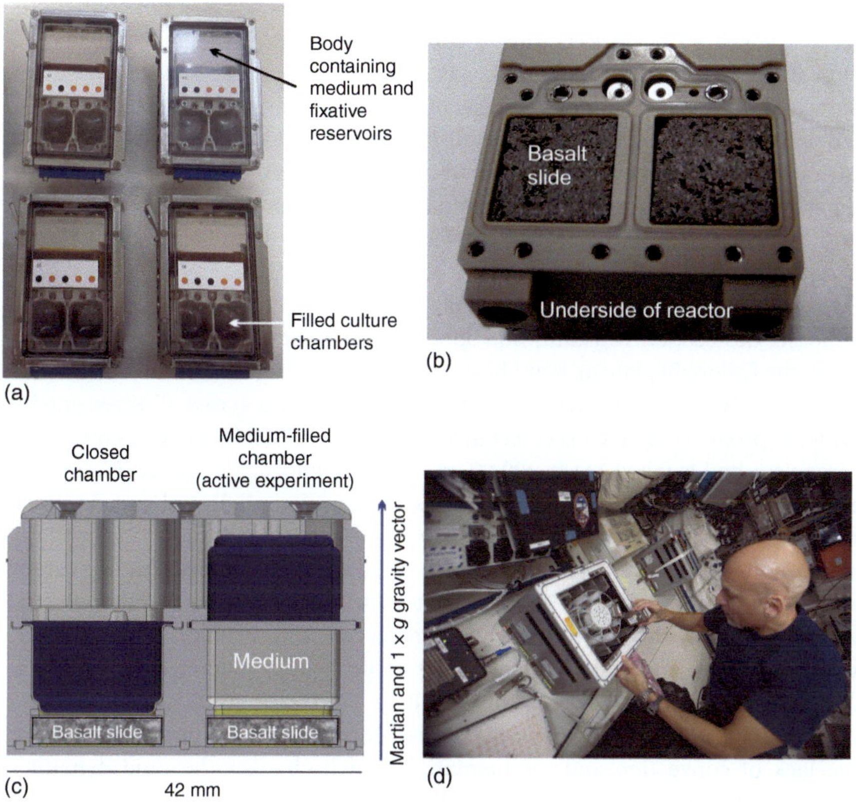

Figure 8.2 A miniature biomining reactor in space. (a) Top-down view of four mining reactors, each with two culture chambers for two separate experiments. (b) Rear side image of reactor showing two basalt slides inserted into the bottom of the culture chambers. (c) A cross-section through the culture chambers showing the position of the basalt slide at the back of the chamber and the operation of the experiment (medium injection and inversion of the membrane: in blue; left side closed, right side inflated with medium). A scale bar shows the size of the unit, which also applies to the images in (a) and (b). (d) ESA Astronaut Luca Parmitano inserts a reactor into a KUBIK incubator on board the International Space Station (credit: ESA) (image adapted from Cockell et al. (2021)). Source: Charles S Cockell / ESA image.

screens, as well as in catalysis, metal alloy and magnet production, and many other high-technology applications.

The European Space Agency (ESA) BioRock experiment was designed to test hypotheses on the effects of gravity on microbe–mineral interactions, biofilm formation, and the implications of these findings for space manufacturing and processing.

Bioleaching experiments were carried out in a miniature biomining reactor (Figure 8.2) in which basalt slices (rectangles of 1.5 × 1.6 cm dimensions and 3 mm thickness) were exposed to three microorganisms for 21 days in 50% R2A medium (Reasoner and Geldreich 1985). The experiment was not optimized for bioleaching since it was also used to investigate the growth of biofilms across a flat contiguous surface of basalt rock, necessitating a compromise in experimental design. The yield of REEs extracted from the rock was generally less than $5 \times 10^{-2}\%$ in all the gravity conditions. These leaching rates would likely be higher with crushed rocks, which on Earth have been shown to result in leaching yields of REEs of $8.0 \times 10^{-3}\%$ to several tens of percent under optimized conditions. Furthermore, the bioreactor volume ($\sim 6\,cm^3$) is also clearly much smaller than a commercial operation and we did not employ mixing because we wanted to study the effects of gravity on microbial growth.

The experiment demonstrated that *Sphingomonas desiccabilis*, a Gram-negative, non-motile, desiccation resistant, non-spore-forming bacterium isolated from soil crusts in the Colorado plateau, was able to enhance leaching of REEs and vanadium from the basaltic substrate. Across all individual REEs and across all three gravity conditions (produced by means of a centrifuge in the KUBIK incubator) on the ISS, the organism had leached 111.9–429.2% of the non-biological controls (Figure 8.3, Cockell et al. 2020). The experiment demonstrated preferentially enhanced release of heavy compared to light REEs.

Intriguingly, there were no statistically significant differences in the absolute final concentrations of REEs in the biological experiments in microgravity, simulated Martian gravity (produced by means of a centrifuge in the KUBIK incubator on ISS), and the simulated $1 \times g$ control (also generated in a KUBIK centrifuge on ISS). We attributed this result to the fact that although microbial cell growth is known to be potentially altered in microgravity, primarily because of the lack of convection and sedimentation which changes the fluid dynamics regime typically observed on ISS, the cell number at the end of the experiment was the same in all gravity conditions (Santomartino et al. 2020). The 21-day length of the experiment likely allowed the cells to reach their final "carrying capacity" (i.e. stationary phase) in the cell fluid after a few days and this may account for the fact that once this was achieved, their leaching capacity was the same regardless of gravity condition produced on the ISS. One implication of this finding could be that provided cells are left long enough to reach stationary phase, the efficacy of biomining extraction of elements would be the same regardless of gravity conditions (e.g. microgravity conditions on small asteroids or higher gravity conditions such as the Moon or Mars). However, caution must be taken in this interpretation because in the BioRock experiment the cell culture volume was small and larger volumes would be expected in a more realistic biomining set up, where convective and sedimentation effects could be more prominent.

REEs are not economically attractive to mine in space and return to Earth (McLeod and Krekeler 2017). Even the REE-enriched KREEP rocks on the Moon have two to three orders of magnitude lower REE concentrations than high-quality REE ores on Earth, such as the Bayan Obo Reserves in China (Table 8.4). Yet they might be extracted and used locally. The biologically enhanced extraction from rock we demonstrated at the low concentrations in basalt also provides a demonstration of the potential of biomining for the extraction and concentration of other elements of industrial interest in a silicate economy.

The BioRock experiment also demonstrated the biologically enhanced extraction of vanadium from basalt. Results showed that both *S. desiccabilis* and *Bacillus subtilis*

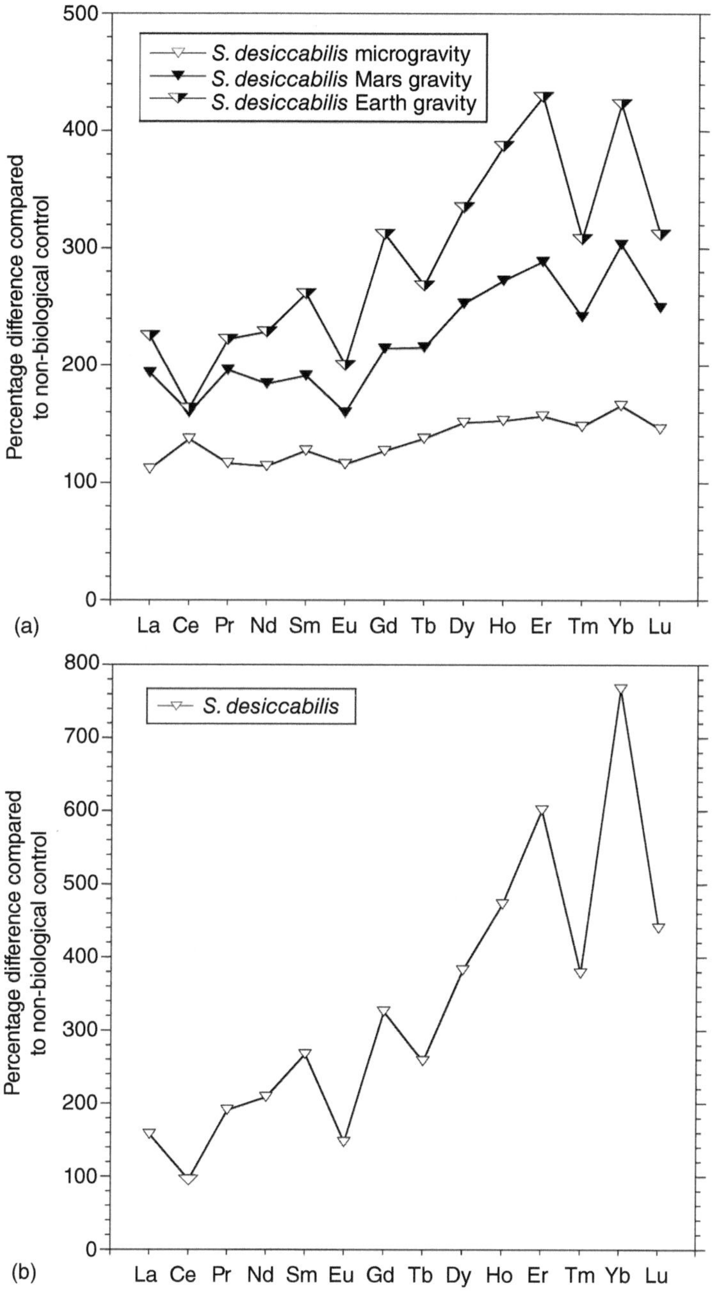

Figure 8.3 Enrichment of rare-earth elements from basalt rock using *Sphingomonas desiccabilis* under microgravity, simulated Martian gravity ($0.38 \times g$), simulated Earth gravity ($1 \times g$) control on board the International Space Station (ISS). Panel (a) shows the mean enrichment as a percentage of the non-biological control for the three gravity conditions on the ISS based on concentrations in the bulk fluid ($n = 3$, $n = 2$ for non-biological microgravity experiment). Panel (b) shows the enrichment data for the ground-based (true) $1 \times g$ control ($n = 3$ for biology, $n = 2$ for non-biological control). Source: Adapted from Cockell et al. (2020). Statistics and details can be found in Cockell et al. (2020).

Table 8.4 Rare-earth elements on Earth and the Moon. Representative concentrations of REEs in lunar and REE ore rocks (ppm) compared to the basalt used for biomining on the International Space Station.

REE	Basalt rock (Table 8.2)	Lunar KREEP rock 15382 (Apollo 15)[a]	Bayan Obo, China, Sample 3. Type II apatite
Lanthanum (La)	6.8 ± 0.5	79.5	7,875
Cerium (Ce)	13.5 ± 1.7	212	24,305
Praseodymium (Pr)	2.3 ± 0.2	—	4,050
Neodymium (Nd)	11.6 ± 0.8	127	20,519
Samarium (Sm)	3.0 ± 0.2	35.2	3,553
Europium (Eu)	1.1 ± 0.1	2.77	1,067
Gadolinium (Gd)	3.7 ± 0.1	42.9	2,835
Terbium (Tb)	0.6 ± 0.6	6.2	345
Dysprosium (Dy)	3.9 ± 0.2	45.7	1,500
Holmium (Ho)	0.8 ± 0.0	—	212
Erbium (Er)	2.4 ± 0.1	28.1	351
Thulium (Tm)	0.3 ± 0.0	—	29.0
Ytterbium (Yb)	2.1 ± 0.2	24	—
Lutetium (Lu)	0.3 ± 0.0	3.43	—

a) LPI (Lunar and Planetary Institute) Lunar samples compendium.

(a Gram-positive, motile, spore- and biofilm-forming bacterium naturally found in a range of environments, including rocks) enhanced the leaching of vanadium under the three gravity conditions compared to sterile controls by 216.32–283.22%, respectively (Cockell et al. 2021). Vanadium is an element of interest because it can be used to make ferrous and nonferrous alloys to enhance hardness, tensile strength, and fatigue resistance (Moskalyk and Alfantazi 2003). On Earth, these alloys are used in structural materials, rails, metallurgical tools, applications that beyond Earth could apply to the fabrication of rovers, pressurized buildings, and particularly objects exposed to lunar and Martian dust and its corrosive, abrasive properties (Ferguson et al. 1999; Stubbs et al. 2005). The element is also used in catalysis in a variety of industrial applications, in the production of superconducting materials and in batteries. The low neutron-absorbing capacity of vanadium makes it a useful additive in the fabrication of alloys for nuclear reactors, including fusion reactors.

We did not attempt to purify the elements from the solutions obtained. Our experiment was focused on the effects of gravity on microbe–mineral interactions, not the practicalities of downstream processing. Nevertheless, the BioRock experiment demonstrates that microorganisms can be used to enhance elemental release from non-ore-quality basaltic substrates under diverse gravity regimes, providing the first proof of concept in space of access to the silicate/basalt economy.

8.4.4 Using Synthetic Biology to Enhance Access to the Silicate Economy

One of the central challenges in accessing the silicate or basalt economy is that no matter whether chemical and biological processes are used to extract the element(s) of interest, many will end up at low leached concentrations by the very nature of the non-ore-quality material that we wish to process. Therefore, in addition to achieving bioleaching, another challenge is to achieve concentration.

Chemical methods might be used to precipitate elements preferentially from leachate. An alternative approach suggested by Cockell (2011) is to use synthetic biology/genetic engineering to achieve biologically mediated concentration. Synthetic biology has been recognized to have huge potential in space manufacturing (Montague et al. 2012; Rothschild 2016).

Bioaccumulation is the process of uptake of metal ions and sequestration within the intercellular space. Some wildtype microbial species can be used to treat mine wastewaters and recover useful and environmentally toxic metals. Genetically engineered microorganisms (GEMs) have been investigated, which will have improved and faster performances, as well as an enhanced tolerance to metal toxicity (Diep et al. 2018). With knowledge of the genetic pathways of elemental extraction and sequestration from raw materials such as basalt, it should be possible to engineer microorganisms to preferentially sequester and bioaccumulate desired elements from basalt within the cell structures, in a type of "directed biomining." This speculative process is shown in Figure 8.4 as an illustrative idea of the type of direction one could pursue. The approach is to engineer an organism that has a genetically engineered preference, and resistance, for bioaccumulating an element of industrial use and grow it in the presence of crushed silicate rock, such as lunar or Martian basalt. When applied to space biomining, bioengineering approaches could also be used to enhance the fitness under space conditions (e.g. radiation, low gravity, toxic compounds) in order to obtain an economically viable biomass. The organism would extract and sequester the element from the rock. The rock leachate with organisms would then be collected, the organisms removed by centrifugation, and the element recovered from the bacteria and the leachate. This possibility might have particular value for very low concentration elements such as REEs, and the range of trace, but economically important, metals shown in Table 8.2.

The capacity to sequester and easily recover many elements at the same time might be made possible by engineering each organism to express a specific color of fluorescent protein, i.e. each element-sequestering bacterium fluoresces a different color. The organisms are added to the ground basalt; weathering and biomining proceeds. The leachate is then collected and passed through an industrial-scale flow cytometer (cell sorter), which sorts the cells based on fluorescent protein color, i.e. according to different elements. Each separate solution is collected and the relevant element extracted from the organisms (illustrated in Figure 8.4). Either the organisms are ground up and the element recovered by some chemical means or the organisms are engineered so that the addition of a compound in their medium induces them to secrete the elements that they have sequestered (this approach

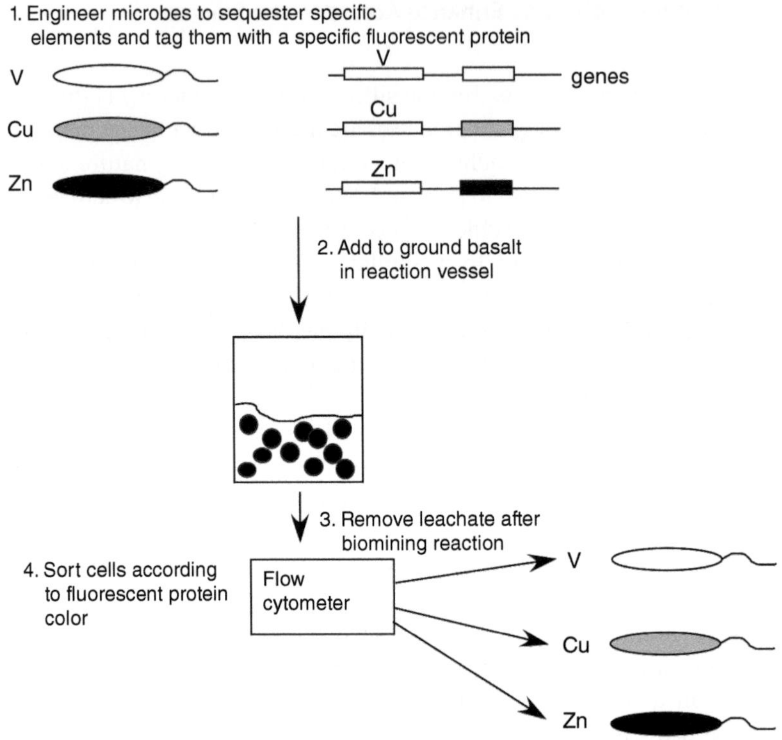

Figure 8.4 A speculative concept for the sequestration and concentration of elements from basalt rock using microorganisms that are engineered to sequester certain elements and tagged to allow their sorting. This illustrates the use of genetic engineering and synthetic biology to access the silicate economy.

would depend on the solubility of the element that they have sequestered). Notably, since bioaccumulation technologies are used for bioremediation, these approaches could be also used for amelioration of the planetary environment (e.g. bioremediation of toxic extraterrestrial compounds, stabilization of regolith, soil formation). Compartments could be added to biological life support systems (BLSSs), enhancing a circular economy approach to space exploration.

The industrial efficacy and practicality of this approach to low concentration element harvesting is open to further analysis. However, it is raised here as an illustration of the way in which novel technologies and approaches, including synthetic biology, might be used to overcome the challenges of opening up the Solar System silicate economy, particularly the processing of what on Earth would be regarded as poor-quality materials.

8.4.5 Some Conceptual Novelties in Extraterrestrial (Bio)mining

The extraterrestrial environment poses, or presents, some novelties with respect to mining and particularly biological mining. There are different physical conditions, for example a harsher radiation environment and the presence of ions such as perchlorates in the Martian soil compared to terrestrial biomining, which must be considered. To illustrate the unfamiliar conditions that must be considered for extraterrestrial biomining, two of these differences are outlined below:

(1) *Fracturing*. On Earth, plate tectonics has erased the long history of impact bombardment of the planetary surface, but this is not the case for other planetary bodies such as the Moon, Mars, and large asteroids. Impact gardening in the near surface of these bodies has fractured and cracked rocks, potentially generating large networks of fractures whose characteristics are not fully understood or characterized. These impact-induced fracture networks could potentially allow for easier physical disruption and movement of rock, but they also open up the possibility of the injection of fluids into fractures to achieve leaching with chemical or biological means, with removal of fluids containing the enriched elements of interest. Challenges associated with the use of these industrially beneficial fracture networks include: (i) connectivity of any given network with the surface environment and thus low pressure-induced volatilization and/or freezing of injected fluids. Freezing could also be caused by low temperatures in the material itself. This may be mitigated in the Martian deep subsurface by the geothermal gradient of that planet, (ii) planetary protection concerns about the application of chemical and/or biological approaches to *in situ* extraction.
(2) *Gravity*. Mining beyond Earth will require operating in altered gravity regimes. Although the experiments by Cockell et al. (2020, 2021) showed no statistically significant effect of gravity on elemental leaching, their volumes were small and we can expect that on low-gravity objects such as asteroids, fluids will form surface tension layers around rock fragments at the small and large scale with little movement unless mixing is introduced. Clearly gravity-driven leaching, similar to "heap leaching" operations on Earth, cannot work on such bodies and may be reduced in efficiency on bodies such as the Moon and Mars.

    There are potentially two ways around this problem. The first is to introduce mixing. In the case of biological processes, the mixing must be sufficient to achieve fluid flow. The second approach could be biological. For example, naturally motile microorganisms (microorganisms that can actively move by propelling themselves forward with flagella or pili), or microorganisms with engineered improved motility, might be used to improve microbial distribution in static fluids under low gravity (Benoit and Klaus 2007). Microbial motility may also improve access of the organisms into small fractures in rocks against static low-gravity fluids. Further investigations are required on the characteristics of organisms that could be used or augmented to improve elemental extraction processes under low-gravity regimes.

8.5 The Circular Silicate Economy

As illustrated in this paper, the silicate economy presents several serious challenges that distinguishes it from traditional ore-based processing on Earth. Two of the most notable are: first, the need to crush and grind substantial quantities of material to access given requirements for elements because of the low concentrations of most elements compared to traditional ores. This imposes significant energy demands. Even if highly effective chemical and biological means of dissolving silicates and releasing elements can be found and improved, crushing and grinding is mandated by the simple physical principle of the need to increase the surface area to volume ratio prior to processing. This is often one of the most energy-intensive steps in the mining process. Second, the need to find ways to improve element sequestration from leachate and their concentration.

Regardless of how well these challenges can be overcome, it is clearly beneficial to be able to recycle valuable elements in space settlements to avoid de novo extraction from local rock to the maximum extent possible. Thus, in addition to developing technologies for the silicate economy, parallel methods to achieve extraction and concentration of elements from recycled materials, waste and disused technology must be accomplished. These approaches may be similar to the extraction and concentration methods used for rock and the bioleaching approaches used on Earth to extract elements from industrial waste (e.g. Hong and Valix 2014; Gu et al. 2018; Johnson 2014; Reed et al. 2016).

8.6 Conclusion

Mining in space will require confronting the need to extract, concentrate, and process economically useful elements and compounds from materials that traditionally have not been regarded as economically useful on Earth, i.e. low-grade crustal rocks. In this chapter, we have referred to this as the "silicate economy" and it has various sub-divisions, noticeably the basalt economy. Mobilizing the silicate economy will require new techniques to process rocks, extract the elements, and achieve high levels of concentration. Here, we have emphasized the use of biology to address the kinetic and energetic challenges of processing these types of materials. A mixture of mining processes with biology, including synthetic biology, could allow for the successful mobilization of the Solar System silicate economy for the long-term human presence in space.

References

Barker, D.C. (2020). Lunar and off Earth resource drivers, estimations and the development conumdrum. *Advances in Space Research* 66: 359–377.

Benoit, M.R. and Klaus, D.M. (2007). Microgravity, bacteria, and the influence of motility. *Advances in Space Research* 39: 1225–1232.

Bibring, J.-P., Langevin, Y., Mustard, J.F. et al. (2006). Global mineralogical and aqueous Mars history derived from OMEGA/Mars express data. *Science* 312: 400–404.

Bosecker, K. (1997). Bioleaching: metal solubilization by microorganisms. *FEMS Microbiology Reviews* 20: 591–604.

Castelein, S.M., Aarts, T.F., Schleppi, J. et al. (2021). Iron can be microbially extracted from Lunar and Martian regolith simulants and 3D printed into tough structural materials. *PLoS One* 16: e0249962.

Clark, C.A. and Norris, P.R. (1996). Oxidation of mineral sulphides by thermophilic microorganisms. *Minerals Engineering* 9: 1119–1125.

Cockell, C.S. (2010). Geomicrobiology beyond Earth—Microbe-mineral interactions in space exploration and settlement. *Trends in Microbiology* 18: 308–314.

Cockell, C.S. (2011). Synthetic geomicrobiology: engineering microbe-mineral interactions for space exploration and settlement. *International Journal of Astrobiology* 10: 315–324.

Cockell, C.S., Santomartino, R., Finster, K. et al. (2020). Space station biomining experiment demonstrates rare earth element extraction in microgravity and Mars gravity. *Nature Communications* 11: 5523.

Cockell, C.S., Santomartino, R., Finster, K. et al. (2021). Microbially-enhanced vanadium mining and bioremediation under micro- and Mars gravity on the International Space Station. *Frontiers in Microbiology* 11: 5523.

Das, T., Ayyappan, S., and Chaudhury, G.R. (1999). Factors affecting bioleaching kinetics of sulfide ores using acidophilic micro-organisms. *Biometals* 12: 1–10.

Diep, P., Mahadevan, R., and Yakunin, A.F. (2018). Heavy metal removal by bioaccumulation using genetically engineered microorganisms. *Frontiers in Bioengineering and Biotechnology* 6: 157.

Elvis, M. (2013). How many ore-bearing asteroids? *Planetary and Space Science* 91: 20–26.

Fathollahzadeh, H., Eksteen, J.J., Kaksonen, A.H., and Watkin, E.L.J. (2019). Role of microorganisms in bioleaching of rare earth elements from primary and secondary resources. *Applied Microbiology and Biotechnology* 103: 1043–1057.

Ferguson, D.C., Kolecki, J.C., Siebert, M.W. et al. (1999). Evidence for Martian electrostatic charging and abrasive wheel wear from the Wheel Abrasion Experiment on the Pathfinder Sojourner rover. *Journal of Geophysical Research* 104: 8747–8759.

Gu, T., Rastegar, S.O., Mousavi, S.M. et al. (2018). Advances in bioleaching for recovery of metals and bioremediation of fuel ash and sewage sludge. *Bioresource Technology* 261: 428–440.

Hong, Y. and Valix, M. (2014, 2014). Bioleaching of electronic waste using acidophilic sulfur oxidising bacteria. *Journal of Cleaner Production* 65: 564–472.

Johnson, D.B. (2014). Biomining: biotechnologies for extracting and recovering metals from ores and waste materials. *Current Opinion in Biotechnology* 30: 24–31.

Mautner, M.N. (1997). Biological potential of extraterrestrial materials—1. Nutrients in carbonaceous meteorites, and effects on biological growth. *Planetary and Space Science* 45: 653–661.

McGuire, M.M., Edwards, K.J., Banfield, J.F., and Hamers, R.J. (2001). Kinetics, surface chemistry, and structural evolution of microbially mediated sulphide mineral dissolution. *Geochimica et Cosmochimica Acta* 65: 1243–1258.

McLeod, C.L. and Krekeler, M.P.S. (2017). Sources of extraterrestrial rare earth alements: to the Moon and beyond. *Resources* 6: 40.

McSween, H.Y., Taylor, G.J., and Wyatt, M.B. (2009). Elemental composition of the martian crust. *Science* 324: 736–739.

Montague, M., McArthur, G.H., Cockell, C.S. et al. (2012). The role of synthetic biology for in situ resource utilization (ISRU). *Astrobiology* 12: 1135–1142.

Moskalyk, R.R. and Alfantazi, A.N. (2003). Processing of vanadium: a review. *Minerals Engineering* 16: 793–805.

Norris, P.R., Burton, N.P., and Foulis, N.A.M. (2000). Acidophiles in bioreactor mineral processing. *Extremophiles* 4: 71–76.

Osinski, G.R., Tornabene, L.L., Banerjee, N.R. et al. (2013). Impact-generated hydrothermal systems on Earth and Mars. *Icarus* 224: 347–363.

Osterloo, M.M., Anderson, F.S., Hamilton, V.E., and Hynek, B.M. (2010). Geologic context of proposed chloride-bearing materials on Mars. *Journal of Geophysical Research* 115: E10012.

Pronk, J.T. and Johnson, D.B. (1992). Oxidation and reduction of iron by acidophilic bacteria. *Geomicrobiology Journal* 10: 153–171.

Rasoulnia, P., Barthen, R., and Lakaniemi, A.-M. (2021). A critical review of bioleaching of rare earth elements: the mechanisms and effect of process parameters. *Critical Reviews in Environmental Science and Technology* 51: 378–427.

Rawlings, D.E. (2005). Characteristics and adaptability of iron- and sulphur-oxidising microorganisms used for the recovery of metals from minerals and their concentrates. *Microbial Cell Factories* 4: 1–15.

Rawlings, D.E. and Silver, S. (1995). Mining with microbes. *Nature Biotechnology* 13: 773–778.

Reasoner, D.J. and Geldreich, E.E. (1985). A new medium for the enumeration and subculture of bacteria from potable water. *Applied and Environmental Microbiology* 49: 1–7.

Reddy, V., Dunn, T.L., Thomas, C.A. et al. (2015). Mineralogy and surface composition of asteroids. In: *Asteroids* (ed. P. Michel et al.), 43–63. Tucson: University of Arizona Press.

Reed, D.W., Fujita, Y., Daubaras, D.L. et al. (2016). Bioleaching of rare earth elements from waste phosphors and cracking catalysts. *Hydrometallurgy* 166: 34–40.

Rezza, I., Salinas, E., Elorza, M. et al. (2001). Mechanisms involved in bioleaching of an aluminosilicate by heterotrophic microorganisms. *Process Biochemistry* 36: 495–500.

Rothschild, L.J. (2016). Synthetic biology meets bioprinting: enabling technologies for humans on Mars (and Earth). *Biochemical Society Transactions* 44: 1158–1164.

Ruzicka, A., Snyder, G.A., and Taylor, L.A. (2001). Comparative geochemistry of basalts from the Moon, Earth, HED asteroid, and Mars: implications for the origin of the Moon. *Geochimica et Cosmochimica Acta* 65: 979–997.

Santomartino, R., Waajen, A.C., de Wit, W. et al. (2020). No effect of microgravity and simulated Mars gravity on final bacterial cell concentrations on the International Space Station: applications to space bioproduction. *Frontiers in Microbiology* 11: 579156.

Schroeter, A.W. and Sand, W. (1993). Estimations on the degradability of ores and bacterial leaching activity using short-time microcalorimetric tests. *FEMS Microbiology Reviews* 11: 79–86.

Skinner, B.J. (1986). *Earth Resources*. Hoboken, New Jersey: Prentice-Hall.

Solisio, C., Lodi, A., and Veglio, F. (2002). Bioleaching of zinc and aluminium from industrial waste sludges by means of *Thiobacillus ferrooxidans*. *Waste Management* 22: 667–675.

Steen, B. and Borg, G. (2002). An estimation of the cost of sustainable production of metal concentrates from the earth's crust. *Ecological Economics* 42: 401–413.

Stott, M.B., Sutton, D.C., Watling, H.R., and Franzmann, P.D. (2003). Comparative leaching of chalcopyrite by selected acidophilic bacteria and archaea. *Geomicrobiology Journal* 20: 215–230.

Stubbs, T.J., Vondrak, R.R., and Farrell, W.M. (2005). Impact of dust on lunar exploration. *Proceedings of Dust in Planetary Systems* 643: 2007.

Volger, R., Pettersson, G.M., Brouns, S.J.J. et al. (2020). Mining Moon and Mars with microbes: biological approaches to extract iron from lunar and Martian regolith. *Planetary and Space Science* 184: 104850.

Wieczorek, M.A. and Phillips, R.J. (2000). The "Procellarum KREEP Terrane": implications for mare volcanism and lunar evolution. *Journal of Geophysical Research* 105: 20417–20430.

Willscher, S. and Bosecker, K. (2003). Studies on the leaching behaviour of heterotrophic microorganisms isolated from an alkaline slag dump. *Hydrometallurgy* 71: 257–264.

9

Near-Earth Asteroids as Promising Candidates for Space Resources

Hideaki Miyamoto

University of Tokyo, School of Engineering, Department of Systems Innovation, 7-3-1 Hongo, Tokyo 113-8656, Japan

9.1 Introduction

Historically, the exploration and development of frontier regions have been motivated by scientific interests and the desire for resources. Areas of severe environments on Earth, such as those with high elevations and cold annual temperatures, may be targeted by adventurers and scientists purely from their interests. However, once valuable resources are found, a village is developed. Good examples include Potosi in Bolivia and Oymyakon in Russia. These cities, which are in the highest elevation and the coldest inhabited area, respectively, were developed after discoveries of gold and silver mines. Space activities may expand in the same way, with initial motivations stemming from scientific interests, followed by economic and political reasons, including securing resources.

Space agencies and companies have reported more than 6100 successful launches in the history of human space exploration. While most are for Earth orbital missions, spacecraft traveling to other bodies has been primarily motivated by scientific purposes. Such missions include more than 200 successful spacecraft that have traveled from Mercury (Solomon et al. 2008) to the outer solar system (Decker et al. 2012), providing *in situ* observational data about roughly 100 extraterrestrial bodies. Thus, human technology already has the capability to send spacecraft to almost every corner of the solar system. Explorations of extraterrestrial bodies have significantly advanced our understanding of the chemical distributions of materials in the solar system and the surface conditions of celestial bodies. Major bodies in the solar system have been studied in detail regarding their sizes, shapes, orbital parameters, chemical compositions, surface conditions, and surrounding environments.

The utilization of space resources has been discussed for decades (e.g., Billingham et al. 1979; McKay et al. 1992). While many parts of the original concepts remain valid, others have gradually been updated with new knowledge of celestial bodies through scientific exploration. Asteroid mining has also been widely discussed (e.g., Lewis and Gump 2015). This paper summarizes our current understanding of asteroids and shows how asteroid exploration is essential for future resource mining.

In-Space Manufacturing and Resources: Earth and Planetary Exploration Applications, First Edition.
Edited by Volker Hessel, Jana Stoudemire, Hideaki Miyamoto, and Ian D. Fisk.

9.2 Materials in the Solar System

Earth is the sixth-largest celestial body in our solar system. Its mass of 5.97×10^{24} kg, while enormous for humans, is a trivial amount of the overall solar system. Even if we total the masses of Earth and all other terrestrial planets, the Moon, rocky satellites, asteroids, and other small bodies, they comprise less than 0.002% of the solar system. Humans have approached about 10 km in depth of Earth's crust, only the surfaces of the Moon, and some asteroids. This means that the total amount of materials that our contemporary technology has allowed us to reach is only about 0.000001% of what comprises the solar system, which has unlimited materials for humans' future activities.

Almost 99.9% of the solar system's total mass is the Sun, which is mainly composed of H and He. The majority of the remainder includes gas giant planets, such as Jupiter, Saturn, Uranus, and Neptune, formed mainly of H and He. Thus, the composition of the solar system is significantly different from that of Earth's surface: while 99% of Earth's crust comprises O, Si, Al, Fe, Ca, Na, K, and Mg, the solar system mainly comprises H, He, C, N, O, Ne, Mg, Si, S, Fe, and Ni.

The overall distribution of materials in the solar system is roughly understood, as explained by its evolutionary history. The solar system was formed due to the gravitational contraction of a cloud of gas and dust. The resulting formation of a protoplanetary disk allowed planetesimals to grow, and they accreted to form planetary embryos and, eventually, planets. As a result, planets and most other bodies orbit the Sun in the same direction as the Sun's rotation. The eight planets—Mercury, Venus, Earth, Mars, Jupiter, Saturn, Uranus, and Neptune—and numerous bodies in the asteroid belt and the Kuiper Belt orbit the Sun almost within the same disk, which is called the ecliptic plane.

A significant temperature variation existed in the protoplanetary disk during the formational process. At that time, the Sun's vicinity was hot, while areas farther away stayed at a lower temperature. As a result, the solar system has been divided into the inner (predominantly rocky and metallic in composition) and outer (primarily gaseous giant planets) regions. The asteroid belt lies between Mars and Jupiter, the boundary between the inner and outer regions, while the Kuiper Belt lies beyond Neptune.

All terrestrial planets have similar chemical compositions and internal structures, with a silicate shell surrounding a metallic core. Earth's crust is depleted in siderophile elements, but it is enriched both in feldspar (Na, K, Ca, Ba) $(Si, Al)_4O_8$ and in elements whose ion radiuses are larger than Mg (such as U and Th). Earth's surface comprises about 5000 minerals—likely the maximum number of minerals on a on a body in the solar system. This variation may indicate that almost all minerals in the solar system can be found on Earth. The high abundance of mineral variations on Earth comes from the fact that (i) Earth has experienced partial melting and solidification to form granites, which helped incompatible elements become enriched in the crust; (ii) interactions between water and rocks, including hydrothermal processes, extensively existed; (iii) circulations of high-pressure minerals to the surface by plate tectonic processes occurred; and (iv) biological activity has been sustained for more than three billion years, allowing (v) diverse biochemical reactions to continue for a sufficient period.

9.3 Asteroids as Extraterrestrial Resources

Statements similar to "No matter what exists outside Earth, we do not call it a resource because the word 'resource' implies economics" were once often quoted in engineering and economics classes, but that thinking is now outdated. Humans are already using extraterrestrial bodies for exploration, at least indirectly. Swing-by and aerobraking techniques are often used to accelerate or decelerate spacecraft. Humans are now trying to use extraterrestrial materials in space via a concept called *In Situ* Resource Utilization (ISRU), which is commonly discussed regarding future space developments. Notably, ISRU does not intend to bring materials back to Earth; while this may happen in the future, it is currently not economically feasible (Elvis 2014).

The bulk composition or interior materials of celestial bodies are almost meaningless when considering extraterrestrial resources because our current technology does not allow us to obtain materials from a deeper part of a body. The most accessible region in the solar system is still the surface of the body, which is the only interest below.

Mars has experienced the most Earth-like evolutionary processes in the solar system (Ehlmann et al. 2011). Volcanic and hydrothermal activities have been extensive in the past, suggesting that Mars is the only planet where Earth-like ore deposits could be developed. Although Mars currently has an arid environment, with an average temperature below −50°, there was a large amount of water near the surface with active volcanism and hydrothermal circulation billions of years ago (Figure 9.1). Recent discoveries of traces of hydrothermal activities, including evaporite-like rocks containing sulfates and chlorides, support this idea. Even today, water ice is extensively distributed over the shallow subsurface of Mars, making it possible to acquire water depending on the location. Such water-ice could be useful as a propellant or to sustain manned activities, such as by providing drinking water,

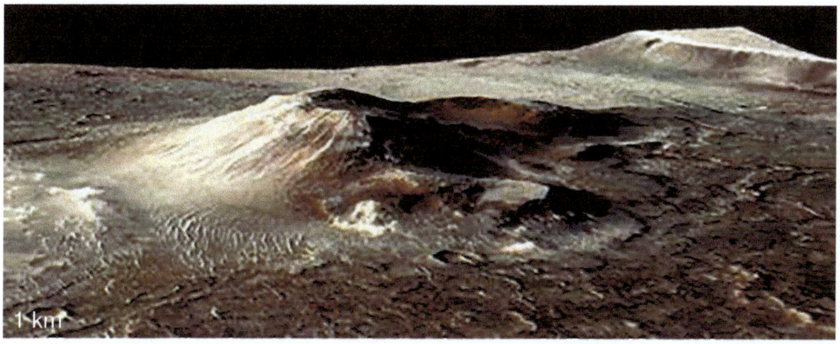

Figure 9.1 Volcanic cone in Nili Patera caldera on Mars showing hydrothermal mineral deposits on the southern flanks (PIA13556). Source: NASA.

an oxygen source, food production, water supplies, etc. Methane has also been found in low concentrations.

The Moon is covered by regolith, fragmental rocks, debris, and glasses. While it is generally a desert of basalts or anorthosites, silicate minerals exist in abundance, especially pyroxene, plagioclase feldspar, and olivine. Several available minerals could be valuable. For example, illuminate has been suggested to be a crucial resource for O_2, Fe, Al, and Ti, while the reduction process for oxides requires high-energy supplies. Recent studies indicate that the regolith of the Moon may help trap exotic materials (Colaprete et al. 2010). For example, volatile materials could be brought by solar wind implantation, endogenic gas release, and the impacts of comets and small bodies and captured by regolith. Enhanced trapping of volatiles in cold regions has been suggested, which may occur widely, predominantly in the polar regions.

Some satellites of giant planets show that they are made almost entirely of ice. Others include those that are thermally active and uniquely evolved due to tidal heating caused by the planet's strong gravitational field. For example, there are massive lakes of liquefied natural gas (LNG) on the surface of Titan (Stofan et al. 2007), Saturn's largest moon (Figure 9.2). Their total amount is estimated to be more than a hundred times the total carbon reserves of Earth from oil, natural gas, etc. The most volcanically active body in the solar system is Jupiter's moon Io (McEwen et al. 1998), which has frequent lava outflows and explosive eruptions (Figure 9.2), resulting in constant surface renewal and significant sulfur concentrations.

The composition of small bodies is predicted mainly by combinations of terrestrial observations and meteorite studies (Nakamura et al. 2011). Although many asteroids have been melted, cooled, or chemically reacted with water, they do not experience significant elemental fractionation, as found on Earth. Thus, most asteroids are believed to be remnants of an earlier stage of the solar system's formation (Bottke et al. 2005). Their relatively pristine nature allowed them to keep many substances not commonly found in Earth's crust. For example, some platinum group elements are tens to thousands of times more abundant in some meteorites than in Earth's crust. Iron meteorites (Figure 9.3) can contain platinum group elements at concentrations of several tens of ppm (Kargel 1994), which is comparable to or sometimes much more than that of platinum group ores on Earth.

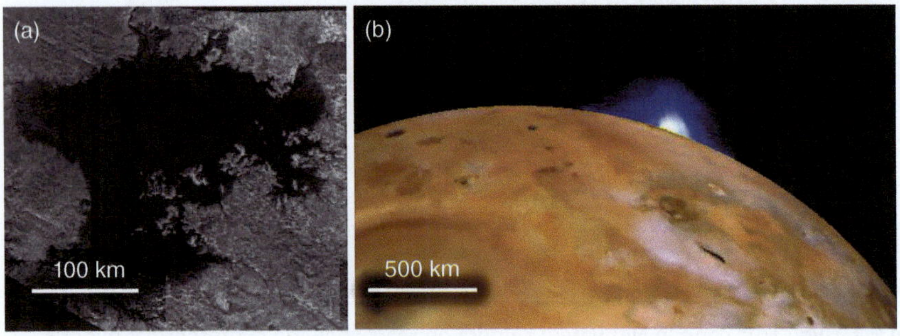

Figure 9.2 Synthetic aperture radar image of a lake of liquified natural gas on Titan (a: PIA20021) and active volcanic eruption on Io (b: PIA00010). Source: NASA.

Figure 9.3 A slice of Gibeon, iron meteorite. Source: University Museum, University of Tokyo.

Importantly, precious materials in a terrestrial market are not necessarily precious in space unless an economic advantage exists. As discussed above, even though we know that various materials exist on celestial bodies in the solar system, their usefulness is limited due to high processing and transportation costs (Elvis 2014). The balance between a material's nature and its accessibility becomes an essential index for evaluating the economic advantage of materials on celestial bodies.

When an object exists at or near a celestial body, its behavior is governed by the body's gravity. To prevent an object from falling to the surface, it must have a relative velocity corresponding to its gravity. If the velocity is fast enough, the object enters an orbit. If it is even faster, the object will enter into a trajectory to another celestial body. The difficulty of moving from one place to another is thus roughly understood by considering the total required velocity change (ΔV). A large ΔV means that more propulsion is required; if this is obtained by chemical propulsion, the fuel requirement increases exponentially.

The altitude of a low-Earth orbit (LEO), often used by Earth-orbiting satellites, is typically about 400 km, which is only about 6% of the radius of Earth (6378 km). However, about 10 km/s of ΔV is needed to insert a spacecraft into the LEO from the surface of Earth. The large ΔV to the LEO is due to Earth's intense gravity. Earth is the biggest and densest rocky body in the solar system, contributing to the strongest gravity field.

Interestingly, the ΔV from LEO to the Moon can only be about 3 km/s, even though the distance between Earth and the Moon is about 1000 times longer than that of Earth's surface to LEO. A smaller asteroid has a weaker gravity field, where spacecraft can explore with zero relative velocity to the asteroid (rendezvous), requiring a smaller ΔV. For example, a round trip to 2000SG344, a small asteroid orbiting far from the Moon, may require only about 3.5 km/s of ΔV because its mass is only about 1/1 000 000 compared to the Moon. Some tiny asteroids need only about 0.1 km/s for round trips from Earth. Such high accessibility is an excellent advantage for utilizing materials on asteroids.

Overall, the difficulty of overcoming Earth's gravity has been the most significant issue in space development. Thus, if we can acquire the materials necessary to support spacecraft activity, especially fuel in space, we can significantly reduce launch costs, which usually comprise the greatest expense of space exploration. If we can obtain fuel in space, we can repeatedly use a spacecraft for multiple purposes. In this way, a new transportation system supported by extraterrestrial materials could be developed. Such a system will be an important goal in near-future space development. Due to the smaller ΔVs for round trips to smaller bodies, asteroids, especially near-Earth asteroids (NEAs), can be primary targets for space resource development.

9.4 Overview of Asteroids

Asteroids are small celestial bodies orbiting the Sun, except for planets and dwarf planets. Their sizes range from a few to 1000 m in diameter, and they are distinguished from comets by their lack of accompanying tails composed of gas and dust particles. Asteroids are considered remnants of planetesimals, which became parts of larger objects, such as planets and satellites, during the solar system's formation.

As of January 2022, about 1.14 million asteroids have been discovered. Most are in the asteroid belt, which lies between Mars and Jupiter, about two and four astronomical units from the Sun, respectively. Asteroids that come close to Earth are classified as NEAs. Their discoveries continue, with orbits of more than 27 000 asteroids being identified. Many other small bodies exist as Trojan asteroids in Jupiter's L4 and L5 Lagrangian points and trans-Neptunian objects (TNOs), where minor planets or dwarf planets exist. Figure 9.4 schematically shows the distribution of asteroids within Jupiter's orbit. The largest asteroid is (1) Ceres, which has a diameter of about 913 km. Note that asteroids are numbered; thus, the full name is written as "(1) Ceres," although simply using Ceres is conventionally accepted. The second largest is Pallas (523 km in diameter). Seven asteroids larger than 300 km in diameter exist, and 81 are larger than 150 km. The cumulative size-frequency distribution of asteroids indicates that smaller asteroids are more abundant (Figure 9.5). Even though the estimated total mass of the bodies in the asteroid belt is only about 0.1% of Earth's mass, numerous undiscovered asteroids of tens of meters in diameter, which are sufficiently large for humans' utilization, are expected to exist.

Asteroids in the main belt are believed to be the remnants of many planetesimals that never accreted into a single large body. The gravitational influence of Jupiter was so strong that it stirred the asteroid population, causing repetitious collisions and breakage. Jupiter still exerts a strong influence on the asteroid belt today; the orbital distributions of asteroids in the main belt show a clear, non-random tendency (e.g. Kirkwood Gaps is deficiencies in the population at values of the semimajor axes corresponding to orbital resonances with Jupiter). The collisional events contribute to putting fragments into orbits much closer to the Sun, explaining why some asteroids near inner planets have unstable orbits and cannot survive for more than a few tens of thousands of years.

Such collisional history would be the key to understanding the significant variety in asteroids' composition. Asteroids are classified into several taxonomic classes (Bus and Binzel 2002), such as bluish, opaque-rich, probably volatile-rich types (C, P, D, and B), mafic-silicate-rich types (S, V, A, and R), and miscellaneous

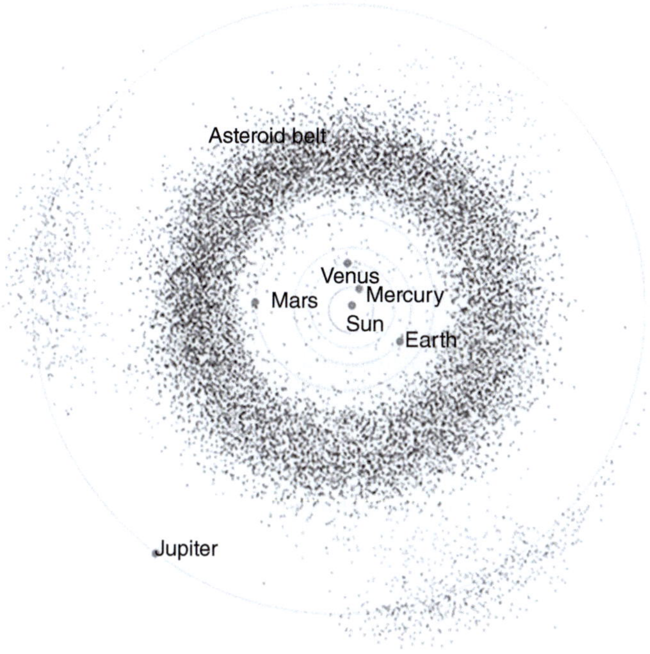

Figure 9.4 Schematic illustration of the orbits of small bodies in the solar system. Source: Mdf, 2006/Wikimedia Commons/Public Domain/Wikimedia Commons/Public Domain.

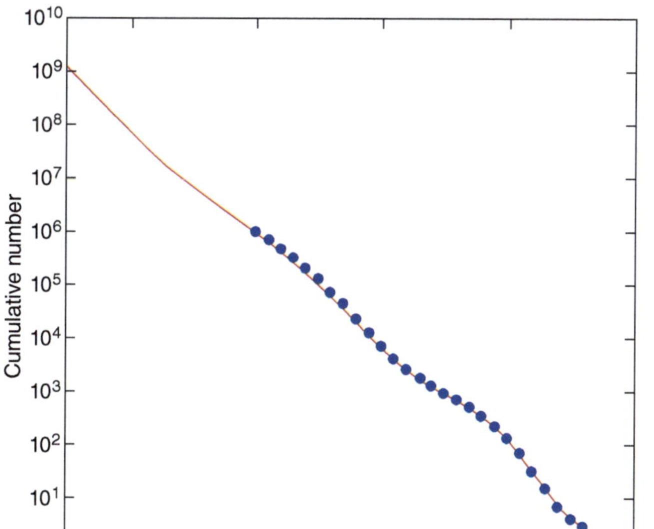

Figure 9.5 The size-frequency distribution of asteroids in the main belt based on observations (dots) and the estimation (line). Source: Walsh 2018/Annual Reviews.

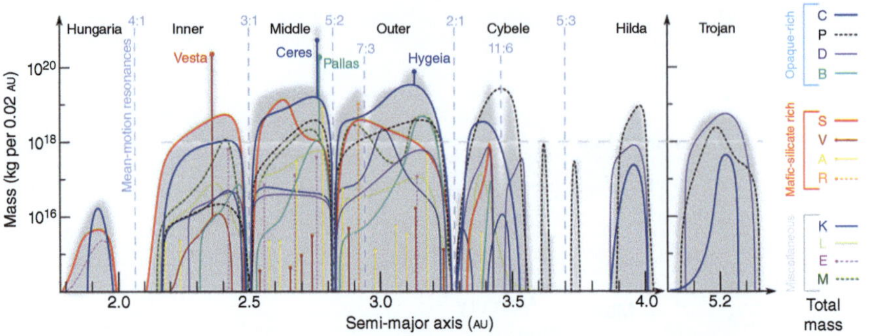

Figure 9.6 The compositional distributions of asteroids and their mass in the asteroid belt and Torojans. Source: DeMeo and Carry 2014, permission conveyed through Copyright Clearance Center, Inc. Republished with permission of Springer Nature.

(K, L, E, and M). Classification is made primarily from the reflectance efficiency of electromagnetic waves and density. The former includes albedo and the reflectance spectrum, which is the relative efficiency of sunlight as a function of wavelength. The overall relative fraction of asteroid classes could be interpreted as a function of distance; a traditional view was that inner main belt asteroids appear redder than outer ones, which are bluer, as remnants of the thermal gradient. However, we now know that rarer asteroid types, such as the crust and mantle remnants of heated and melted bodies, exist in all regions of the main belt, and bluer D-type (or Trojan-like) asteroids appear to exist in the inner belt (Figure 9.6) (DeMeo and Carry 2014). The compositions of asteroids are more diverse with size and distance from the Sun than previously known, implying substantial mixing through various processes, such as planetary migration and subsequent dynamical processes (DeMeo and Carry 2014). Importantly, precise determinations of the compositions of asteroids, especially their inner parts, are challenging because asteroids are generally too small to observe from Earth. Detailed studies of each asteroid can only be done through space missions.

To date, only 13 small bodies have been explored using spacecraft. Remote-sensing missions include Galileo (Veverka et al. 1994), Near Earth Asteroid Rendezvous – Shoemaker (NEAR Shoemaker) (Veverka et al. 2001), Rosetta (Sierks et al. 2011), and Dawn (Jaumann et al. 2012), and sample-return missions include Stardust (Zolensky et al. 2006), Hayabusa (Nakamura et al. 2011), and Hayabusa-2 (Watanabe et al. 2019). These missions confirmed that meteorites are fragments of asteroids. Thus, identifying the corresponding meteorite types of the targeted bodies is among the crucial goals of these asteroid missions. We created the world's largest meteorite composition database (Miyamoto et al. 2016). Based on statistical comparisons of the database and orbital information, reflection spectra, and other observational information, we are developing an economic list for each asteroid by incorporating the accessibility and surface conditions of asteroids.

Volatiles could be the most valuable materials because they can be used for propulsion and solvents. Water is helpful for life support, while its use as fuel, a solvent, or a source of separated hydrogen and oxygen is valuable due to its extensive application possibilities. Finding volatiles in extraterrestrial bodies will not be that complex because the solar system is, as discussed above, enriched in H, He, O, and C, and the same is true for their simple combinations, such as water (ice), methane, and organic matter. A typical carbonaceous meteorite has 10–20% water and around 3% hydrocarbons—the same amount as shales, where oil sand and shale gas are produced. Thus, the parent bodies of carbonaceous meteorites can supply volatile components. This idea has been confirmed by observations of the carbonaceous asteroids Ryugu and Bennu by Hayabusa-2 and Origins, Spectral Interpretation, Resource Identification, Security, Regolith Explorer (OSIRIS-REx), respectively.

Even if volatile components exist far away from Earth, it is unlikely that we can immediately establish a system to utilize them. For example, LNG is useful on Earth but not on Titan because of the lack of oxidizing agents. At the initial stage of space resource utilization, water extractions for ISRU purposes are likely to occur, where NEAs are promising in terms of their optimal accessibility. Comets originating from the outer solar system that fly into the inner solar system could be good water sources because their volatile components may remain inside, even in the vicinity of Earth. However, their typical orbital parameters often require higher ΔVs for a round trip from Earth. Comets are distinguished from asteroids by their solar-influenced tails. When the volatile components on their surfaces disappear, it becomes impossible to determine whether they are depleted cometary nuclei or rocky asteroids. Still, they are expected to retain abundant volatile components in the interior. Such objects are generally difficult to detect by terrestrial observations because of their low albedo (i.e. they are very black). In other words, many easy-to-use small objects may remain undiscovered.

The Wide-field Infrared Survey Satellite (WISE) has discovered many small objects (Wright et al. 2010), and discoveries are expected to continue. With many small celestial objects being discovered every year, it will not be surprising if a highly accessible asteroid with valuable materials for human use is soon found in the vicinity of Earth. There is a theoretical expectation that objects larger than 1 m can be captured in the region closer than the Moon for a few hundred days at a frequency of about one per decade. Many of these objects, called Temporarily Captured Objects (TCOs) (Granvik et al. 2012), are expected to have a ΔV of only 1–3 km/s from LEO. They could contain carbonaceous meteorite-like substances, making them promising candidates for volatile acquisition.

9.5 Findings Through *In Situ* Explorations of Small Asteroids

Mining operations on small asteroids would be different from most terrestrial mining projects. On Earth, precious materials are usually found as ore, from which the best portion is excavated for mining. However, no ordinary ore would be found on an asteroid because the lack of thermal, hydrous, or biological processes does not allow the formation of terrestrial-style concentrations of some chemicals. Additionally, the

Figure 9.7 Itokawa, Bennu, and Ryugu at the same scale (upper) and their close-up images at the same scale (lower).

surface conditions are significantly different in terms of extremely low gravity and zero atmospheric pressure.

Only three small asteroids have been closely observed by spacecraft: the S-type asteroid Itokawa by Hayabusa (Fujiwara et al. 2006) and the C-type asteroids Ryugu (Watanabe et al. 2019) and Bennu (Lauretta et al. 2019) by Hayabusa-2 and OSIRIS-Rex, respectively (Figure 9.7). These missions revealed that the surfaces of such small bodies are different from those of other larger asteroids; specifically, they are covered by numerous boulders (Figure 9.7) and are highly porous. For example, Ryugu's estimated porosity is higher than 50%, indicating a rubble-pile structure, which was surprising because small asteroids were considered monolithic before the observations of these missions (the sample collection devices were developed with the capability of sampling even a monolithic target). Rubble-pile structures are believed to be common among other NEAs. Thus, methods for harvesting resources on such asteroids will not be drilling or breaking to excavate from the surface but will simply involve gathering, sieving, and crushing particles at the surface.

Asteroids and comets were likely major water sources for early Earth, as indicated by returned samples and proximity observations. Water is abundant in a comet as water vapor is the main species observed in a coma. Thus, water is believed to be the major constituent of cometary nuclei, as confirmed by the exposed water-ice region on the surface of the nucleus of comet 67P Churyumov-Gerasimenko observed by the Rosetta mission (Filacchione et al. 2016). Water was found in carbonaceous chondrites and C and C-series asteroids, with most in phyllosilicates. For example, Bennu contains widespread, abundant hydrated phyllosilicates, and its surface has about 0.71 wt% water (Praet et al. 2021). This is in the range of aqueously altered meteorites, such as CMs (e.g. C2 Tagish Lake). Traces of water can be found in fine carbonaceous dust. Observations from Eros, Ganymede, Bennu, and Itokawa suggest that 30–300 ppm of water content is possible for S-type asteroids, which was confirmed by the return sample of Itokawa showing 160–510 ppm for the entire asteroid (Chan et al. 2021). Nevertheless, such water found on an S-type asteroid is water of constitution (crystallized water), which requires significant energy to extract. Ice, absorbed water, or interlayer water in clay minerals would be preferred, which can be more easily found on comets and C-type and similar asteroids.

Rubble-pile asteroids are likely formed through catastrophic collisions, disruptions, and reaccumulations (Michel et al. 2020), whose resulting fragments are held together by their weak gravities, making the surface states easy to modify by later processes (Walsh 2018). The properties of gravel on the surface (regolith) can differ from one object to the next, depending on many factors, including the asteroid's mineralogy and gravity field. The mechanical properties of regolith depend largely on chemistry and processes for producing the particles, which are controlled by impacts and variation histories. The evolutionary processes of the regolith include impact cratering (Schmedemann et al. 2014), earlier internal aqueous and thermal alterations (Wilson et al. 1999), mass movements (Miyamoto et al. 2007), thermal fatigue (Delbo et al. 2014), and space weathering (Hiroi et al. 2006).

The resultant surfaces from such processes on small bodies show a specific homogeneity/heterogeneity in chemistry, mechanical variations, and a range of particle sizes, from smaller than centimeter to tens of meters and larger. For example, while both Bennu and Ryugu are C-type asteroids (among the darkest objects in the solar system), Ryugu seems to be more homogeneous (Kitazato et al. 2019) than Bennu, which has significant albedo heterogeneity (Lauretta et al. 2019). Ryugu has an almost continuous equatorial ridge, although Itokawa does not have a large-scale topographic feature. Instead, Itokawa is homogeneous in albedo and color, although it is geomorphologically inhomogeneous. Itokawa shows two distinctive areas of smooth and rough terrains resulting from the global segregation of fine materials (Miyamoto et al. 2007). The top shape of Ryugu and Bennu is believed to have resulted from rotationally induced deformation or accumulation, while Itokawa could be a contact binary asteroid.

Knowledge of regolith properties is essential for science missions to design scientific objectives, secure landing safety, and select landing sites. The same is true for resource acquisition. Specifically, segregations and potential differentiation through the granular process (Miyamoto et al. 2008) can be significant for resource utilization. A small asteroid's weak gravitational field may further enhance the complex behaviors of smaller particles, especially dust, on the surface. Powdery materials could have significant advantages for radiation shielding in terms of ease of handling. Fine particles are preferred for developing concrete. Ferrous metals can be used for structural materials in space, and separations and extractions of ferrous metals also favor powdery materials.

Dust particles could be released from the surface due to outgassing from the subsurface, shaking invoked by impact or gravitational interaction with other celestial bodies, cracking due to thermal fatigue, or electrostatic repulsing invoked by solar EUV irradiation. The asteroid 433 Eros was precisely observed by NEAR, which found several craters filled with dust grains. This characteristic structure is called a "pond," which could be formed by electrostatic dust levitation or its combination with external triggers, such as impact. The typical size of a "pond" structure on Eros is smaller than 100 m, and it is not clear whether these are common structures. Surface rocks may be broken by repetitious small impacts, as observed on the Moon. The median survival time of rock fragments >2 m in diameter is estimated to be about 40–80 Ma on the Moon. Diurnal temperature cycling and associated stress may also contribute to the destruction of rocks. In this sense, asteroids' surfaces may often contain plentiful fine particles, as observed on the Moon.

9.6 Mars, the Moon, and Asteroids

Space agencies in various countries have declared the growth of their exploration programs in the future. Most importantly, NASA has indicated increased activity in three primary domains: LEO, the Moon, and Mars. As such, the utilization of these domains will expand significantly rather soon. The resultant scientific findings will further enhance our scientific understanding of the formation and evolution of the Moon, Earth, and the solar system.

The surface of Mars is undoubtedly the most Earth-like; previous missions have confirmed the existence of vast water and easy-to-use minerals on the surface of Mars, making it the best place to acquire the materials necessary to develop bases and support life. The current environment of Mars, while severe for humans, is the least harsh among extraterrestrial bodies. Technical difficulties, including relatively restricted accessibility due to the intense gravity and the extended distance from Earth and the Sun, would be critical factors for the extensive development of Mars. Debates exist about when to start, but NASA and other major space agencies have suggested that a sustained long-term presence on the surface of the Moon should come earlier than a further voyage to Mars. Thus, the Moon will be the center of near-future space explorations.

NASA has announced the Artemis program to build a lunar-orbiting space station and land the first woman and first astronaut of color on the Moon through international collaborations. This program will significantly encourage activities to obtain space resources on the Moon. NASA's concept includes using the Moon as a gateway to the solar system; thus, the proposed Artemis program provides the foundation for a sustained long-term presence on the Moon's surface, which can provide a stable platform for extensive infrastructures and serve as feedstocks for its vast materials. In addition, the pros of the Moon include its proximity to Earth and the possible existence of volatiles.

NASA's plans for continuous exploration of and activities on the Moon include extended manned missions and enhanced robotic activities, including deliveries by Commercial Lunar Payload Services (CLPS). Sustainable operations on and near the Moon would require decreasing supply needs from Earth. Perhaps the

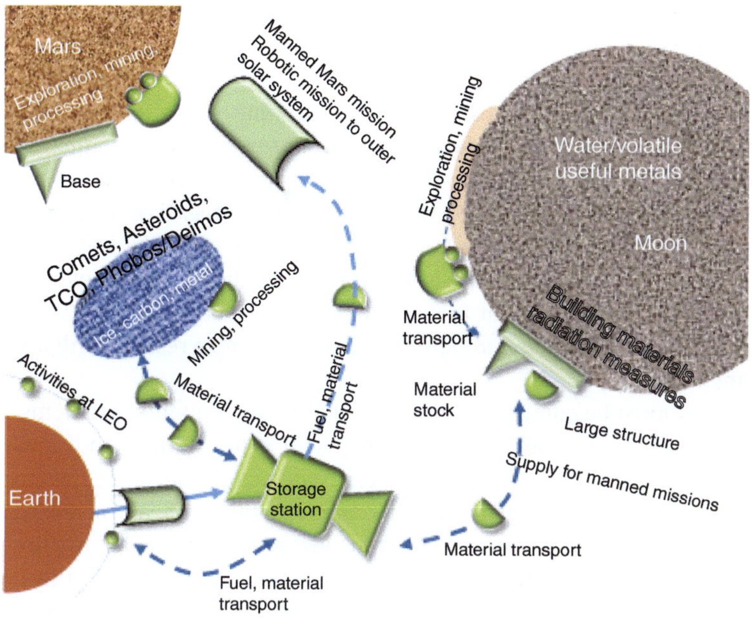

Figure 9.8 Concept of space-resource utilization in the future. Regolith on the Moon and small bodies, including asteroids, will be collected for manufacturing. Volatiles will be excavated and used for solvents or fuels stored in a station to support transportation.

availability of volatiles on the Moon could impact the approach to near-future operations. If a valuable amount of water exists in an easy-to-use form, such as absorbed or cold-trapped ice, activities on the Moon could be extensively supported by lunar materials. If not, other sources of volatiles would be the primary targets.

The cons of utilizing materials on the Moon include not only its monotonous mineral compositions but also its relatively large gravity. If materials cannot be consumed near the area harvested, transportation costs could prohibit extensive utilization because of still relatively large gravity field of the Moon. Other accessible bodies may provide competing merits against lunar materials. NEAs could be the most reasonable target for this purpose, while Phobos and Deimos could also be significant targets for resources due to their accessibility from Earth (Figure 9.8). Currently, Japan Aerospace Exploration Agency (JAXA) is working on a Mars satellite exploration project (Martian Moons eXploration (MMX), scheduled for launch in 2024), which is positioned as a successor to Hayabusa-2 and is trying to acquire samples from the Martian satellite Phobos. MMX will also perform remote-sensing observations of both Phobos and Demos, both of which are potential targets for resource acquisition.

9.7 Future View

The current total population of Earth is more than 7.5 billion, which means that more than 7% of the people born in the history of Earth are alive on its surface today.

Since the Industrial Revolution, the energy and resources consumed by each person have significantly increased, and each person now consumes 6L of oil equivalent and 35 kg of resources every day. As a result, non-renewable, easy-to-process resources, such as oil and sulfides, have been extensively consumed. Some metals, such as lead, nickel, tungsten, and molybdenum, are feared to have short reserves-to-production ratios (RPRs) of less than 100 years, and copper, lead, and zinc are predicted to have 20–30 years, faster than that of crude oil. Although discussions based only on RPR can be misleading because the sizes of reserves depend on the technological level, the production lifespan of an average mine is far shorter than the timescales of mineral deposit formation, suggesting that metal mining is unsustainable for human timescales. No matter how much technology advances, recoverable reserves will not increase permanently.

Another issue that must be considered is balance with environmental problems. When cultivating bacteria in the laboratory, they will grow if the conditions are right. However, the number of viable bacteria can rapidly decrease due to environmental degradation caused by their own emissions. Similarly, an increased number of humans use massive resources, including low-grade ores, which require more energy and have heavier environmental impacts, potentially resulting in a deterioration in the population. Even if recycling is carried out with high efficiency, there will still be impurities in the materials used to make the next product, and it is theoretically impossible to reuse 100% of all materials due to energy costs. Thus, recycling resources cannot extend materials' lifetimes indefinitely. Ultimately, humans face complicated problems regarding resources, energy, and the environment, which are mutually connected. Therefore, we must develop the technological capability to solve these problems and extend the sustainability of our civilization.

Earth's total population is expected to reach 10 billion by around 2050, which will require more road networks, buildings, and machinery, including in developing countries. There is likely no single solution to issues regarding resources, energy, and the environment. A matured society may curb the consumption of resources and energy through economic efficiency, and a super-efficient society could be driven by new technologies, such as superconductivity, advanced elemental strategies, and nuclear fusion. Nevertheless, it is still important to secure more resources and energy sources, especially from other than Earth's surface. Historically, humans have always solved resource and environmental issues by expanding their territories. This does not imply simply extracting from greater depths; even deep subsurface development cannot exclude environmental impacts. Similarly, deep-sea mining could be expensive due to the high environmental costs, as recent discoveries of the marine biosphere and its role on Earth imply. After an era of using exhaustible resources on Earth, the only choice is to acquire resources from outside Earth's surface.

9.8 Conclusion

Exploration of the solar system is no longer limited to traditional space agencies. China has become the first country to land on the Moon's far side, and India's lunar exploration program is moving forward on its Mars mission. Space-X is contracted to launch almost every two weeks, proving that private companies have already become an integral part of space exploration. Several private companies have announced plans to land on the Moon using only private capital. NEAs are also being targeted by several private companies, which is reasonable given their accessibility and chemical variety.

Space resources will soon be utilized in space. Regolith soil will be used for radiation shielding, and powdery metals in regolith soils on the Moon and asteroids will be collected for manufacturing. Volatile components, such as carbonaceous chondrite-like materials on asteroids and putative ice on the Moon, will be used for propellants or solvents. Volatile components acquired in space can be used to support multiple flights between celestial bodies, which will significantly enhance the use of space resources in the construction of artificial structures (Figure 9.8).

Humanity has developed its civilization by consuming large amounts of easy-to-use resources on Earth. However, we now face interconnected issues of resources, energy, and the environment. We need to rapidly promote research to realize a sustainable society, including promoting recycling and alternative element strategies, yet we still must seek resources outside our living environment in the future. Obtaining extraterrestrial resources is more advantageous than mining those located deep underground or under the ocean. Once multiple spacecraft flights are supported by materials obtained in space, the use of extraterrestrial resources will rapidly expand. The era of large-scale space development, which will bring enormous wealth to mankind, is approaching, and it would not be surprising if a major shift in human civilization, comparable to the Industrial Revolution, occurred within the next 100 years. However, the initial costs will be high, and it is important to consider these costs based on the scientific knowledge of the solar system acquired through scientific exploration.

Acknowledgment

This work is supported in part by JAXA's LUPEX mission, SSG-SST activities of the MMX mission, and TSUKIMI mission by the Ministry of Internal Affairs and Communications (0155-0236; jpmi00316), Japan.

References

Billingham, J., Gilbreath, W., and O'Leary, B., and NASA(1979). *Space Resources and Space Settlements*. Washington, DC: NASA.

Bottke, W.F., Durda, D.D., Nesvorny, D. et al. (2005). The fossilized size distribution of the main asteroid belt. *Icarus* 175 (1): 111–140.

Bus, S.J. and Binzel, R.P. (2002). Phase II of the small main-belt asteroid spectroscopic survey – a feature-based taxonomy. *Icarus* 158 (1): 146–177.

Chan, Q.H.S., Stephant, A., Franchi, I.A. et al. (2021). Organic matter and water from asteroid Itokawa. *Scientific Reports* 11 (1): 5125. https://doi.org/10.1038/s41598-021-84517-x.

Colaprete, A., Schultz, P., Heldmann, J. et al. (2010). Detection of water in the LCROSS ejecta plume. *Science* 330 (6003): 463–468.

Decker, R.B., Krimigis, S.M., Roelof, E.C., and Hill, M.E. (2012). No meridional plasma flow in the heliosheath transition region. *Nature* 489 (7414): 124–127.

Delbo, M., Libourel, G., Wilkerson, J. et al. (2014). Thermal fatigue as the origin of regolith on small asteroids. *Nature* 508 (7495): 233.

DeMeo, F.E. and Carry, B. (2014). Solar system evolution from compositional mapping of the asteroid belt. *Nature* 505 (7485): 629–634.

Ehlmann, B.L., Mustard, J.F., Murchie, S.L. et al. (2011). Subsurface water and clay mineral formation during the early history of Mars. *Nature* 479 (7371): 53–60.

Elvis, M. (2014). How many ore-bearing asteroids? *Planetary and Space Science* 91: 20–26.

Filacchione, G., De Sanctis, M.C., Capaccioni, F. et al. (2016). Exposed water ice on the nucleus of comet 67P/Churyumov-Gerasimenko. *Nature* 529 (7586): 368.

Fujiwara, A., Kawaguchi, J., Yeomans, D.K. et al. (2006). The rubble-pile asteroid Itokawa as observed by Hayabusa. *Science* 312 (5778): 1330–1334.

Granvik, M., Vaubaillon, J., and Jedicke, R. (2012). The population of natural Earth satellites. *Icarus* 218 (1): 262–277.

Hiroi, T., Abe, M., Kitazato, K. et al. (2006). Developing space weathering on the asteroid 25143 Itokawa. *Nature* 443 (7107): 56–58.

Jaumann, R., Williams, D.A., Buczkowski, D.L. et al. (2012). Vesta's shape and morphology. *Science* 336 (6082): 687–690.

Kargel, J.S. (1994). Metalliferous asteroids as potential sources of precious metals. *Journal of Geophysical Research, Planets* 99 (E10): 21129–21141.

Kitazato, K., Milliken, R.E., Iwata, T. et al. (2019). The surface composition of asteroid 162173 Ryugu from Hayabusa2 near-infrared spectroscopy. *Science* 364 (6437): 272.

Lauretta, D.S., DellaGiustina, D.N., Bennett, C.A. et al. (2019). The unexpected surface of asteroid (101955) Bennu. *Nature* 568 (7750): 55.

Lewis, J.S. and Gump, D. (2015). *Asteroid Mining 101 : Wealth for the New Space Economy*. Moffett Field, CA: Deep Space Industries, Inc.

McEwen, A.S., Keszthelyi, L., Geissler, P. et al. (1998). Active volcanism on to as seen by Galileo SSI. *Icarus* 135 (1): 181–219.

McKay, M.F., McKay, D.S., Duke, M.B. et al. (1992). *Space Resources : Overview*. Washington, DC: NASA.

Michel, P., Ballouz, R.L., Barnouin, O.S. et al. (2020). Collisional formation of top-shaped asteroids and implications for the origins of Ryugu and Bennu. *Nature Communications* 11 (1): 2655. https://doi.org/10.1038/s41467-020-16433-z.

Miyamoto, H., Yano, H., Scheeres, D.J. et al. (2007). Regolith migration and sorting on asteroid Itokawa. *Science* 316 (5827): 1011–1014.

Miyamoto, H., Kargel, J.S., Fink, W., and Furfaro, R. (2008). *Granular Processes on Itokawa, a Small Near-Earth Asteroid: Implications for Resource Utilization*, vol. 6960, 69600I. Space Exploration Technologies https://www.spiedigitallibrary.org/conference-proceedings-of-spie/6960/69600I/Granularprocesses-on-Itokawa-a-small-near-Earth-asteroid/10.1117/12.784634.short?SSO=1.

Miyamoto, H., Niihara, T., Kuritani, T. et al. (2016). Cluster analysis on the bulk elemental compositions of antarctic stony meteorites. *Meteoritics and Planetary Science* 51 (5): 906–919.

Nakamura, T., Noguchi, T., Tanaka, M. et al. (2011). Itokawa dust particles: a direct link between S-type asteroids and ordinary chondrites. *Science* 333 (6046): 1113–1116.

Praet, A., Barucci, M.A., Clark, B.E. et al. (2021). Hydrogen abundance estimation and distribution on (101955) Bennu. *Icarus* 363: 114427. https://doi.org/10.1016/j.icarus.2021.114427.

Schmedemann, N., Kneissl, T., Ivanov, B.A. et al. (2014). The cratering record, chronology and surface ages of (4) Vesta in comparison to smaller asteroids and the ages of HED meteorites. *Planetary and Space Science* 103: 104–130.

Sierks, H., Lamy, P., Barbieri, C. et al. (2011). Images of asteroid 21 lutetia: a remnant planetesimal from the early solar system. *Science* 334 (6055): 487–490.

Solomon, S.C., McNutt, R.L., Watters, T.R. et al. (2008). Return to mercury: a global perspective on MESSENGER's first mercury flyby. *Science* 321 (5885): 59–62.

Stofan, E.R., Elachi, C., Lunine, J.I. et al. (2007). The lakes of Titan. *Nature* 445 (7123): 61–64.

Veverka, J., Belton, M., Klaasen, K., and Chapman, C. (1994). Galileo encounter with 951-gaspra – overview. *Icarus* 107 (1): 2–17.

Veverka, J., Farquhar, B., Robinson, M. et al. (2001). The landing of the NEAR-Shoemaker spacecraft on asteroid 433 Eros. *Nature* 413 (6854): 390–393.

Walsh, K.J. (2018). Rubble pile asteroids. *Annual Review of Astronomy and Astrophysics* 56 (56): 593–624.

Watanabe, S., Hirabayashi, M., Hirata, N. et al. (2019). Hayabusa2 arrives at the carbonaceous asteroid 162173 Ryugu-A spinning top-shaped rubble pile. *Science* 364 (6437): 268.

Wilson, L., Keil, K., and Love, S.J. (1999). The internal structures and densities of asteroids. *Meteoritics and Planetary Science* 34 (3): 479–483.

Wright, E.L., Eisenhardt, P.R.M., Mainzer, A.K. et al. (2010). The wide-field infrared survey explorer (wise): mission description and initial on-orbit performance. *Astronomical Journal* 140 (6): 1868–1881.

Zolensky, M.E., Zega, T.J., Yano, H. et al. (2006). Report - mineralogy and petrology of comet 81P/Wild 2 nucleus samples. *Science* 314 (5806): 1735–1739.

10

Mining of Phosphorus from Moon Crust – Spillover of Learning from Earth

Changping Zhuang[1], Daniel Kinasz[1], Nam N. Tran[1,2], and Volker Hessel[1,2]

[1] *University of Adelaide, School of Chemical Engineering and Advanced Materials, North Terrace campus, Adelaide, SA 5005, Australia*
[2] *University of Adelaide, Andy Thomas Centre for Space Resources, North Terrace campus, Adelaide, SA 5005, Australia*

10.1 Introduction

The global food production is now highly dependent on the continuing use of phosphates, which account for 50–60% of all phosphorous (P) supply [1]. The aim of this chapter is to mirror Earth needs for phosphorus as critical nutrient and element, the Moon's respective resource opportunities, and the relevance of new kinds of circular bioprocess technologies to solve extraterrestrial nutrient supply. Thus, this chapter provides a spillover between Earth problems and space solutions to stimulate an anticipatory technology development. In particular, this chapter gives a perspective view of a lunar-suited processing technologies from the portfolio of Earth best-practice minerals processing technologies for phosphorus. In view of the limitations of those, a bio-inspired proprietary minerals processing is proposed for the exploitation of extraterrestrial resources, and demands connected to their mining systems for space farming in a recycle fashion. Experimental details to justify the approach taken, providing experimental results on the concept of lunar phosphorus mining proposed here, are provided.

10.2 Phosphorus Geomineral Sources

10.2.1 Sources of Phosphorus on Earth

Looking back in history, farmers in places such as England and US used bones as a source of fertilizer in the seventeenth to eighteenth centuries [2], by dissolving bones in sulfuric acid to make a liquid fertilizer [3]. Also, guano (bird and bat droppings) were discovered as concentrated mineral sources of phosphorus, but it could not meet the demand of harvest crops [2].

Today, rock phosphate is the only practicable raw material resource to feed the demand of P fertilizers [4]. Usually, the available P fertilizer can be produced by

In-Space Manufacturing and Resources: Earth and Planetary Exploration Applications, First Edition.
Edited by Volker Hessel, Jana Stoudemire, Hideaki Miyamoto, and Ian D. Fisk.

extraction from the rock phosphate. However, it is conservatively estimated that the Earth will run out of phosphorus within about 150 years with the continuous increase of the global population [1]. Hence, it is necessary to find solutions or alternatives in advance. Two main paths could provide rational responses to the potential phosphorus shortage. One is that terrestrial phosphorus usage needs to become more efficient, for instance reducing agricultural run off, using agricultural waste products and even phosphorus recovery from water. The other is that extraterrestrial phosphorus resource might be an option. For the latter, phosphorus extraction in space needs to be developed.

Lunar P Production Learning

Lunar phosphorous (P) rock materials are the first choice for the mining of phosphorus. Yet, lunar rock materials containing P do not exactly match those on Earth. Examples of lunar P rock materials are mainly whitlockite [ideally $Ca_2(PO_4)_2$] and apatite [ideally $Ca_5(PO_4)_3(OH,F,Cl)$] [5]. Commonly, those phosphates are associated with metal elements and/or rare earth elements [5, 6]. They are simple in structure due to the lack of water and chemical weathering on the moon [6]. However, phosphates are minor accessories in meteorites and lunar rock, which is usually less than 1 vol% [6]. Thus, their mining (extraction in the sense of comminution and flotation) and minerals processing (isolation and purification of P) might be different to Earth P minerals and is largely unknown.

10.3 Phosphorus Mining Sites

10.3.1 Mining Sites of Phosphorus on Earth

Table 10.1 shows the supply of mined phosphate rock in the world from the year 2019 to 2020 and also the reserves. A steady increase is visible, which continues until today. It also shows that we use about 0.3% of the world resources annually; meaning all reserves will be extinguished in about 300 years. Three countries, Morocco, China, and Egypt, hold about 80% of all Earth reserves. Thus, alike on the moon (Section 10.3.2), the distribution of phosphorus is distinct at a very concentrated locations. Local qualities of phosphorus make a global phosphorus supply non-resilient, and the terrestrial extraction techniques are not sustainable if used on the moon.

10.3.2 Mining Sites of Phosphorus on the Moon

As for the source of P in the space, the Moon is considered as relevant source. The Moon crust contains substantial amounts of phosphorus. The sample rocks and dust from the moon by Apollo program have identified the Moon as a KREEP-rich place [8]. KREEP materials are rich in potassium (chemical symbol K), rare-earth elements (REE), and phosphorus (chemical symbol P), which composes its name [9]. The P sources available on the Moon include both the KREEP material in igneous rocks and phosphides distributed among the lunar fines [9]. Moon crust is widespread across the lunar surface [9].

Table 10.1 Global phosphorus production in 2019 and 2020 and current reserves (in 1000 metric tonnes Note: 1 metric tonne = 1.102 US tons).

	Mine production		
	2019	**2020[a)]**	**Reverses**
Algeria	1,300	1,300	2,200,000
Australia	2,700	2,700	1,100,000
Brazil	4,700	5,500	1,600,000
China	95,000	90,000	3,200,000
Egypt	5,000	5,000	2,800,000
Finland	995	1,000	1,000,000
India	1,480	1,500	46,000
Israel	2,810	2,800	57,000
Jordan	9,220	9,200	800,000
Kazakhstan	1,500	1,500	260,000
Mexico	558	600	30,000
Morocco and Western Sahara	35,500	37,000	50,000,000
Peru	4,000	4,000	210,000
Russia	13,100	13,000	600,000
Saudi Arabia	6,500	6,500	1,400,000
Senegal	3,420	3,500	50,000
South Africa	2,100	2,100	1,400,000
Syria	2,000	360	1,800,000
Togo	800	800	30,000
Tunisia	4,110	4,000	100,000
United States	23,300	24,000	1,000,000
Uzbekistan	900	900	100,000
Vietnam	4,650	4,700	30,000
Other countries	1,140	1,100	840,000
World total (rounded)	227,000	223,000	71,000,000

a) Estimated.
Source: Summaries [7]/U.S. Geological Survey / Public Domain.

Through the Lunar Prospector mission's Gamma-Ray Spectrometer (GRS), a map of thorium abundance on the Moon (Figure 10.1) was made, which is considered as a marker of KREEP [11]. It is demonstrated that the Procellarum and Imbrium regions of the Moon make up a unique geochemical crustal province, which is full of KREEP basalt [12]. Studies of the composition of KREEP basalt 15386 (the largest pristine sample of KREEP basalt in the Apollo collection) show that KREEP basalt 15386 contains 13–50% plagioclase, 30–50% olivine, 10–15% cristobalite/potassium feldspar, 3–11% ilmenite, and minor amounts of sulfide, iron metal, zirconolite, and phosphates [13], which provide useful information of the KREEP compounds.

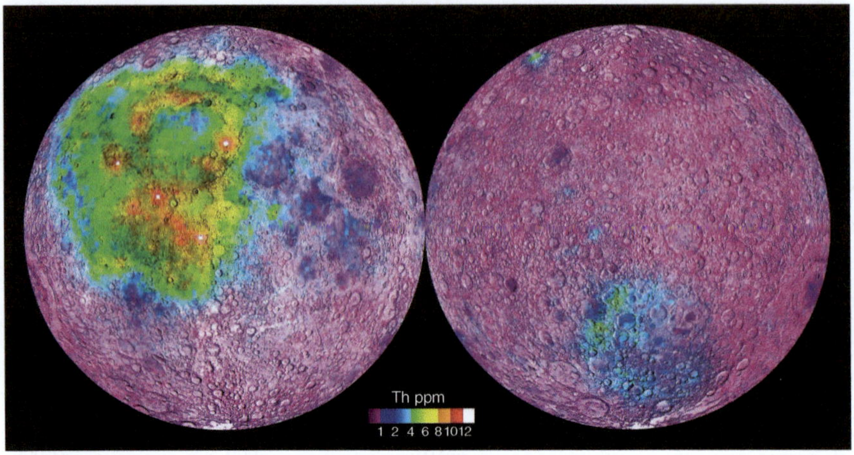

Figure 10.1 Thorium map of the moon as a marker for phosphorus. Procellarum-Imbrium region is left blue, green yellow, orange and red area; right upper is Feldspathic Highland Terrane and lower is South Pole-Aitken Terrane. Source: Reprinted from Jolliff et al. [10] with permission of Wiley.

Lunar P Production Learning

The learning for a lunar production is that the distribution of phosphorus is as geographically concentrated as on Earth. While this gives advantage from the view of mineral production, especially in terms of supply chains and their role in the integration of production process, it raises issues of (Moon) resilience in form of space policy and law, which are addressed further in Chapters 19 and 20. At this point and with positive view, it can only be stated that this offers a new chance to achieve a higher resource resilience and sustainability.

10.4 Phosphorus Industrial Production from Mined Resources

10.4.1 Industrial Production of Phosphorus on Earth

Currently, rock phosphate is the basic raw material used in the manufacture of phosphatic fertilizers like single superphosphate (SSP), diammonium phosphate, and nitro phosphates. Useable rock phosphate occurs in nature as stores of apatite together with other minerals such as quartz, silicates, carbonates, sulfates, and sesquioxides. Four common types of rock phosphate minerals are carbonate apatite [$3Ca_3(PO_4)^2 \cdot CaCO_3$], fluorapatite [$3Ca_3(PO_4)_2 \cdot CaF_2$], hydroxyapatite [$3Ca_3(PO_4)_2 \cdot Ca(OH)_2$], and sulfa apatite [$3Ca_3(PO_4)_2 \cdot CaSO_4$]. The apatite has igneous and metamorphic origin with less reactivity [4].

Rock phosphate can be used to produce phosphoric acid by the wet process. The production of phosphoric acid is made using the steps shown below (Figure 10.2). In the grinding mill, the rock phosphate is ground to a level of fine powder. Finely ground rock powder is mixed with sulfuric acid in a digester at a temperature about 70–80 °C to form phosphoric acid and gypsum as the chemical equation:

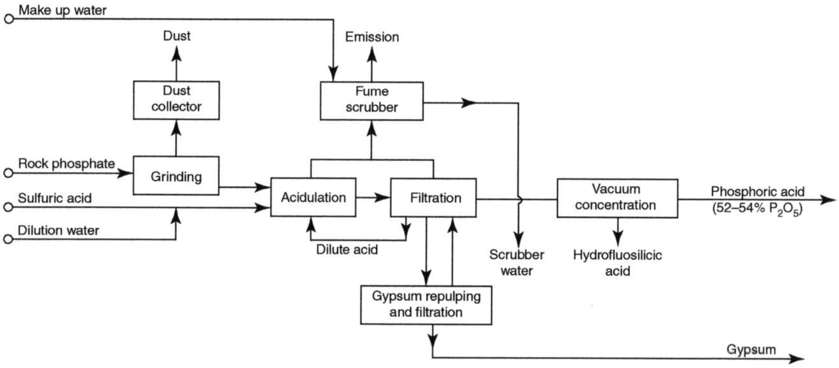

Figure 10.2 Phosphoric acid production process. Source: Cheremisinoff [14]/Reproduced with permission of Elsevier.

$$Ca_{10}(PO_4)_6F_2 + 10H_2SO_4 + 20H_2O \rightarrow 10(CaSO_4 \cdot 2H_2O) + 6H_3PO_4 + 2HF.$$

The obtained gypsum is separated from phosphoric acid in the pan filters. Gypsum is then washed to make it free from phosphoric acid and is processed into gypsum slurry or gypsum cake.

To produce single super phosphate (SSP), ground rock phosphate requires a 90% passaging through a 100-mesh sieve before being mixed with sulfuric acid (concentration of 55–75%) in a specially designed mixer which unloads the product to a wide belt conveyor. The reaction below is completed in the belt conveyor:

$$Ca_{10}(PO_4)_6F_2 + 7H_2SO_4 \rightarrow 3Ca(H_2PO_4)_2 + 7CaSO_4 + 2HF$$

The reacted product is then sent to a curing shed where the product is stored for three to four weeks for curing and drying. The cured product is dried, milled, and screened to obtain the desired SSP. The P fertilizer product can be obtained after granulation.

To produce triple superphosphate (TSP), the rock phosphate is ground to a fine mesh and then reacted with concentrated phosphoric acid (concentration of 52–54% P_2O_5) in a cone mixer:

$$Ca_{10}(PO_4)F_2 + 14H_3PO_4 \rightarrow 10Ca(H_2PO_4)_2 + 2HF$$

The resulting product is then fed to a continuous den and piled for curing. After curing, the material can be sized and used as a P fertilizer. This material can also be granulated if needed.

Lunar P Production Learning

The lunar production of pure elemental phosphorus and any relevant P molecules, such as fertilizers, is assumed to be the same or similar as on Earth, based on present evidence. Likely, when more is known from actual Moon sample extraction and processing, this view will have to be adapted and changed. Yet, as now, the mining and minerals processing are the technologies expected to change most, considering the difference of processing conditions between the Earth and the space such as gravity, atmospheric environment, ore location and structure, etc.

10.5 Phosphorus Cycle to Ensure Circularity of P Provision vs. Consumption

10.5.1 Land-Based Cycle of Phosphate

Phosphorus (P) has an extraordinarily slow cycle in nature, unlike the circulation of carbon (C), nitrogen (N), and sulfur (S) [1]. With human interference, the cycle of these elements has become more rapid than by nature. Three kinds of the P cycle are known, land-based cycle, water-based cycle, and geological sediment cycle [15]. The land-based cycle is shown in Figure 10.3, which shows the processes of P cycle both by nature and by human additions. The residual P goes to the river and takes part in the water-based cycle. It is notable that human activities have roughly intensified releases of P from the nature cycle [1].

Lunar P Production Learning

The learning for a lunar production is that human interaction in the farming use of nutrients like P fertilizers needs to be minimized and rather an automated production is mandatory. On the other hand, the slow speed of the cycle favors a lunar production and recycling of P, as the underlying processes of minerals extraction and farming-based recycling might be slower when fulfilled on the moon, and therefore more efficient, as well.

10.6 Phosphorus Extraction Technologies

10.6.1 Total Process Technology Overview

Chen and Graedel [16] made a conceptual framework of the Dynamic Phosphorus Flows Analysis (DPFA) model (Figure 10.4), indicating that the first step to deal with phosphate rocks is the extraction process. As moon crust can be regarded as

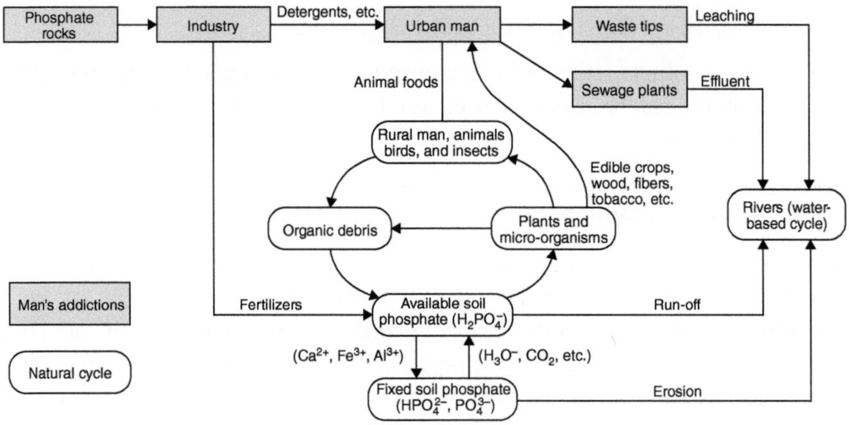

Figure 10.3 Land-based cycle of P. Source: Greenwood and Earnshaw [15]/Reproduced with permission of Elsevier.

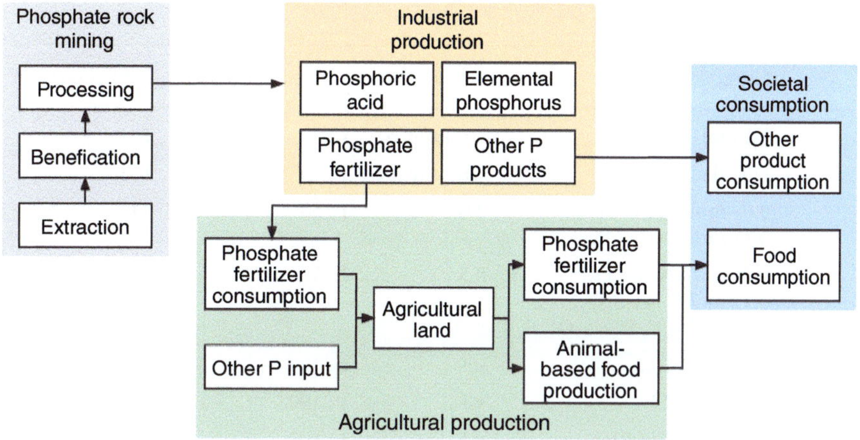

Figure 10.4 Block diagram of the Dynamic Phosphorus Flows Analysis (DPFA) model. Source: Chen and Graedel [16] / Reproduced with permission of Elsevier.

a new kind of phosphate rock, it stands to reason that the DPFA analysis has to be revised for a lunar phosphorus production. Thus, a potential phosphorus extraction applicable on the Moon crust has to be worked out, suitable for lunar rocks or soil.

Question is if the lunar process can be chosen from Earth-established phosphorus extraction processes, which are available and much diversified with a broad choice of chemical extraction and ion-sink extraction methods. It would be ideal that circular extraction systems that have no recurring reagent input can be used to allow for more efficient and potentially automated extraction on the Moon. Sixteen phosphorus extraction methods that might be applied to extract phosphorus from P raw materials are at hand. The extractants and extracting solutions of each method are shown in Table 10.2.

Table 10.2 Overview of the soil phosphorus (P) extraction methods [17–19].

Method	Extracting solution	Solution PH	Soil- solution ratio	Extraction time
H_2O	Distilled H_2O	Unbuffered	1 : 0.3–1 : 1.2	12 h
$CaCl_2$	0.01 mol/L $CaCl_2$	Unbuffered	1 : 10	2 h
LiCl	0.4 mol/L LiCl	Unbuffered	1 : 1.8–1 : 4.0	2 × 2 h
Olsen	0.5 mol/L $NaHCO_3$	8.5	1 : 20	30 min
Bray II	0.03 mol/L NH_4F, 0.1 mol/L HCL, 0.2 mol/L CH_3COOH, 0.25 mol/L NH_4NO_3	1.0	1 : 7	40 s
Mehlich 3 [19]	0.015 mol/L NH_4F, 0.013 mol/L HNO_3, 0.001 mol/LEDTA	2.5	1 : 10	5 min
CAL	0.3 mol/L CH_3COOH, 0.05 mol/L $C_6H_{10}CaO_6$, 0.05 mol/lL $(CH_3COO)_2Ca$	4.0	1 : 20	2 h

(Continued)

Table 10.2 (Continued)

Method	Extracting solution	Solution PH	Soil- solution ratio	Extraction time
Fe-oxide P_i	0.1 mol/L $CaCL_2$	Unbuffered	1 : 40	16 h
CAEM	Distilled H_2O	Unbuffered	1 : 40	16 h
Oxalate	0.08 mol/L $(COOH)_2$, 0.11 mol/L $(COONH_4)_2$	3.0	1 : 50	4 h
Dithionite	0.2 mol/L $NaHCO_3$, 0.12 mol/L $Na_2S_2O_4$, 0.24 mol/L $C_6H_5Na_3O_7$	8.5	1 : 50	16 h
HCL	0.5 mol/L HCl	<1	1 : 10	2 h
Organic P	1 mol/L HCl	<1	1 : 100	16 h
Total P	1 mol/L HCl (after calcination)	<1	1 : 100	16 h
Mehlich 3 [17]	1.5 mol/L NH_4F, 0.1 mol/L EDTA	<7	1 : 5	30 min
Miller-Axley	0.03 mol/L NH_4F, 0.015 mol/L H_2SO_4	<7	1 : 5	30 min
Modified Kelowna	0.015 mol/L NH_4F, 0.25 mol/L NH_4OAc	<7	1 : 5	30 min

CAL: calcium-acetate-lactate; Fe-oxide P_i: iron-oxide-impregnated filter paper; CAEM: cation- and anion-exchange membranes; Organic P was determined as the difference of P extracted with 1 mol/L HCl from an ignited (550 °C; 1 h) and unignited sample [19].

Lunar P Production Learning

It is concluded that the iron-oxide-impregnated filter paper (Fe-oxide P_i) method suits lunar processing – with its low P concentration and its low P extraction rate – because it is based on an infinite P sink approach. Ion-sink methods are considered to be the best methods of plant-available P terrestrially [20] because the exchange membranes provide similar functionality as does the root system in the soil [19]. There is a second lunar aspect which favors this technology. As given later in the chapter, we propose to integrate lunar minerals processing and farming, i.e. to leverage auto-accumulative "silent" bioprocesses. Indeed, Fe oxide filter papers serve as a P sink, providing a similar functionality alike plant roots do (Figure 10.5). Through absorption of P from the solution to the filter paper, further desorption and re-supply from the soil are induced, using $CaCl_2$ [19]. The extraction time is extremely long, reaching 16 hours.

As an alternative to the above concept, another sink concept may be utilized by the cation- and anion-exchange membranes (CAEM) method [19], using anion strips and cation strips and distilled H_2O under shaking. The extraction time is also long, being 16 hours.

Figure 10.5 Iron oxide impregnated filter papers. Source: Reprinted from Wünscher et al. [18] with permission of Wünscher et al, University of Natural Resources and Life Sciences (BOKU).

For the first 14 methods by [19], the amount of extracted P increased in the order $H_2O < CaCl_2 < LiCl <$ Fe-oxide $P_i <$ Olsen < CAL < CAEM < Mehlich 3 < Bray II < dithionite < organic P < HCl < oxalate < total P. Also, by Wünscher et al. [18], the average rate of extraction increased along the order of $H_2O < CaCl_2 < LiCl <$ Fe-oxide $P_i <$ Olsen < CAL < CAEM < Mehlich 3 < Bray II < Dithionite < Organic P < HCl < Oxalate < Total P (Figure 10.6).

From these results, it is found that the physical extraction method based on the dissolution of P forms generally may not suit the P extraction from Moon crust due to the low dissolution of the compounds in the KREEP. The methods of Fe-oxide P_i and CAEM can have a good extraction performance, but their implementation on the Moon would need to refrain from the use of filter paper and membranes, and development of another lunar-suited concept would be required. Otherwise, the iron impregnated filter paper or resin is able to be internally regenerated or recycled in a way of ideal resupply. Chemical extractants show a higher extraction efficiency than those methods only relying on physical dissolution. In Section 10.7, we propose a biochemical reactive extraction process using plant-derived oxalic acid.

(Continued)

(Continued)

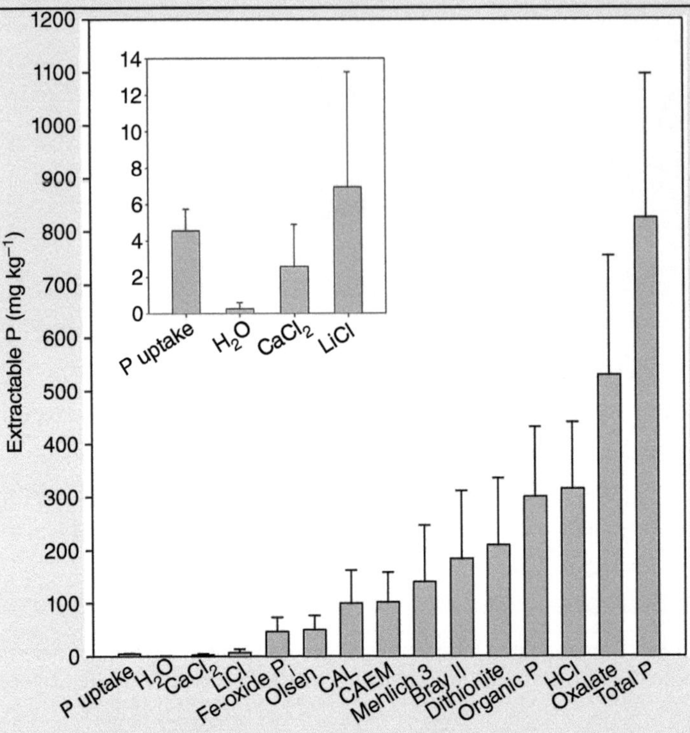

Figure 10.6 Extractable P per weight amount of soil for different P leaching techniques. Source: Wünscher et al. [18].

Reviewing the components in the KREEP basalt 15386, chemical extractants can be qualified to extract phosphorus, particularly the acids. However, some extractants such as inorganic acids, ammonium and sodium bicarbonate, or even EDTA cannot be obtained directly on the moon or in the space by simple processes. Transportation of those from Earth to Moon is too costly. In this sense, a regenerative method of oxalate, grown in space farms on the moon, hints at a suitable extractant. For example, oxalic acid commonly exists in the leaf or root of plants such as spinach, soursob flower, etc. and citric acid can be generally found in the fruit of plants such as lemon, orange, etc. Malic acid and tartaric acid can also be usually found in the plants. More importantly, plants have been successfully grown in space [21], so that it is very likely to produce organic acids directly on the Moon by growing ideal plants. In fact, the studies of phosphorus extraction by useful lixiviant from plants such as soursob have been attempted, and the results showed that extract of the plants (i.e. soursob) can be used as a useful lixiviant for phosphate ore [22], which reveal a good prospect.

10.7 Lunar Biomining of Phosphorus as Integrating Minerals Processing and Space Horticulture

10.7.1 Bio-Extractants for Phosphorus Minerals Processing

From above analysis, the use of oxalic acid as P extractant for lunar soils and on lunar ground seems to be favorable, both for reasons of high extraction efficiency and the fact that this extractant can be sourced from growing plants, such as from a lunar space farm. Soursobs are known to be rich in oxalic acid and is considered as plant source in the following analysis.

Soursob flower is regarded as a sort of weed, whose scientific name is *Oxalis pes-caprae*. It can be commonly found in the wild (Figure 10.7) since the soursob can grow on a wide range of soil types in both heavy shade and exposed places, and it can also endure both rainy and dry seasons [24]. Thus, soursob seemingly has potential to grow fast with less management in the moon base. Besides, Millerd et al. [25] researched the oxalic acid synthesis in *oxalis pes-caprae* and found that the organic acids in the extracts of oxalis shoot contained glyoxylic acid, glycolic acid, oxalic acid, malic acid, citric acid, isocitric acid and tartaric acid, and oxalic accounts for most, shown in Table 10.3.

There are many more green plants, which are also rich in oxalic acid, and Table 10.4 provides an overview of their respective performance. This study makes clear that the choice of the final green plant and vegetable to be used in space farming depends on its oxalic acid content, but also on its growth rate, nutrition

Figure 10.7 Oxalis pes-caprae. Source: Reprinted from Evenor and Knight [23] with permission of Zachi Evenor and MathKnight, Flickr / CC BY SA 4.0.

Table 10.3 Acid contents in soursob [25].

Plant	Type of acid	Volume percentage (%) recovered as specific acids
Soursob	Glyoxylic acid	1.29
	Glycolic acid	5.85
	Oxalic acid	81.16
	Malic acid	1.78
	Citric acid	1.68
	Isocitric acid	0.99
	Tartaric acid	1.19

value, and ease of harvesting under space farming conditions. Table 10.4 provides lots of opportunities for further refined space phosphate accumulation studies.

10.8 Moon Phosphate Circular Process Based on Microfluidic Biomining

We propose here an anticipatory integrated process for lunar bioleaching of P and its "silent" accumulation in green plants, which drives the circular process by constantly providing a concentration gradient for the extraction. As the soursob flowers are unlikely to suit this function, lettuce might instead be used, as it is known to accumulate high phosphorus loads. Lettuce will take up the phosphorus in its roots, in a manner of automatic harvesting. In this way, a circular lunar process can be designed (Figure 10.8). Soursob can be used to accumulate the phosphorus, yet it is not useful as a food product and therefore would only be a processing aid and not a food product. Using soursob as the collecting/concentrating plant means producing reagent and accumulating phosphorus and other elements useful to plants in one step.

The first, fast step, phosphorus extraction will be undertaken in an intensified micro-reactor, suited for solid–liquid (KREEP-water) operation. Bioagents will therein be used for intensification of phosphorus extraction from lunar soil. Those bioagents come from soursob flowers, which produce an abundance of oxalic acid. It is hypothesized that the soursobs are grown from nutrients extracted from the Moon regolith. The soursobs are then juiced in a perforated basket to extract the phosphate at about 40 tonnes of pressure. For space application, small scale squeezing machines like a screw press or a bag press (commonly used in the wine industry) may possibly be suited that will have less reliance on gravity. Heating or the addition of salt/sugar is known to help to shift the extraction at lower pressures. Our early-stage work, to be published, shows that the oxalic-acid juice can be derived from flowers and the thick stem, that can dissolve considerably

Table 10.4 Oxalic acid content in common vegetables.

Vegetables					
1–4 mg oxalate			**Over 10 mg oxalate**		
Item	Portion	Oxalate (mg)	Item	Portion	Oxalate (mg)
Alfalfa sprouts	½ cup	0	Bamboo shoots	1 cup	35
Bok choy (raw)	1 cup	1	Beans (fava)	½ cup	20
Broccoli	½ cup	1	Beans (kidney)	½ cup	15
Brussel sprouts	½ cup	2	Beans (navy)	½ cup	76
Cabbage	½ cup	1	Beans (refried)	½ cup	16
Cauliflower (cooked)	½ cup	1	Beets	½ cup	76
Celery (raw)	1 stalk	3	French fries	½ cup	51
Corn	½ cup	1	Olives	10	18
Cucumber	¼	1	Okra	½ cup	57
Endive	½ cup	0	Parsnip	½ cup	15
Green pepper	1 ring	1	Potato (baked with skin)	1 medium	97
Kale (chopped)	1 cup	2	Potato (chips)	1 ounce	21
Lettuce	1 cup	0	Potato (mashed)	1 cup	29
Mung beans	½ cup	3	Potato (salad)	½ cup	25
Mushrooms	1	0	Potato (sweet)	1 cup	28
Onions	1 small	0	Rhubarb	½ cup	541
Peas	½ cup	1	Rutabaga (mashed)	½ cup	31
Radish	10	0	Spinach (cooked)	½ cup	755
Rapini (chopped)	1 cup	4	Spinach (raw)	1 cup	656
Squash	½ cup	1	Tomato sauce	½ cup	17
Water chestnuts	4	0	Turnip	½ cup	30
Zucchini	½ cup	1	Yam	½ cup	40
5 to 10 mg oxalate					
Artichoke	1 small bud	5			
Asparagus	4 spears	6			
Carrot (cooked)	½ cup	7			
Carrot (raw)	½ large	10			
Celery (cooked)	1 cup	10			
Collards	1 cup	10			
Mixed vegetables	½ cup	5			
Conversion table for amounts					
½ cup	125 ml		1 ounce	28 g	
1 cup	250 ml				

Source: Hamilton SJsH [26]/Reproduced with permission of St. Joseph's Healthcare Hamilton.

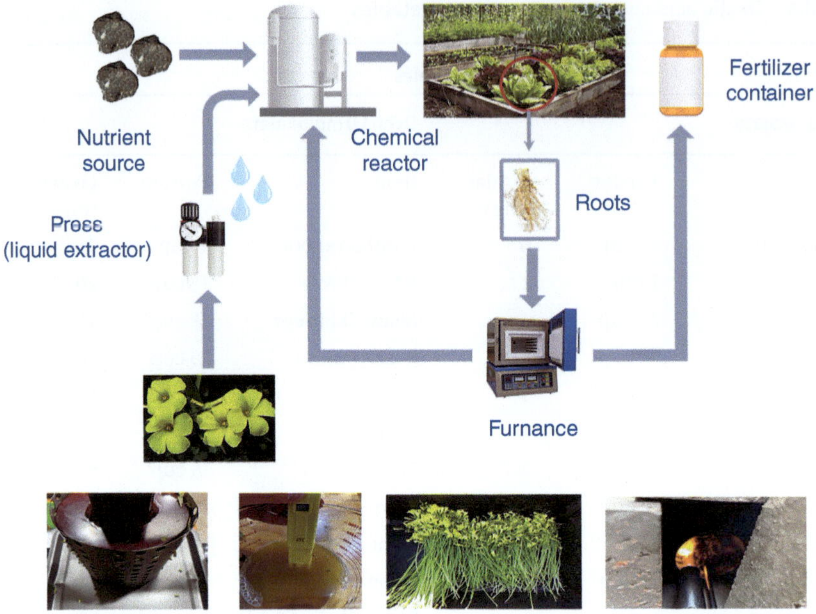

Figure 10.8 Anticipatory lunar process of silent P auto-accumulation, using a continuous-flow reactor for KREEP nioleaching, using oxalic acid from soursobs, and recycling part of the P from ashes of roots of lettuce plants, grown in the P nutrient stream.

more hydroxyapatite than the juice from the leaves, stems, and roots. To purify the P nutrient stream (from other leached component), a second continuous-flow extraction step might be considered; done in a liquid–liquid (water–oil) fashion; using a scavenger. Such extracted P nutrients feeds lettuce growth in a lunar space farm; this vegetable is known to be a phosphate (and nutrient)-rich green plant, which also grows fast and in a reliable manner. The plant will then be dried by heat, to take phosphorus from the ash, again by extraction.

10.9 Scale-up for First Lunar Outpost and Lunar P Life-Support Systems

For the purpose of calculating phosphate accumulation capacity per lettuce head and per lunar greenhouse container, we have taken from web the following plant growth data.

- Mature time (normal earth farming) = 45–55 days.
- Mature time (advanced earth farming: hydroponics) = 7 days (one week).
- Basically, four different kinds of lettuce are known, which differ in weight of the head, growth rate, nutrient content, and taste. Our preferred type at this point of time is the large-headed iceberg lettuce [27].
- Weight of one iceberg lettuce head = 539 g.

For the calculation of the demand of an adult with lettuce in space, Earth recommendations on (general) vegetable healthy consumption were taken. In this sense, it is assumed:

- Demand of lettuce for each adult = 500 g/day.
- Phosphate content in lettuce: 10 mg phosphate in 28 g (1 ounce) lettuce.
- P content in 1 head of lettuce = 539÷28 × 10 = 193 mg phosphate/head.

There is no current space greenhouse on a pilot- or production level; to our best knowledge, there is not even a design sketch reported in the open public. Thus, the further estimations given here rely on the best-reported Earth productivity from the most advanced greenhouses.

To provide an example, the Greenery hydroponic farm container of Freight Farms Company is presumably among the best such systems on Earth, and it is specifically made to operate under harsh conditions, where green plants are unable to grow (such as Northern Canada, near the Arctic). It is also made for farming in large cities, so-called Vertical Farming. The temperature inside is always warm, and irradiation by LEDs is done continuously, with no dark phases. The plants obtain all their nutrition from water in a hydroponic fashion. This provides independence from land, climate, and season.

The following specifications of the lunar farm container are relevant for this study.

- Container weight: 8 tonnes.
- Container volume = 15.24 m (l) × 3.5 m (w) × 2.96 m (h) = 158 m^3.
- Demand of water: 5 gallons water/day = 15.4 l water/day.

With those specifications and the claim from the web to be able to produce 800 lettuce heads per week in one container, the following conclusion on phosphate accumulation can be drawn.

- Total lettuce-moderated phosphate accumulation in one Greenery container; assuming 193 mg phosphate/head and 800 heads/week = 154 g phosphate/(container × week)

Literature states that lettuce needs an advanced phosphate fertilizer P use of 225 kg/hectare (225 kg/10,000 m^2) [28] (Hoque et al. 2010). Taking this value, the demand of phosphate fertilizer for lettuce in a lunar container sums up to 1.2 kg phosphate fertilizer/(container × 7.7 weeks). The accumulated phosphate value in the container then is 0.154 kg phosphate/(week × container).

Thus, the annual total need for phosphate fertilizer for a lunar container plant is 6.7 kg/acre; assuming ~7.7 weeks use of 1.2 kg. With the above assumptions and literature information, it can be calculated that the production of lettuce is 431 kg lettuce/(week × container). Bringing data together, the lunar space farm could provide sustainable food production for 123 people. This meets NASA's vision on the first steps of colonization [29]. Thus, the current best-earth technology can meet the demands of space colonization if it would work as effectively as on Earth.

With the above assumptions and literature information, it can be calculated that the needed microfluidic KREEP mineral-leaching chemical setup will have to have a capacity of 80 L/h. Our current laboratory systems, homemade and from Corning Company, have 36 and 72 L/h.

10.10 Conclusions

Overviewing lunar and Earth P reserves and their distribution sites, as well as relevant Earth process technologies and what might be their lunar counterpart, we propose herein an anticipatory integrated process for lunar bioleaching of P and its "silent" accumulation in green plants. This shall drive the circular process by constantly providing a concentration gradient for the extraction. Bioleaching is provided from soursob flowers, and lettuce is proposed to accumulate high phosphorus loads. Burning the roots of lettuce and any unused soursob biomass to ash, and extraction the phosphorus closes the circle. This includes the waste from the pressing of the sour sobs and the inedible parts of the lettuce, such as the flowering stem used to produce seed to grow more lettuce.

By comparing the various Earth phosphorus extraction methods, it is proposed to use organic acids, such as oxalic acid, from the green plants which can be grown directly on the Moon. Microfluidic or continuous-flow bioleaching, done in a solid–liquid fashion, is proposed as means to achieve the process intensification needed. Through calculation and comparison, the current home-made and commercial microfluidic extraction and leaching systems seem to be sufficient to provide an adequate phosphate supply for the vegetable feed of a population of >100 Moon or potentially even Mars colonialists assuming similar conditions and techniques producing equivalent quantities. The results indicate that the bottleneck is on the side of the lettuce plant growth, and not on the minerals processing.

References

1 Smil, V. (2000). Phosphorus in the environment: natural flows and human interferences. *Annual Review of Energy and the Environment* 25 (1): 53–88.

2 Ashley, K., Cordell, D., and Mavinic, D. (2011). A brief history of phosphorus: from the philosopher's stone to nutrient recovery and reuse. *Chemosphere* 84 (6): 737–746.

3 Liu, Y. (2005). *Phosphorus Flows in China: Physical Profiles and Environmental Regulation*. Wageningen University and Research.

4 Samreen, S. and Kausar, S. (2019). Phosphorus fertilizer: the original and commercial sources. In: *Phosphorus-Recovery and Recycling* (ed. T. Zhang), 1–14. London: IntechOpen.

5 Papike, J., Taylor, L., and Simon, S. (1991). Lunar minerals. In: *Lunar Sourcebook: A User's Guide to the Moon*, 121–181. Cambridge University Press.

6 Buchwald, V.F. (1984). Phosphate minerals in meteorites and lunar rocks. In: *Phosphate Minerals*, 199–214. Springer.

7 Summaries, M.C (2021). Mineral commodity summaries. Report Reston, VA: 200. p. 123. https://doi.org/10.3133/mcs2021

8 Crawford, I.A. (2015). Lunar resources: a review. *Progress in Physical Geography* 39 (2): 137–167.

9 Pasek, M.A. (2015). Phosphorus as a lunar volatile. *Icarus* 255: 18–23.

10 Jolliff, B.L., Gillis, J.J., Haskin, L.A. et al. (2000). Major lunar crustal terranes: surface expressions and crust-mantle origins. *Journal of Geophysical Research, Planets* 105 (E2): 4197–4216.

11 McLeod, C.L. and Krekeler, M. (2017). Sources of extraterrestrial rare earth elements: to the moon and beyond. *Resources* 6 (3): 40.

12 Wieczorek, M.A. and Phillips, R.J. (2000). The "Procellarum KREEP Terrane": implications for mare volcanism and lunar evolution. *Journal of Geophysical Research, Planets* 105 (E8): 20417–20430.

13 Neal, C. and Kramer, G. (eds.) (2003). The composition of KREEP: a detailed study of KREEP basalt 15386. *Lunar and Planetary Science Conference*, The Woodlands, Texas.

14 Cheremisinoff, P.N. (1995). Industry profile—fertilizers. In: *Waste Minimization and Cost Reduction for the Process Industries*, 222–284. Elsevier.

15 Greenwood, N.N. and Earnshaw, A. (2012). *Chemistry of the Elements*. Elsevier.

16 Chen, M. and Graedel, T. (2016). A half-century of global phosphorus flows, stocks, production, consumption, recycling, and environmental impacts. *Global Environmental Change* 36: 139–152.

17 Zhang, M., Wright, D., Heaney, D., and Vanderwel, D. (2004). Comparison of different phosphorus extraction and determination methods using manured soils. *Canadian Journal of Soil Science* 84: 469–475.

18 Wünscher R, Unterfrauner H, Peticzka R, Zehetner F (2013). A comparison of different phosphorus extraction methods with the phosphorus uptake of wheat: na, University of Natural Resources and Life Sciences, Vienna, Austria.

19 Wuenscher, R., Unterfrauner, H., Peticzka, R., and Zehetner, F. (2015). A comparison of 14 soil phosphorus extraction methods applied to 50 agricultural soils from Central Europe. *Plant, Soil and Environment* 61 (2): 86–96.

20 Nafiu, A. (2006). Soil-phosphorus extraction methodologies: A review. *African Journal of Agricultural Research* 1 (5): 159–161.

21 Ferl, R., Wheeler, R., Levine, H.G., and Paul, A.-L. (2002). Plants in space. *Current Opinion in Plant Biology* 5 (3): 258–263.

22 Kisnasz, D. (2019). *Your Hardware Store in Space*. South Australia: The University of Adelaide.

23 Evenor. Z. and Knight, M. (2009). *Oxalis pes-caprae* 16: Flickr. https://www.flickr.com/photos/zachievenor/3755827278/in/set-72157623475255073/.

24 Hunter, I. (2015). Plant policy: soursob (*Oxalis pes-caprae*). In: *Sustainability EaC*, 1–2. Australia: Government of South Australia.

25 Millerd, A., Morton, R., and Wells, J. (1963). Oxalic acid synthesis in shoots of *Oxalis pes-caprae* (L.). *Biochemical Journal* 86 (1): 57–62.

26 Hamilton SJsH (2016). Oxalate in Food: St. Joseph's Healthcare Hamilton. https://www.stjoes.ca/patients-visitors/patient-education/patient-education-k-o/pd-9447-oxalate-in-food.pdf.

27 Picard, C. (2019). 11 Types of Lettuce to Try So You'll Never Have a Boring Salad Again: Hearst Digital Media. https://www.goodhousekeeping.com/food-recipes/cooking/g532/types-of-lettuce/.

28 Tiesenhausen, G. (1970). Fifty-man space base population organization.

29 Hoque, M.M., Ajwa, H., Othman, M., Smith, R. and Cahn, M. (2010). Yield and postharvest quality of lettuce in response to nitrogen, phosphorus, and potassium fertilizers. HortScience 45 (10). https://doi.org/10.21273/HORTSCI.45.10.1539.

11

Lunar Resources in Support of Human Interplanetary Settlement and Limitations

Zoe Nunn[1], *Samantha Smith*[1], *and John Culton*[2]

[1] *Scotch College Adelaide, Carruth Road, Torrens Park, Adelaide, SA 5062, Australia*
[2] *University of Adelaide and Andy Thomas Centre for Space Resources, N104 Engineering North, North Terrace Campus, Adelaide, SA 5005, Australia*

11.1 Introduction

Exploring deep space is something that many space organizations (both government and private) have been pursuing to advance humanity to becoming an interplanetary species. The costs associated with taking a kilogram of payload into low-Earth orbit is around US$10,000/kg, meaning launching a crewed settlement mission from Earth is not economically viable (Spudis 2010). "The Moon is not a hostile, barren rock in space – it is humanity's stepping stone into the Solar System" (Spudis 2006). The Moon is the only celestial object in deep space that is both close and constantly accessible (European Space Agency 2018). Research from various missions in the early-mid-2000s uncovered the presence of resources such as lunar water, regolith, and sunlight, all of which can be utilized to sustain a crewed mission farther and for longer than humanity has previously achieved (NASA 2020a). Lunar *In Situ* Resource Utilisation (Lunar ISRU) is the process of using the materials found on the Moon to advance humanities space capabilities (Sanders 2018). The extent to which these resources can be used depends on their abundance, ease of extraction, and how effectively they can achieve their theorized use. If resources for human sustainment in prolonged spaceflight and interplanetary habitation can be found on the Moon, it will open the door for constant and recurring missions that enable the human settlement of planetary bodies like Mars and beyond.

In-Space Manufacturing and Resources: Earth and Planetary Exploration Applications, First Edition.
Edited by Volker Hessel, Jana Stoudemire, Hideaki Miyamoto, and Ian D. Fisk.

The table below depicts resources and their potential uses:

Resource	Where it is found	Primary use(s)
Silicates (Ilmenite, Pyroxene, Olivine, Anorthite, etc.)	Regolith	Power/Energy + Life Support
Lanthanides/REE (Uranium, Thorium, etc.)	Regolith	Power/Energy
Helium-3	Regolith	Power/Energy
Carbon	Regolith	Production Processes/Infrastructure
Metals (Aluminum, Titanium, etc.)	Regolith	Production Processes/Infrastructure
Hydrogen	Regolith/Cold traps	Power/Energy
Lunar Ice (Polar Water)	Cold Traps	Power/Energy + Life Support
Sunlight	Lunar Surface	Power/Energy

The three resources that have been analyzed are lunar ice, sunlight, and regolith. The specific resources in the regolith have been combined as many of the individual resources have similar uses and potential extraction processes and hence can be discussed together.

11.2 Lunar Ice

Several international lunar missions (e.g. NASA's Clementine and India's Chandrayaan-1) have returned data that support the idea that water molecules can reside in the lunar poles of the Moon (Figure 11.1) (Spudis 2010; Siddiqi 2019; Crawford 2015).

Due to the Moon's spin axis being a 1.5° angle to the solar ecliptic plane (the Earth's is 23°), sun exposure for different lunar areas varies greatly (Bussey, Spudis, and Robinson 1999). Permanently shadowed regions, located in polar areas, are usually craters where there has been no sun exposure for billions of years (Crawford 2015; Haine et al. 2014). Temperatures in these areas are measured below 40 K (−233.15 °C) meaning lunar ice will be stable and trapped as there is not enough energy for it to escape (Paige et al. 2010). NASA's Lunar Crater Observation and Sensing Satellite (LCROSS) impactor mission in 2009 found lunar ice at a wt% of 5.6 ± 2.9 at the poles within the first meter of regolith, hence 100 million to 1 billion tonnes is theorized to be found (Crawford 2015; Colaprete et al. 2010; Taylor 2019).

11.2.1 Settlement Uses

Lunar Ice is the Key Resource says Professor Phil Bland, Curtin University (personal communication 2021). When colonizing other planets, lunar water can rehydrate food, and sustain human life by providing drinking water and oxygen,

Figure 11.1 Presence of water and hydroxyl (blue) on lunar surface. Source: Siddiqi (2019).

which crewmembers can use for air stores in this oxygen-depleted environment (Spudis 2010; Roger et al. 2018). However, the most beneficial purpose of lunar ice is the potential for creating rocket fuel (from electrolyzing H_2O molecules) to use the Moon as a fuel station to reach other planetary bodies (Moses *personal communication* 2021; Crawford 2015). Head of the Rocket Propulsion Technology Department of the German Aerospace Centre, Dr. Justin Hardi, highlights, "hydrogen and oxygen are the most efficient propellant combination of oxidizer and fuel" (Justin Hardi: via Skype on 16th February 2021, A research engineer at the German Aerospace Centre. Head of Rocket Propulsion Technology Department). Hydrogen is a very energy-dense molecule and when combusting with an oxidizer (oxygen) a large amount of energy is released: $H_{2(l)} + \frac{1}{2}O_{2(l)} \rightarrow H_2O_{(g)}$ $\Delta H_f^\circ = -285.8\,\text{kJ mol}^{-1}$(Çakanyıldırım 2015). This production of energy burns hydrogen and oxygen together, creating gaseous water that can be used to thrust the spacecraft forwards (Manoharan et al. 2019).

To allow for the combustion reaction to take place using lunar ice in the form H_2O, it must be electrolyzed to separate the hydrogen and oxygen components. This is completed through the oxidation of the oxygen molecules and the reduction of hydrogen molecules in an electrolytic cell (WHA International 2018). The process requires a large amount of energy input to allow for the electrodes to separate these molecules out (Office of Energy Efficiency and Renewable Energy 2019). When separated, to maximize the quantity stored in the fuel tanks and hence the energy efficiency of the rocket, oxygen must be liquified at −180 °C and hydrogen at −253 °C to increase their densities (European Space Agency 2020a). Additionally, launching the rocket from the Moon allows for more fuel to be stored in the tank as less would be consumed escaping the Moon's gravity. One option being considered is launching fuel generated from lunar materials to a fuel depot in lunar orbit.

Therefore, exploration vehicles launched from Earth would only need enough fuel to reach lunar orbit, significantly reducing the costs of interplanetary travel (Crawford 2015; WIRED 2019; Bland *personal communication* 2021).

11.2.2 Limitations

Albeit there is a high probability of lunar ice being present in the permanently shadowed regions, it is unconfirmed as they have not been prospected. As researchers do not currently understand ice as a resource, its exact amounts and ease of extraction need to be determined (Anand 2012; European Space Agency 2020a; Spudis 2010). Furthermore, even if confirmed to be present, researchers theorize the extraction of the lunar ice is difficult due to the extreme 40 K temperatures of the environment where these ice stores are located (Taylor 2019; Anand 2012).

11.3 Regolith

Regolith has varied potential uses to further deep space exploration. The composition of the Moon's surface is broadly split into two categories: highland and mare (Figure 11.2). Highland regolith is iron poor anorthic material that dates back to the early lunar crust. Mare material is volcanic in nature, associated with the lava that flooded the large lunar basins. It has a higher iron content than highland material (Crawford 2015; Robinson et al. 2007). The regolith on the Moon is further divided into subsections depending on the TiO_2 concentration by mass (above 6% = high Ti,

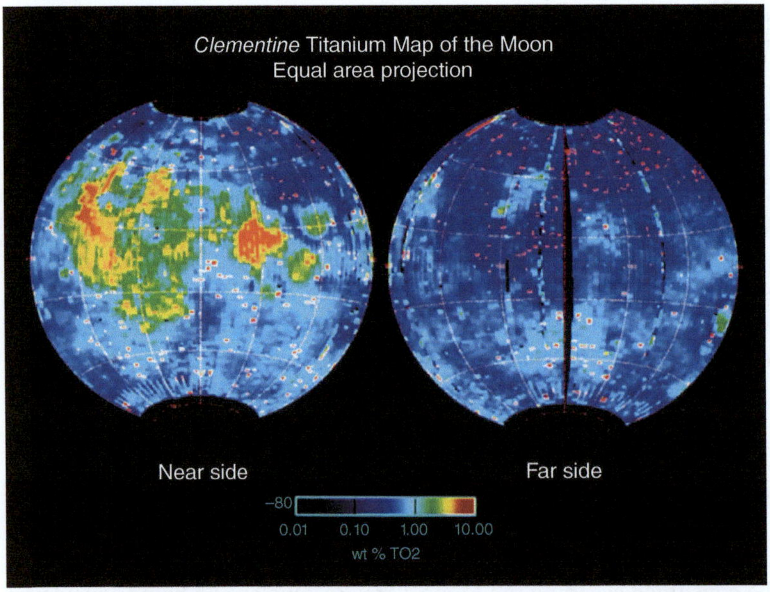

Figure 11.2 TiO_2 concentration on lunar surface. Source: ACT Cooperation et al. (n.d.).

below 6% = low Ti) (see Figure 11.2) (Robinson et al. 2007; Anand 2012, Crawford 2015; Taylor 2000). Research indicates that when extracted, these compounds can be used for a variety of purposes (Crawford 2015; Taylor 2019; Anand 2012; Sanders et al. 2019).

11.3.1 Settlement Uses

Regolith's raw state can be used to build and reinforce lunar surface structures that can help advance interplanetary settlement (Taylor 2019; European Space Agency 2020a; Sanders et al. 2019). It can also act as a radiation shielding, protecting the astronauts inside their habitat. Associate Professor John Culton, Director of the Andy Thomas Centre for Space Resources, is currently supporting research utilizing regolith to form materials that can build lunar infrastructure for habitats and interplanetary settlement (Spudis 2010; Roger et al. 2018; European Space Agency 2018; Taylor 2019). Other researchers such as the late Professor Paul Spudis also support the use of this 3D-printing method (European Space Agency 2018).

Potentially useful rare and abundant Earth metals may aid in building technologies and structures and creating tools for interplanetary settlement, surveying, and exploration (Taylor 2019; Crawford 2015; Astrum 2021; Sanders 2018). Currently, Professor Volker Hessel of University of Adelaide utilizes the abundance of TiO_2 in regolith to create pain-relief medicine for astronauts that does not degrade while in spaceflight (Savage 2020; Cairns and Short 2018). Furthermore, lunar regolith is approximately 42% oxygen due to a high abundance of oxides in the lunar regolith. Releasing this oxygen by cracking these minerals means it can be used for life support and fuel supplies necessary for deep space settlement (European Space Agency 2018; Taylor 2019; Burke 2021).

Lastly, Helium-3 found in the regolith is the ideal nuclear power source to power technologies on and off the lunar surface (Astrum 2021; European Space Agency 2015; Li and Wittenberg 1992). In the regolith, although it can be found in minimal concentrations due to implantation from the solar wind (Figure 11.3), helium-3 can be gathered with any mining operation as stores are found all over the lunar surface (Astrum 2021; Burke 2021; Fa and Jin 2007). Helium-3, along with deuterium (another commonly found resource), can be used in a nuclear fusion reactor as an efficient source of power without any waste or radiation (Barnatt 2016; European Space Agency 2015). When the deuterium and helium-3 undergo nuclear fusion, a helium molecule and a proton are produced as well as 12.86 MeV of energy that can be transformed into electrical energy when one helium-3 molecule is fused. This could potentially power operations on the lunar surface as well as support lunar bases on the Moon's surface (Barnatt 2016).

11.3.2 Limitations

The composition of the lunar surface can be theorized; however, as the area has not been fully surveyed, this composition cannot be proven. Furthermore, research focusing on 3D printing using raw regolith is completed using simulants, potentially

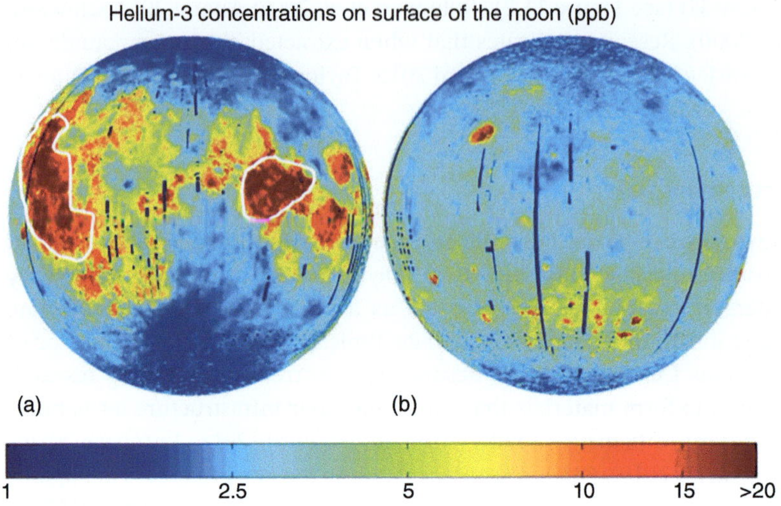

Figure 11.3 3He concentrations on lunar surface. Source: Fa and Jin (2007)/with permission of Elsevier.

reducing its validity when reproduced on the lunar surface (A/Prof John Culton, Director Andy Thomas Centre for Space Resources, personal interview conducted on-campus 18 Feb 2021; Crawford 2015).

The production of specific compounds is also a limitation. Due to the harsh, low atmospheric environment, certain industrial processes used on Earth would not be possible on other planetary bodies without increased infrastructure, which may not be economically viable (Crawford 2015). Furthermore, regolith extraction processes are very energy intensive, especially helium-3 and oxygen extraction, so for this to be a viable practice, an energy source must be found to provide consistent power for these processes (Anand 2012; European Space Agency 2019).

11.4 Sunlight

As mentioned above, the lunar axis allows for varying sun exposure, most drastically at the lunar poles. Consequently, there are permanently shadowed craters, while other regions experience high sun exposure during the lunar day. A study completed during the Clementine mission found that some areas were exposed to the sun over 80% of the lunar day (approximately 591 of 708 hours) (Figure 11.4) (Bussey, Spudis, and Robinson 1999).

These areas are usually situated on lunar crater rims due to them being the highest terrain (European Space Agency 2018; Bussey, Spudis, and Robinson 1999). Thus, there is potential for solar panels to be used to capture this sunlight and convert it into electrical energy (Anand 2012; Taylor 2019). Dr. Taylor, emeritus research professor in the Hawai'i Institute of Geophysics and Planetology, supports this theory by stating, "an average of 1360 watts per square meter (W/m^2) of electrical energy can be obtained from the sunlight" (Taylor 2019; Bussey, Spudis, and Robinson 1999).

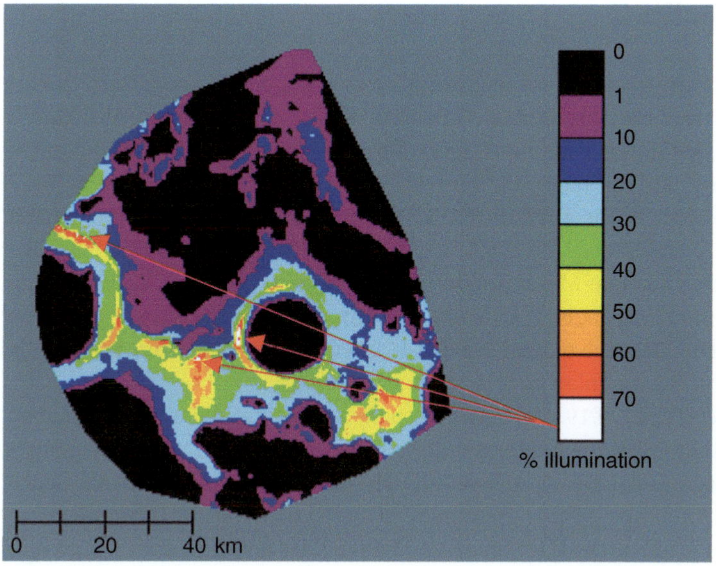

Figure 11.4 Lunar pole illumination %/lunar day. Source: Bussey, Spudis, and Robinson (1999), permission conveyed through Copyright Clearance Center, Inc. Republished with permission of John Wiley & Sons.

11.4.1 Settlement Uses

Sunlight, when converted into energy, can be utilized to progress interplanetary settlement further. Solar energy can be used as a reliable power source for habituated lunar bases, allowing humans to live, operate machinery, and test samples on the surface (European Space Agency 2018, Bussey, Spudis, and Robinson 1999; Burke 2021; European Space Agency 2020a). It can also allow for energy-intensive industrial processes (like the pyrolysis process or regolith sintering) to be completed which can extract the compounds needed to progress humanity further (NASA 2020a). Furthermore, Adrian Stoica, NASA and Caltech Jet Propulsion Laboratory, theorizes that the sunlight can be utilized by directing beams of sunlight using heliostats (highly reflective curved surfaces that move with the sun and rover positioning) reflecting sunlight to power the rover in permanently shadowed craters to help it extract lunar ice (Taylor 2019; Hall 2017). These operations will allow the above materials to be extracted and utilized for planetary settlement.

11.4.2 Limitations

A significant shortcoming is the large amount of purpose-built infrastructure required to harness, transport, and utilize this sunlight for power, resulting in greater costs of production (Taylor 2019; Anand 2012). Also, insufficient energy could be produced, not meeting the demand of resource utilization processes (Bland *personal communication* 2021; European Space Agency 2020a).

11.5 ISRU Policies

Like on Earth, activities undergone in space need to be regulated to ensure peaceful and ethical practices are undertaken (Malcolm Davis: via Zoom on 4th March 2021 , an ASPI Senior Analyst in Defence Strategy and Capability). Currently, many space-faring nations have set their sights on the Moon due to the IRSU potential. The growth of the current $US300-billion space industry is stimulated by research and development of technologies (Lisk 2019). Hence, missions like Artemis, Heracles, and Blue Moon are being created, with countries collaborating to help meet the mission's goals (NASA 2020b; European Space Agency 2020b; Blue Origin). Various treaties have been created to govern these activities, potentially impacting Lunar ISRU and planetary settlement.

The Outer Space Treaty of 1967 governs all space activities and has been signed by 137 countries. Many missions have created their accords to align with this treaty's articles (NASA 2020a): Article 1[1], Article 2[2] (United Nations Office for Outer Space Affairs 1966). A limitation of this document is that it is broad, focusing more on activities in space itself, rather than on the celestial bodies and thus does not go into depth around ownership and use of *in situ* materials (Fuentes 2015; Parkinson 2019).

Another noteworthy document is the Moon Agreement which was implemented in 1979 as space agencies started to theorize how the Moon could impact the future of space exploration. However, this treaty has only been signed by 18 countries, only 11 of which are ratified, and hence is not widely accepted as customary international space law (United Nations Office for Outer Space Affairs 1979). Hence, there is no prescriptive application outside of the 18 signatory nations meaning it is an active treaty, but only needs to be abided by for these 18 countries. Despite the lack of countries that have signed and ratified this treaty, the section that highlights the need for a resource's regime when exploitation of them is feasible is Article 11 which would be a valuable contribution to sustainable use of space resources.[3] Although this exploitation is now feasible, no regime has been created to govern the use of these materials. This could cause tension between countries around usage and property rights of lunar areas and resources (Fuentes 2015; Parkinson 2019).

The Artemis Accords created in 2020 by NASA is a document that outlines a political commitment upheld by all countries participating throughout NASA's Artemis Mission. This document highlights the necessary restrictions of the Artemis Program when it is conducted on the Moon. It gives evidence why the procedures are formulated the way they are and justifies the decisions NASA has made during this program. The Artemis Accords are governed by the key principles that exploration will be completed in a peaceful manner with a high degree of transparency to the public by releasing information, plans, and scientific data collected. It also ensures that the mission will work toward preserving outer space

1 *No celestial bodies can be claimed by sovereignty* by *any countries.*

2 *Exploration should be carried out for the benefit of humankind*

3 *All resources on the surface or subsurface of the moon are the property of mankind and thus cannot be claimed by one nation* and *when exploitation of resources is feasible countries shall establish an international regime to ensure to govern the exploitation of said resources*

heritage by removing and safely disposing of space debris as well as preventing harmful interference of other space missions. Safety precautions are also taken to ensure the protection of humans working throughout the Artemis mission, but also any other space activities occurring. The Artemis Accords draw heavily on the Outer Space Treaty throughout the key principles of the document, especially when discussing the safety precautions and the purpose of exploration on the lunar surface. This document has been signed and ratified by 11 countries including Australia and the United Kingdom; however, it should also be noted that many large space-faring countries (China and Russia for example) have not signed the Artemis Accords and thus may have a conflict when Lunar ISRU begins on the Moon at scale for planetary settlement. The Artemis Accords are the first step toward using space resources in cooperation with other countries to explorer deep space (NASA 2020b).

11.6 Conclusion

With humankind poised to return to deep space via NASA's Artemis Program, which promises long-term human presence on the Moon and Mars, we are confronted with the unrelenting reality of the rocket equation: bringing everything we need to live for long periods in deep space from the surface of the Earth is not feasible. However, based on continuous advances in our understanding of the solar system, we now know that we don't need to. Both the Moon and Mars possess abundant resources, such as water, oxygen, titanium, iron, aluminum, and many more. With these resources, astronauts could make air to breathe, fuel for their spacecraft, water for themselves and their gardens, and construction materials for their habitats. Researchers all over the world are examining how we might empower exploration by learning to become a truly space-faring species, learning how to live and work in space using the resources we find there rather than bringing them from the Earth. Research is also being done with regard to the legal regime needed to ensure this is done in a sustainable and responsible manner.

References

Anand, M. (2012). A brief review of chemical and mineralogical resources on the Moon and likely initial in situ resource utilization (ISRU) applications. *Planetary and Space Science* 74 (1): 43. https://www.sciencedirect.com/science/article/abs/pii/S0032063312002498?casa_token=D_mfAO8hqEYAAAAA:cLY8-0O9_aK_A5IY1R3cOJuOUrUKl532_C6wwm1fMBTeVp7anHfcjqN2D4ll9TUyae89xPMswg#bib14.

Astrum (2021). *The Real Reason Countries Are Going Back To The Moon*, https://www.youtube.com/watch?v=q8aqVh-LJYA.

Barnatt, C. (2016). Helium-3 Power Generation, Future Technologies, https://www.explainingthefuture.com/helium3.html.

Bland, P. (2021). Personal communication, 25 February.

Blue Moon (n.d.). Blue Origin, Back to the Moon – To Stay. https://www.blueorigin.com/blue-moon/lunar-transport.

Burke, J. (2021). Lunar resources. https://www.britannica.com/biography/Eugene-Andrew-Cernan.

Bussey, B. (2021). Principal Professional Staff Scientist at The Johns Hopkins Applied Physics Laboratory. Former NASA Chief Exploration Scientist.

Bussey, B., Spudis, P., and Robinson, M. (1999). Illumination conditions at the lunar south pole. *Geophysical Research Letters* 26 (9): 1187–1190.

Cairns, I. and Short, W. (2018). *Australian Space Research Conference*. http://www.nssa.com.au/18asrc/resources/18ASRC-A5-BOOKLET-full.pdf.

Çakanyıldırım, M. (2015). Science of Energy II. https://personal.utdallas.edu/~metin/Merit/MyNotes/energyScience2.pdf.

Colaprete, A., Schultz, P., Heldmann, J. et al. (2010). Detection of water in the LCROSS ejecta plume. *Science* 330 (6003): 463–468, American Association for the Advancement of Science.

Crawford, I. (2015). Lunar resources: a review. *Progress in Physical Geography* 39 (2): 137–167, SAGE, SagePub.

Dowling, J (2021). Toyota Mirai: Australia's first hydrogen car on sale to the public 'within a couple of years', NEWS. , https://www.caradvice.com.au/942563/2022-toyota-mirai-australias-first-hydrogen-car-on-sale-to-the-public-within-a-couple-of-years/.

European Space Agency (2015). Helium-3 mining on the lunar surface. https://www.esa.int/Enabling_Support/Preparing_for_the_Future/Space_for_Earth/Energy/Helium-3_mining_on_the_lunar_surface.

European Space Agency (2018). Space Bites: The Value of the Moon|Paul D. Spudis, online video, 12 January. https://www.youtube.com/watch?v=Ne7MLSVKSf4&t=2s.

European Space Agency (2019). Helium-3 mining on the lunar surface, Enabling & Support. http://www.esa.int/Enabling_Support/Preparing_for_the_Future/Space_for_Earth/Energy/Helium-3_mining_on_the_lunar_surface.

European Space Agency (2020a). Meet the ESA experts – Resources on the Moon, online video, 21 February. https://www.youtube.com/watch?v=Ne7MLSVKSf4&t=2s.

European Space Agency (2020b). Landing on the Moon and returning home: Heracles, Science and Exploration. https://www.esa.int/Science_Exploration/Human_and_Robotic_Exploration/Exploration/Landing_on_the_Moon_and_returning_home_Heracles.

Explaining The Future (2014). Mining the Moon, online video, May 16. https://www.youtube.com/watch?v=n9EMcyfGMDA.

Fa, W. and Jin, Y. (2007). Quantitative estimation of helium-3 spatial distribution in the lunar regolith layer. *Icarus* 190 (1): 15–23, Elsevier, ScienceDirect.

Fuentes, U. (2015). Understanding the legal status of the Moon, Space News. https://www.thespacereview.com/article/2703/1.

Haine, P., Ingersoll, A. and Paige, D. (2014). New Approaches to Lunar Ice Detection and Mapping, Keck Institute for Space Studies. https://www.kiss.caltech.edu/final_reports/Lunar_Ice_final_report.pdf.

Hall, L. (2017). Trans-Formers for Lunar Extreme Environments: Ensuring Long-Term Operations in Regions of Darkness and Low Temperatures. https://www.nasa.gov/feature/trans-formers-for-lunar-extreme-environments-ensuring-long-term-operations-in-regions-of/.

Li, Y. and Wittenburg, L. (1992). Lunar surface mining for automated acquisition of helium-3: methods, processes, and equipment. In: *The Second Conference on Lunar Bases and Space Activities of the 21st Century*, vol. 2, 609–617. Johnson Space Center.

Lisk, J. (2019). Space Law. Its a real thing', Blog post, 19 July. https://www.cowellclarke.com.au/news/space-law-its-a-real-thing.

Manoharan, Y., Hosseini, S., Butler, B. et al. (2019). Hydrogen fuel cell vehicles: current status and future prospect. *Applied Sciences* 9 (11): 2296. https://www.mdpi.com/2076-3417/9/11/2296#cite.

Moses, R. (2021). Personal communication, 9 March.

NASA (2020a). NASA's Lunar Exploration Program Overview 2020. https://www.nasa.gov/sites/default/files/atoms/files/artemis_plan-20200921.pdf.

NASA (2020b). The Artemis Accords. https://www.nasa.gov/specials/artemis-accords/img/Artemis-Accords-signed-13Oct2020.pdf.

Office of Energy Efficiency and Renewable Energy (2019). Hydrogen Production: Electrolysis, U.S Department of Energy. https://www.energy.gov/eere/fuelcells/hydrogen-production-electrolysis.

Paige, D., Siegler, M., Zhang, J. et al. (2010). Diviner lunar radiometer observations of cold traps in the Moon's south polar region. *Science* 330 (6003): 479–482, American Association for the Advancement of Science.

Parkinson, J. (2019). Can anyone 'own' the Moon?, BBC, Science. https://www.bbc.com/news/science-environment-46877417.

Robinson, M., Hapke, B., Garvin, J. et al. (2007). High resolution mapping of TiO_2 abundances on the Moon using the hubble space telescope. *Geophysical Research Letters* 34 (13): 1–4, AGU, John Wiley and Sons.

Roger, H., Charisma, J., Pesquira, D. et al. (2018). *Mahina*. Honolulu, Hawai'i: Charisma Industries.

Sanders, G. (2018). Overview of Past Lunar In Situ Resource Utilization (ISRU) Development by NASA. https://ntrs.nasa.gov/citations/20190001035.

Savage, C. (2020). Testing Time for Pills in Space, Newsroom. https://www.adelaide.edu.au/newsroom/news/list/2020/09/21/testing-time-for-pills-in-space.

Siddiqi, A. (2019). Chandrayaan-1 / Moon Impact Probe, Diagram, NASA. https://solarsystem.nasa.gov/missions/chandrayaan-1/in-depth/.

Spudis, P. (2006). Ice on the Moon. https://www.thespacereview.com/about.html.

Spudis, P. (2010). The Moon: Creating Capability in Space and Getting Value for our Money, Smithsonian Institution. https://www.airspacemag.com/daily-planet/the-moon-creating-capability-in-space-and-getting-value-for-our-money-156159408/.

Taylor, G. (2000). The titanium contents of lunar mare basalts. *Meteoritics and Planetary Science* 35 (35): 193–200.

Taylor, G. (2019). Using the Resources of the Moon to Expand Earth's Economic Sphere, Planetary Science Research Discoveries, Headline Articles. http://www.psrd.hawaii.edu/Nov19/lunar-isru.html.

Sanders, G., Linne, D., Kleinhenz, J., et al. (2019). NASA Oxygen and Water Production Architectures for Early Reusable Lander, Powerpoint Presentation, Explore Moon to Mars, Lunar ISRU Workshop.

United Nations Office for Outer Space Affairs (1966). Treaty on Principles Governing the Activities of States in the Exploration and Use of Outer Space, including the Moon and Other Celestial Bodies, Space Law. https://www.unoosa.org/oosa/en/ourwork/spacelaw/treaties/outerspacetreaty.html.

United Nations Office for Outer Space Affairs (1979). Agreement Governing the Activities of States on the Moon and Other Celestial Bodies, Space Law. https://www.unoosa.org/oosa/en/ourwork/spacelaw/treaties/moon-agreement.html.

WHA International (2018). Will Hydrogen Power The Future Of Aerospace?, Forsythe Creativehttps://wha-international.com/will-hydrogen-power-the-future-of-aerospace/.

WIRED (2019). Scientist Explains How Moon Mining Would Work|WIRED. https://www.youtube.com/watch?v=ix6BxLf1bew.

Part V

Space Farming and Food

12

PONDS: A New Method for Plant Production in Space

Howard G. Levine[1], Jeffrey T. Richards[2], Lawrence L. Koss[2], Mark Weislogel[3], David Reed[4], Anna Nguyen[5], Hector Barea[5], and David Kusuma[5]

[1] *NASA Utilization and Life Sciences Office, Kennedy Space Center, FL 32899, USA*
[2] *AECOM, Kennedy Space Center, FL 32899, USA*
[3] *Portland State University, Portland, OR 97207, USA*
[4] *Techshot, Space Life Sciences Laboratory, Kennedy Space Center, FL 32899, USA*
[5] *Tupperware Corporation, Orlando, FL 32837, USA*

12.1 Vegetable Production System (Veggie)

The vegetable production system, or veggie, was developed by ORBITEC Corporation (Madison, WI; now Sierra Nevada Corporation) to be a simple, high-growth volume facility (planting area of 29 cm × 36 cm and maximum height of 45 cm) capable of producing fresh vegetables on the International Space Station (ISS) (Morrow et al. 2005). It is an open system that uses a fan at its top to draw in air from the adjacent cabin environment and therefore relies upon the ISS ambient temperature (typically around 22 °C) and carbon dioxide levels (typically around 3000 ppm) for plant growth. Portable temperature and relative humidity data loggers (HOBO U12, Onset Computer Corporation, Bourne, MA) are used to collect data for the duration of the experiments.

Veggie has a light bank with red (630 nm), green (530 nm), and blue (455 nm) light-emitting diodes (LEDs) that provide 100–500 $\mu mol\,m^{-2}\,s^{-1}$ of light. Investigator-specified photoperiods are programmed at experiment initiation. (See Khodadad et al. 2020; Massa et al. 2013, 2015, 2016, 2017 for a more complete description of the veggie hardware, protocols that have been developed, and results obtained.)

Veggie's LED light bank can be seen in Figure 12.1a. For the salad crop production studies performed to date, veggie has employed plant pillows (Figure 12.1b,c) with quick disconnect fittings for the addition of water (typically 100–150 ml per pillow at initiation). The pillows are filled with a solid porous ceramic arcillite substrate (Turface Profile Products, Buffalo Grove, IL) that is sieved to 1–2 mm, washed, autoclaved, and mixed with slow-release fertilizer pellets for nutrient provision. One such mixture consists of 5 $g\,l^{-1}$ of 18N-6P-8K Nutricote T100 and 2.5 $g\,l^{-1}$ of $Ca(NO_3)_2$ polymer-coated pellets (Florikan, Sarasota, FL). Dry seeds are inserted

In-Space Manufacturing and Resources: Earth and Planetary Exploration Applications, First Edition.
Edited by Volker Hessel, Jana Stoudemire, Hideaki Miyamoto, and Ian D. Fisk.

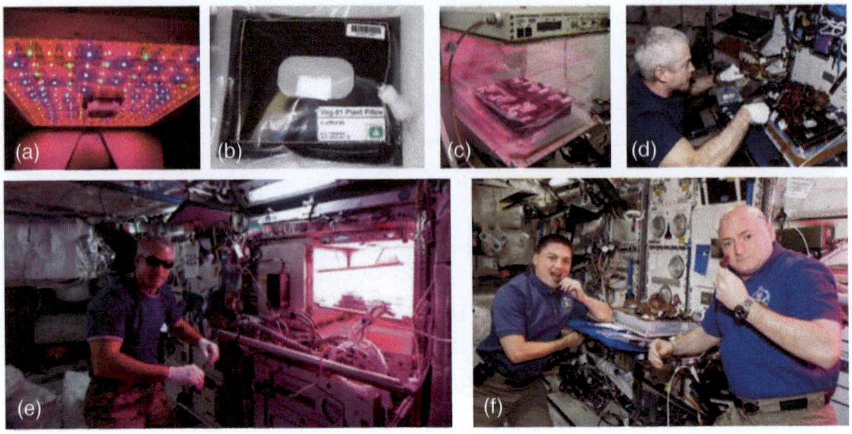

Figure 12.1 Vegetable production system (veggie). (a) Veggie light bank. (b) Veggie plant pillow with quick disconnect fitting for adding water. (c) Veggie with six plant pillows contained within adjustable bellows. (d) Astronaut Steve Swanson harvesting lettuce on ISS. (e) Astronaut Steve Swanson basking in the glow of veggie on ISS. (f) Astronaut Kjell Lingren (left) and Scott Kelly (right) enjoying freshly harvested lettuce on ISS. Source: NASA.

into the plant pillows and sealed in gas-impermeable bags prior to launch to the ISS. Using this approach the system has successfully been used to produce a variety of salad crop species for both crew consumption and research purposes (Figure 12.1d–f). On-orbit, astronauts manually pump water from a storage bag through quick disconnect fittings into either a 2 l root mat reservoir situated below the pillows, or directly into the plant pillows themselves at experiment initiation. The pillow and root mat reservoir each contain a membrane that when mated together are intended to passively supply water through capillary action from the reservoir to the substrate within the pillow. Ground control experiments are conducted at Kennedy Space Center (often with a 1–2 day delay from the flight experiment) in veggie hardware placed within International Space Station Environmental Simulator (ISSES) chambers that control temperature, relative humidity, and CO_2 levels based on data downlinked from the ISS.

Reliable seed germination generally requires an environment that is neither too dry nor too wet. However, water passage in the pillow/mat system can be inconsistent due to an occasional lack of good contact between the pillows and mat reservoir, clogging of the membranes with precipitates, biofilms, etc. Another deficiency of this approach is the inability to determine water content within the pillows. This results in a need to create a watering schedule based upon the use of ground-based empirically derived data, visual cues from on-orbit pictures, or emergency crew actions upon detection of underperforming (or wilting) plants.

Veggie has also supported outreach activities and petri plate-based science experiments mostly with the model plant species *Arabidopsis thaliana*. The plants are typically photographed at predetermined time points, harvested, and preserved within Kennedy Fixation Tubes (Zhou et al. 2019; Califar et al. 2020). In some instances, petri plates have undergone fluorescent microscopy within the ISS light microscopy module (LMM) (Simon Gilroy, personal communication).

12.2 PONDS Prototype Development Efforts

At NASA's Kennedy Space Center (KSC) we have developed another approach for growing plants within veggie called the passive orbital nutrient delivery system (PONDS). PONDS uses a root cylinder made from commercial off-the-shelf PVC fittings (Figure 12.2a) filled with the arcillite/fertilizer pellet mixture (Figure 12.2c) (or alternative substrates) described above for the veggie pillows. It is screwed into an exterior polycarbonate reservoir to which water is added. The reservoir can either be black or covered to minimize light penetration and thus limit algal growth within the system. Reservoirs included a coverable viewing port (with marked gradations) to permit inspection of the water level within and the progression of roots outward from the root cylinder into the exterior reservoir.

To enhance aeration of the root zone, oxygen-permeable membranes (in our prototypes we used TYVEK commercial wrap available from E. I. du Pont de Nemours and Company, Wilmington, DE) are incorporated into both the root cylinder itself (Figure 12.2a) and on the outside of the reservoir (Figure 12.2b).

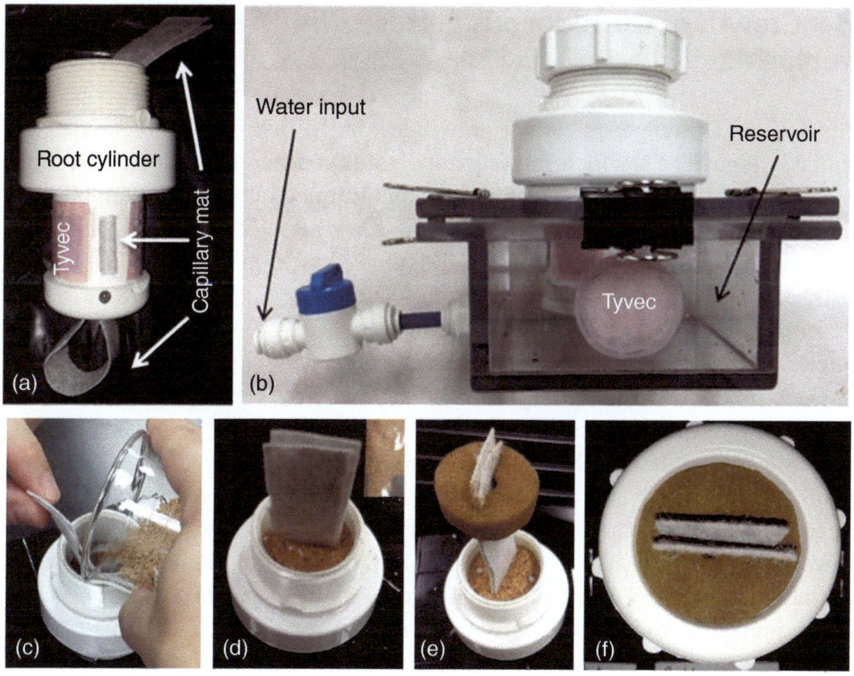

Figure 12.2 KSC PONDS prototype #1. (a) Root cylinder. (b) Side view of prototype #1. (c) Filling root cylinder with 1–2 mm arcillite. (d) Capillary mat wick extending up through the root cylinder. (e) Oasis rootcube foam plug insertion. (f) Top view of root cylinder prior to seed insertion. Source: Levine.

Critical to the functioning of PONDS is the use of hydrophilic wicks composed either of a capillary mat (Cap Mat) material (CAPMAT II, Phytotronics, Inc., Earth City, MO) or commercially available germination paper that passively link the water reservoir to the removable root cylinder. Water is delivered from the reservoir to the substrate contained within the root cylinders both from the bottom and from the sides using these wicking materials. The centrally located wick forms a loop beyond the bottom of root cylinder that extends upward through the root cylinder to its top (Figure 12.2a,d–f) where the seeds are inserted (and most often glued using either guar gum or other biocompatible glues). The system is therefore self-watering and continuously supplies water to the root cylinder.

At the top of the root cylinder a foam plug surrounds the wick at the seed insertion zone to contain the substrate within. In early prototypes the plug utilized ROOTCUBE foam medium (OASIS Grower Solutions, Kent, OH), which is hydrophilic and approximately 98% void. The plug's hydrophilic character facilitates removal of excess water in the vicinity of the seed insertion zone. This can mitigate the occurrence of excess water surrounding the seed that can inhibit germination, and is especially important under microgravity environments where a water film can form around seeds limiting oxygen availability and increasing the risk of microbial proliferation (Figure 12.2e,f).

12.3 Growth of Outredgeous Lettuce Within KSC PONDS Prototype #1

Results from an Outredgeous Lettuce growth test in the first PONDS Prototype are presented in Figure 12.3, which shows lettuce status at different days after planting (DAP) timepoints. Three seeds were planted and thinned down to one established seedling on DAP 6. Here, and in all following growth studies, the growth chamber conditions were: 85% red/15% blue LEDs (plus a low level of green LEDs); light intensity of 200–250 $\mu mol\ m^{-2}\ s^{-1}$; a light/dark cycle of 16/8; temperature of 22 °C; a CO_2 setpoint of 2000 ppm; a relative humidity setpoint of 40–45%. These last three parameters represent typical ambient conditions of the ISS which are not controlled by veggie.

The lettuce plant was harvested for the first time on DAP 39 (Figure 12.3d,e) and yielded 18 leaves with a fresh weight of 35.6 g (1.25 oz). Figure 12.3f shows how the plant looked immediately after this harvest. The second harvest (Figure 12.3h,i) occurred 22 days after the first harvest and yielded 18 lettuce leaves with a fresh weight of 28.8 g (1.02 oz). Figure 12.3j shows how the plant looked immediately after the second harvest and Figure 12.3k,l show when the flowering stalk first appeared and subsequently developed. After the second harvest there was a change in the phenotype of the lettuce leaves with a velvety morphology appearing (Figure 12.3k). Normal seed development occurred, demonstrating "self-pollination" and 95% of the harvested seeds germinated demonstrating "self-compatibility."

Figure 12.3 Outredgeous' lettuce test within PONDS prototype #1. (a) DAP 4: 100% germination of three seeds. (b) DAP 12: View of seedling remaining after thinning. (c) View of plant at DAP 20. (d) View of plant immediately before harvest #1. (e) Harvested lettuce leaves. (f) View of plant immediately after harvest. (g) View of plant at four days after harvest (DAH). (h) View of plant immediately before harvest #2. (i) Harvested lettuce leaves at 22 DAH #1. (j) View of plant immediately after harvest #2. (k) Initiation of flower stalk and appearance of "velvety" leaf morphology. (l) Fully developed flower stalk. DAH = days after harvest, DAP = days after planting. Source: Levine.

12.4 Growth of Red Robin Tomato Within KSC PONDS Prototype #2

Results from a Red Robin Tomato growth test in the second PONDS Prototype are presented in Figure 12.4. Flowers started opening on DAP 34, fruit formation was observed on DAP 41 (Figure 12.4c), and ripe tomatoes were obtained by DAP 86 (Figure 12.4f). The first harvest occurred on DAP 92 (Figure 12.4g), a second harvest occurred on DAP 185 (Figure 12.4i,j), and a third harvest occurred on DAP 283 (Figure 12.4k). For these extended-duration growth tests, MiracleGro™ nutrient supplements were added (at a dilution of 1:1000) into the exterior reservoir given that the fertilizer pellets are eventually exhausted. These supplements started at the time of flower initiation and were first performed once a week and subsequently progressed up to three times a week. This nutrient supplement approach using a commercially

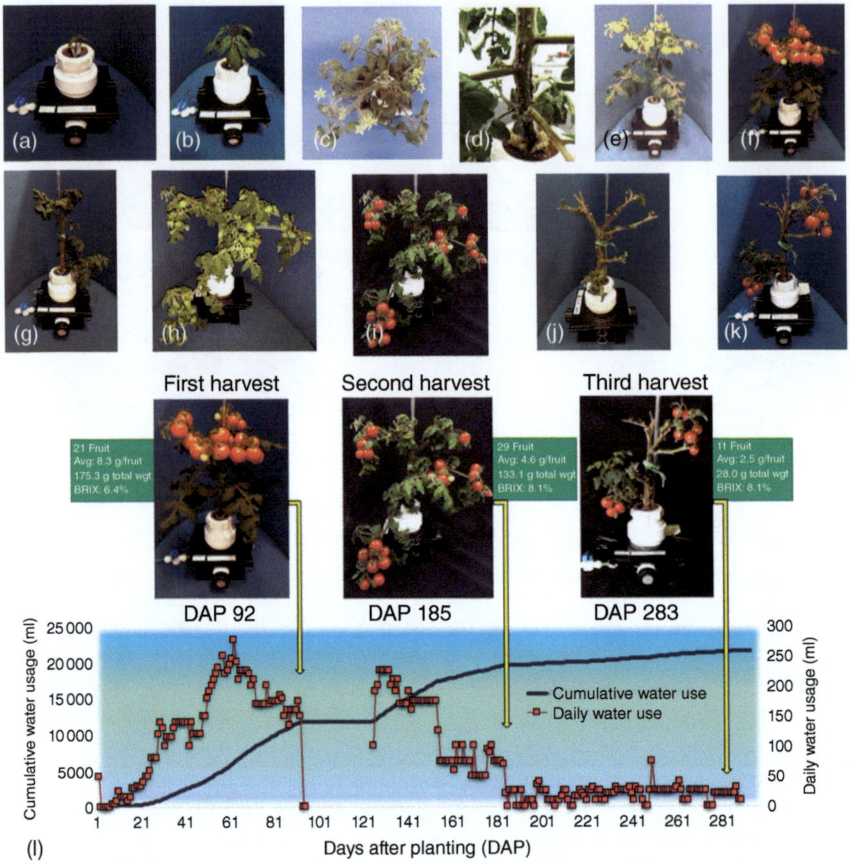

Figure 12.4 Red robin tomato test in PONDS prototype #2. (a) View of plant at DAP 13. (b) View of plant at DAP 24. (c) View of plant at DAP 41 with 25 flowers and appearance of first fruit. (d) View of thickened stem at DAP 48. (e) Plant at DAP 62 with immature fruit. (f) Plant at DAP 86 with ripened fruit. (g) Plant after first harvest on DAP 92. (h) Newly developed immature fruit. (i) Prior to second harvest at DAP 185. (j) Plant immediately after harvest #2. (k) Plant prior to third harvest at DAP 283. (l) Cumulative and daily tomato water usage over the course of the experiment. Source: Levine.

available product has great promise for the provision of nutrients to plants grown within PONDS units over prolonged periods of time. In Figure 12.4l the maximum water usage rate was found to approach a level of 300 ml per day, which PONDS was able to accommodate with crew water injections once every three days.

12.5 Growth of Zinnia and *Arabidopsis thaliana* Within KSC PONDS Prototypes #2 and #3

A veggie ISS experiment using the pillow approach resulted in the production of numerous "Profusion" Zinnia flowers (*Zinnia hybrida* cv. "Profusion"), much to

Figure 12.5 Zinnia PONDS prototype #2 growth test. (a) Single Zinnia seedling after thinning at DAP 8. (b) View of Zinnia plant at DAP 14. (c) View of Zinnia plant at DAP 35. (d) View of root cylinder lifted out from the reservoir section with roots growing outward into the reservoir visible. (e) View of multiple Zinnia flowers at DAP 144. Source: Levine.

Figure 12.6 *Arabidopsis thaliana* PONDS prototype #3 growth test. (a) 100% germination of three seeds at DAP 7. (b) Single seedling remaining after thinning on DAP 11. (c) Arabidopsis rosette stage at DAP 27. (d) After stem bolting and flower formation on DAP 41. (e) View of mature Arabidopsis plant with siliques containing seeds at DAP 71. Source: Levine.

the delight of the crew. We consequently undertook a Zinnia grow-out experiment within the KSC PONDS Prototype #2 unit (Figure 12.5) and demonstrated excellent Zinnia flower formation. Again, we were able to maintain prolonged growth and flower formation for close to a year using the MiracleGro nutrient supplementation procedure as was used for the tomato experiment.

The plant *Arabidopsis thaliana* has become the primary model organism for basic research into the physiology of plants both on Earth and in space. We therefore performed an *A. thaliana* grow-out experiment within the PONDS Prototype #3 unit to demonstrate to the plant research community that PONDS can support basic research studies with Arabidopsis. As can be seen in Figure 12.6, PONDS was able to support excellent Arabidopsis growth all the way to flowering, silique (seed pod) formation, and seed production.

12.6 Development of the PONDS Flight-Rated Units

To transition from the KSC PONDS prototype designs to a flight-rated version of PONDS (Figure 12.7a,b,d), a partnership was formed with the Tupperware and Techshot Corporations. The objective was to produce an easily fabricated and reasonably priced version of the PONDS design that was suitable for operation under microgravity conditions utilizing Tupperware's knowledge of advanced injection molding technology and food-safe materials. Critical to the success of this effort is the inclusion of Techshot as an experienced spaceflight hardware provider that guided Tupperware through the requirements and regulations required to

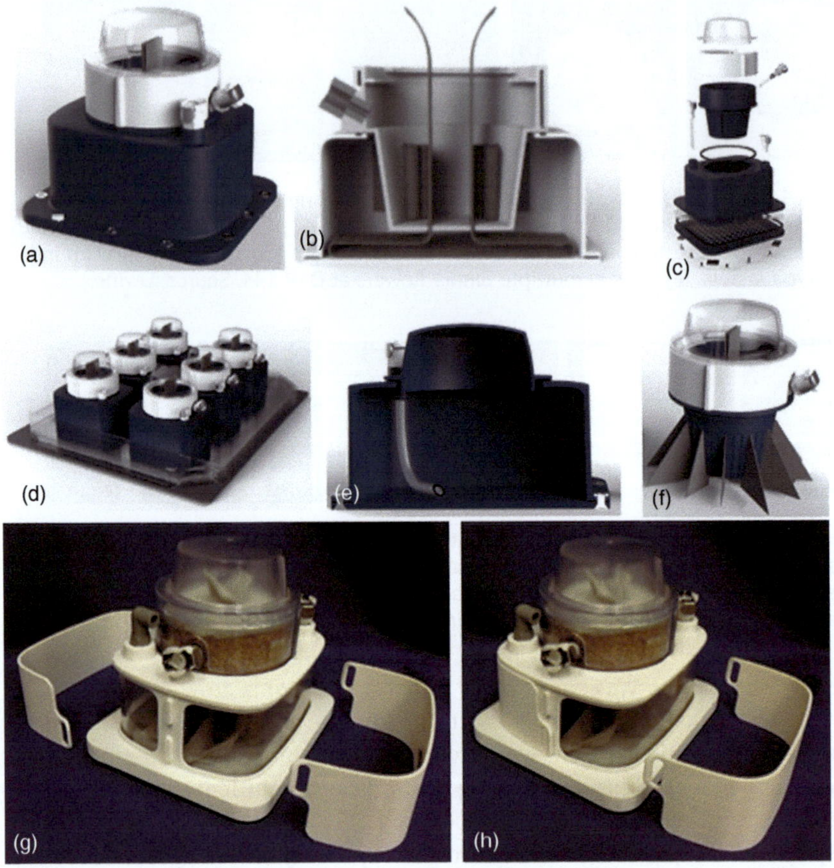

Figure 12.7 Flight-rated PONDS design as developed by Tupperware and Techshot Corporations. (a) Fully assembled flight design. (b) PONDS cross sectional view. (c) PONDS exploded view. (d) Conceptual image of six PONDS units prior to insertion into veggie. (e) Diagram showing placement of reservoir water introduction tube. (f) Diagram showing triangular Cap Mat fins extending outward from the root cylinder into the reservoir area. (g) PONDS flight version with transparent exterior reservoir and two thermal shroud covers removed to facilitate visualization of interior conditions. (h) PONDS flight version with one of two thermal shroud covers snapped in place to prevent light penetration. Source: Levine.

produce flight-rated hardware suitable for use on ISS. Both companies contributed significant engineering expertise to develop the final design, which included several iterations. Low-gravity fluid physics experts from NASA's Glenn Research Center and Portland State University were consulted for strategies to control water within the PONDS.

In the flight version of PONDS, the geometry, wettability of materials, and capillary action within the exterior reservoir, root cylinder, apertures and wicking elements were designed in a manner to ensure that surface tension causes water in the reservoir to fill and stay in the proper areas to facilitate water flow to the growing plant. Instead of screwing into the reservoir (as was done with the KSC prototypes), the root cylinder is now attached and removed using a bayonet mount connection. Also employing a bayonet mount approach is the Germination Cap (GC) cover that is secured above the root cylinder. The GC may be either opaque to prevent light penetration or clear to allow entry of light during germination and early seedling development. GCs ensure a high humidity environment during the seed germination phase and are removed and stowed once the seedlings are established.

Triangle-shaped fins composed of Cap Mat extend outward from the root cylinder (Figure 12.7f) and wick in water from the reservoir into the root cylinder. Their outer corners are closely adjacent or touch the sides of exterior reservoir. This provides access to water that may be situated distant from the root cylinder and helps to limit inertial sloshing of free water due to crew manipulations. A cylinder pad (also composed of Cap Mat) is inserted onto the bottom of the reservoir prior to closure to aid in the even initial distribution of water within the reservoir upon filling, and the bottom cover of the reservoir is secured with tank bolts. A gasket is inserted to prevent water leakage (Figure 12.7c).

Using quick disconnect ports, water (or nutrient solutions) can now be introduced either directly into the root cylinder (to initiate germination) or directly into the exterior reservoir. In the latter case, the quick disconnect port is attached to a tube that discharges at the base of the root cylinder (Figure 12.7e), which is domed to create an initial meniscus as the reservoir fills. To ensure this tube placement, the root cylinder includes an alignment tab having a notch to receive the free end of this water introduction tube.

The current design of the flight-rated PONDS units incorporates a white light reflective thermal shroud (Figure 12.7g,h) to minimize heating up of the unit while under high-intensity LEDs. In this implementation, the exterior reservoir has been made transparent so that the operator can look inside and assess how much water remains, where it is located, observe root growth out from the root cylinder, etc.

Water behaves differently in microgravity, which can cause problems especially if it blocks the reservoir's air vents, in which case there can be a pressure build-up as more water is introduced. The internal pressure could prevent further water flow into the reservoir or force water through the air vents and into the cabin. To prevent this, structural features of PONDS use the premise that water in microgravity is drawn to sharp geometric corners and is not drawn to curved smooth corners. The design of PONDS arranges these features in a way that causes the water within the reservoir to mimic a filling pattern similar to that obtained under normal gravity, even while under microgravity conditions.

It is critical that the speed with which filling operations occur is below a certain threshold such that it does not exceed the point where the water's inertia will be greater than surface tension bonds resulting in the water filling the reservoir in a chaotic manner. Moderately slow, steady manual operation of the filling syringe provides an acceptable flow rate to achieve the desired fill pattern. As the torus of water under the bottom end of the root cylinder (where it is introduced by the water input tube) expands radially outward, it encounters the radially inner lower edges of the Cap Mat fins and begins to wick up the fins to enter the root cylinder and supply water within. The result is that the PONDS design facilitates filling of water in a microgravity environment with fill characteristics similar to that seen under 1 g conditions.

12.7 Science Verification Test Using the Flight Version of PONDS

We performed a high-fidelity 29-day Science Verification Test (SVT) using the flight-rated PONDS units at KSC (Figures 12.8–12.10). Three lettuce seeds were planted in each PONDS unit. On DAP 0, 120 ml was added to the root cylinder to initiate germination. By DAP 5, 88% of the planted seeds had germinated (Figure 12.8a–c) and 100 ml was added to the exterior reservoirs to prevent the drying out of the seedlings. On DAP 8 (Figure 12.8d–f) the GCs were removed, seedlings were thinned to one per PONDS unit, and 100 ml added to the exterior reservoirs. Additions of 200 ml to the exterior reservoirs were made on DAP 15 (Figure 12.8g–i), DAP 19 (Figure 12.8j–l), DAP 23, and DAP 27. Figure 12.8m shows the PONDS exterior reservoir on DAP 15 before the addition of 200 ml and Figure 12.8n shows the PONDS exterior reservoir after the addition of 200 ml.

The final results of the 29-day SVT with the flight version of PONDS are shown in Figure 12.9. Figure 12.9a is a top view of the lettuce from 6 PONDS units that were situated within a single veggie. In Figure 12.9b a single lettuce plant from a PONDS Unit is shown, and in Figure 12.9c roots that have grown outward from the root cylinder into the exterior reservoir are evident.

A key measurement of plant growth is evapotranspiration (ET), which is the sum of water loss due to both evaporation and plant transpiration. Figure 12.10 shows the evapotranspiration data from 6 PONDS units over the course of the SVT study. Minimal water loss occurred until DAP 8 when the Germination Caps (labelled as RH Caps) were removed. From that point onward, ET increased at each growth interval as the plants grew and transpired water at greater rates based upon their increased biomass. The maximum water loss occurred in the H3 PONDS unit during the DAP 27 → DAP 29 growth interval, reaching an ET rate of 66 ml per day. These data are crucial for estimating the frequency with which the crew will need to supply water to the PONDS units.

12.8 Conclusions

The PONDS plant growth approach has been shown to facilitate water delivery to seeds for germination and subsequent transport of water from the reservoir for plant growth while providing nutrients and aeration to the root zone. PONDS units have successfully grown a variety of plant species, and the use of MiracleGro nutrient supplements added directly into the exterior reservoir was demonstrated to be capable of supporting plant growth within PONDS units over prolonged periods of time. Once validated under microgravity conditions, PONDS will facilitate a variety of plant growth studies on ISS. It is an entirely passive device with no power requirements

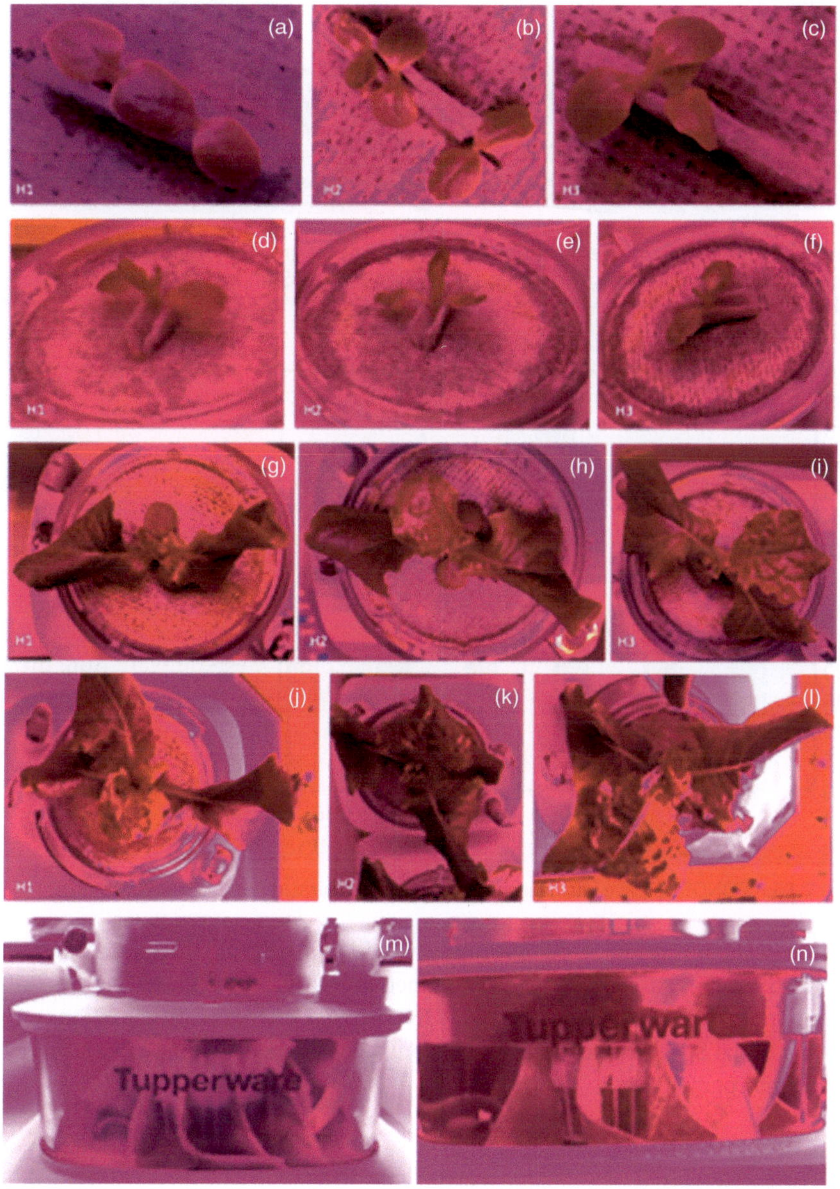

Figure 12.8 PONDS science verification test at Kennedy Space Center. (a–c) Views of germinating seedlings from PONDS units H1, H2, H3 at DAP 5. (d–f) Views of remaining seedlings after thinning from PONDS units H1, H2, H3 at DAP 8. (g–i) Views of plants from PONDS units H1, H2, H3 at DAP 15. (j–l) Views of plants from PONDS units H1, H2, H3 at DAP 19. (m) Side view of PONDS unit on DAP 15 before addition of 200 ml water into the exterior reservoir. (n) Side view of PONDS unit on DAP 15 after addition of 200 ml water into the exterior reservoir. Source: Levine.

Figure 12.9 PONDS science verification test after 29 days of growth. (a) Top view of lettuce grown in 6 PONDS units within veggie. (b) Single PONDS unit on scale at harvest. (c) Side view showing roots in reservoir. Source: Levine.

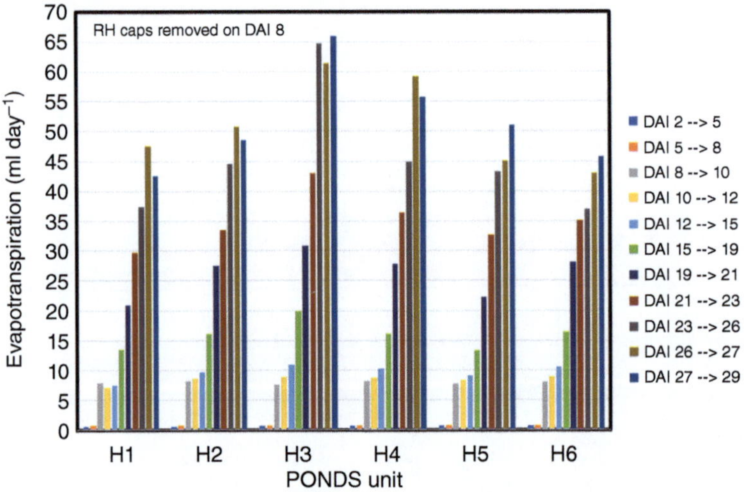

Figure 12.10 PONDS science verification test evapotranspiration rates. Source: Levine.

or moving parts. It is anticipated that the crew will enjoy their participation in the plant growth studies based upon results obtained thus far from veggie ISS studies. In 2022 we hope to validate the operation of PONDS under microgravity conditions on the International Space Station to facilitate continued evaluations of plant and crop growth in microgravity.

Acknowledgments

Many thanks to members of KSC's Utilization and Life Science Office (UB-A), the NASA Biological and Physical Sciences (BPS) Division and the ISS Program Office for support of the PONDS Development Project.

References

Califar, B., Sng, N.J., Zupanska, A. et al. (2020). Root skewing-associated genes impact the spaceflight response of *Arabidopsis thaliana*. *Frontiers in Plant Science* 11: 239.

Khodadad, C.L., Hummerick, M.E., Spencer, L.E. et al. (2020). Microbiological and nutritional analysis of lettuce crops grown on the International Space Station. *Frontiers in Plant Science* 11: 199. https://doi.org/10.3389/fpls.2020.00199.

Massa, G.D., Newsham, G., Hummerick, M.E. et al. (2013). Preliminary species and media selection for the Veggie space hardware. *Gravitational and Space Research* 1 (1): 95–106.

Massa, G.D., Wheeler, R.M., Stutte, G.W., Richards, J.T., Spencer, L.E., Hummerick, M.E., Sirmons, T. (2015). Selection of leafy green vegetable varieties for a pick-and-eat diet supplement on ISS. *ICES-2015-[252], 45th International Conference on Environmental Systems*, Bellevue, WA, 1–16.

Massa, G.D., Wheeler, R.M., Morrow, R.C., and Levine, H.G. (2016). Growth chambers on the International Space Station for large plants. *Acta Horticulturae (ISHS)* 1134: 215–222.

Massa, G.D., Dufour, N.F., Carver, J.A. et al. (2017). VEG-01: Veggie hardware validation testing on the International Space Station. *Open Agriculture* 2: 33–41.

Morrow, R.C., Remiker, R.W., Mischnick, M.J., Tuominen, L.K., Lee, M.C., Crabb, T.M. (2005). A low equivalent system mass plant growth unit for space exploration. *SAE Technical Paper 2005-01-2843*.

Zhou, M., Sng, N.J., LeFrois, C.E. et al. (2019). Epigenomics in an extraterrestrial environment: organ-specific alteration of DNA methylation and gene expression elicited by spaceflight in *Arabidopsis thaliana*. *BMC Genomics* 20: 205. https://doi.org/10.1186/s12864-019-5554-z.

13

Space Food for the Future: Nutritional Challenges and Technological Strategies for Healthy and High-Quality Products

Paola Pittia[1] *and Martina Heer*[2,3]

[1] *University of Teramo, Faculty of Bioscience and Technology for Food, Agriculture and Environment, via Renato Balzarini 1, 64100 Teramo Italy*
[2] *International University of Applied Sciences, Kaiserplatz 1, Bad Reichenhall, Germany*
[3] *University of Bonn, Institute for Nutritional and Food Sciences, Nussallee 9, Bonn, Germany*

13.1 Introduction

The availability of high-quality food with high nutritional value is essential for the humans on Earth and even higher for space travelers during their missions as an adequate energy and nutrients intake counteracts the weight loss and minimizes negative effects on the metabolic and health status that, in turn, could impair the working and mental performances of astronauts. Despite the advancements in the food supply and space food processing, space travelers still lose about 5% of their body mass during six-month space missions because of insufficient energy intake, and this corresponds to a reduced nutrients consumption (e.g. vitamins, minerals, bioactive compounds).

Many potential explanations have been proposed that might lead to less food intake during space missions [1]. Among others, stressful environmental conditions, reduced appetite, lowered palatability of foods, changes in sensory perception of the astronauts, or restricted variability of food options have been considered. The availability of varied and high-quality food with high palatability is one way to improve appetite, increase food intake, and ensure adequate nutrition, but it is known that food variability might decrease for long-duration missions and this may lead to meal fatigue.

Space food systems have also to comply with high-level safety requirements while keeping qualitative properties and sensory acceptability for storage times matching the duration of space missions. In general, conventional and intense food technologies such as canning, drying, and freeze-drying are proven technologies used for space foods as they guarantee the safety and long storage time requirements. However, it is known that these processes reduce both the initial nutritional value and qualitative properties (in particular, the sensory ones, e.g. aroma, taste, and color)

In-Space Manufacturing and Resources: Earth and Planetary Exploration Applications, First Edition.
Edited by Volker Hessel, Jana Stoudemire, Hideaki Miyamoto, and Ian D. Fisk.

of the foods when compared to the fresh ones, and trigger degradation reactions that will develop during the space missions, thereby diminishing their acceptability. Long-term stability, quality, and acceptability of space foods could be impaired also by other factors related to the space ecosystem (e.g. radiation, environmental CO_2) that have been studied only to a limited extent so far.

Currently, the interplay of factors of different nature, whose cause and effects have not yet been unraveled, makes rather complex the matching of the nutritional needs to keep the good nutritional status of space travelers and reduce weight loss with the space food systems quality available. On the other side, the implications of reduced dietary intake are not only related to the lower energy intake but also to the loss of lean body and fat mass along with a series of effects on metabolic pathways that could affect directly or indirectly the health status of the crew (Figure 13.1).

Thus, it is important both to better understand the role of the various factors that contribute to the adequate food intake and to find solutions by favoring a close interaction between various disciplines including food sciences, technologies, nutrition, and health.

The application of alternative innovative food processes to produce high-quality and safe products for space missions meeting the nutritional, quality, and safety requirements and higher sensory acceptability is a key solution. Currently, innovative technologies (e.g. high hydrostatic pressure, microwave-assisted sterilization), hurdle technology approaches, and new packaging materials could represent an interesting option. The enrichment of foods with micro- and macro-nutrients as well as of bioactive components (e.g. probiotics, antioxidant compounds) to fulfill the nutritional requirements is a modern strategy to enhance the nutritional status of space foods.

Moreover, recent scientific evidences highlighted the need to provide space foods with an adequate bioavailability and bioaccessibility of nutrients in order to

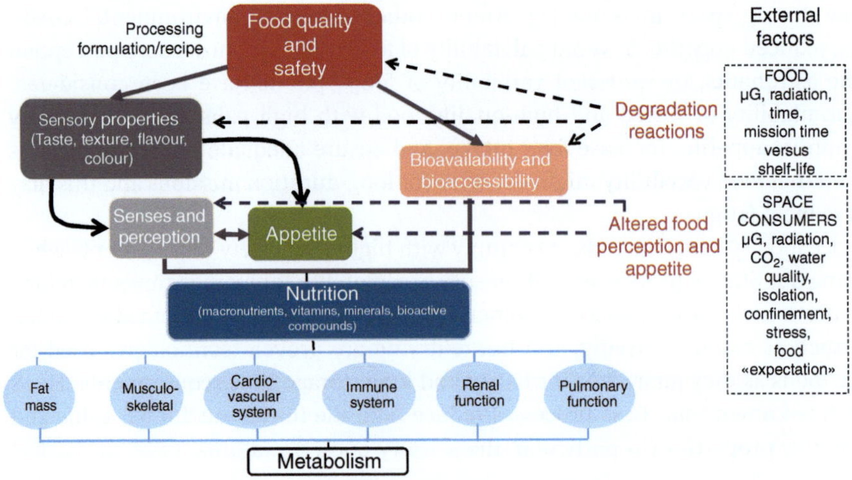

Figure 13.1 The relationship between food properties, nutrition, and metabolism.

maximize the nutritional value of foods, aspects that, thus, need to be taken into account when designing and developing future products for space missions.

Based on the current scenarios and scientific knowledge in the field, this chapter will review the nutritional challenges of space crew, the main space food types, and the related technologies for their manufacturing taking into account critical qualitative and safety aspects, storability, and limits in balancing the energy and metabolic requirements to maintain the health status. Eventually, the perspectives in the design and development of space foods for future missions to better meet the nutritional and metabolic challenges of astronauts will be briefly discussed.

13.2 Nutritional and Metabolic Challenges of Astronauts

Nutrition together with other environmental influences (e.g. microgravity, noise, temperature, air quality) impacts all physiological systems and the resulting physical and behavioral health and performance. Thus, as on Earth, providing an adequate amount of nutrients tailored to the respective demands is mandatory to maintain physical health and performance. Inadequate nutrition during long-duration missions, such as to Moon or Mars or beyond, might lead to serious health problems and eventually to mission failure as seen with some pioneering missions on Earth.

Today, most astronauts lose body mass during their mission (on average six months) on the International Space Station (ISS). More than half of 79 astronauts examined on ISS lost 5–15%, compared to their pre-flight body mass mainly because of low caloric intake [2]. Insufficient dietary energy intake leads to degradation of fat mass, but also of lean body mass as an energy source. The loss of lean body mass during space missions is significant because of both insufficient caloric intake and lower physical activity as well as an increased proteolysis and catabolism related to oxidative stress [3]. Inadequate energy intake, even at a moderate level, leads to bone loss to a similar degree as does microgravity [4]. Results from space analog studies have shown that caloric intake also weakens the cardiovascular system, leading to reduced orthostatic tolerance [5].

13.2.1 Energy Expenditure, Exercise, and the Musculoskeletal System: Find a Good Balance

Total energy expenditure mainly consists of basal metabolic rate (i.e. the energy needed to maintain life) plus the energy needed for physical activity. Lowering mechanical loading such as in microgravity or immobility leads, aside others, to muscle atrophy and bone loss [6]. However, landing on the Moon or Mars or getting back to Earth requires certain muscle and bone strength. During space missions muscle and bone mass therefore should be maintained by an adequate exercise regimen [7]. With higher physical activity levels, energy expenditure, that also tends to increase with the duration of the mission, will rise and more food has to be provided to match energy expenditure. Inadequate caloric intake already at moderate exercise levels [2] could cause muscle mass degradation and muscle and bone loss [8, 9] which could lead to not favorable effects during microgravity life. Finding the right balance between the exercise regimen needed to maintain muscle and bone mass and strength and other physiological systems and a minimum impact on energy expenditure is highly desirable [8].

Maintaining muscle mass as a result of the balance between protein synthesis and breakdown needs an adequate dietary protein intake. Increase in dietary protein intake might positively affect protein synthesis and thereby muscle mass [10, 11]. However, whether a rise in protein intake leads to higher protein synthesis depends on the dietary protein intake level consumed before [12]. In spaceflight the average protein intake already exceeds recommended levels [2], and this might lead to side effects such as a translational and moderate reduction in blood pH levels and, in turn, by an increased osteoclasts activity, to lower bone mineral density [13–15].

13.2.2 Immune

Adequate nutrient supply is extremely important for an optimally functioning immune system. Historically, in spaceflight crews usually are not optimally nourished with an energy intake that is on average 25% below the energy requirements which could play a significant role in host immune response and resistance to infection because of the protein synthesis decreases [16]. Insufficient energy intake is also correlated with increased free radical propagation because of a diminished protein-based antioxidant defense mechanism [17]. Although protein intake is, on average, exceeding the recommendations in space missions, it is important to guarantee the dietary intake of essential amino acids that might enhance immune function. Concentrations of glutathione, the endogenous antioxidant which plays an important role in antioxidant defenses, were lower in limited number of mice flown to ISS suggesting an adaptation to increased oxidative stress [18], and the effect of providing food high in cysteine and glycine, the amino acids from which glutathione is synthesized in the body in supporting the immune function, is unknown . The ISS food system is rather low in fruit and vegetables and rich in meat and meat products. Although there is no scientific evidence demonstrating that higher consumption of fruit and vegetables improves immune function, there are several studies demonstrating a positive impact [19].

13.2.3 Microbiota

The gut microbiota enjoys a symbiotic relationship with its host: the latter provides the microbiota with a stable environment and the former provides functions such as digestion of complex dietary components, production of nutrients and vitamins, and defense against pathogens. The microbiota also supports maintenance of the immune system, and can also affect cognition and behavior [20, 21].

The Astronauts' Microbiome project studied the impact of long-term space travel aspects (e.g. microgravity, g-forces, radiation, stress, and anxiety) on the microbiota of crew members and surrounding ISS environment and the consequences on human health. A change in the microbiota composition of the gastrointestinal tract, skin, nose, and tongue during flight was observed, and it became more similar between astronauts [22]. Space travel alters the intestinal microbiota and, thus, the associated metabolic and immune functions. Different factors can affect astronaut gastrointestinal microbiota including the more sterile space food system than a traditional diet on Earth due to the need to inactivate microorganisms and promote stability [23] and a diet low in fiber [2], which may lead to a decrease in metabolites associated with the intestinal microbiota such as short-chain fatty acids, as well as the immune response against pathogens [24] and the prevention of some diseases with immunological components like cancer [25].

13.3 Space Missions and Food Supply

Foods are primary component of the health and wellbeing of each individual on Earth as well as for crewed space missions, and an adequate food supply is essential to deliver long-term success [1]. From the first space mission of a very short duration (Gagarin, 108 minutes), where the food provided was a pureed meat and chocolate sauce in squeeze-tubes, to the more recent ones, both the spaceflight length and the number of crew members changed along with the evolution of the space food systems with the availability of differently packed and stabilized products (thermostabilized, irradiated), equipment to heat or to keep at chilling or freezing temperatures and spaces to store foods in the spaceship.

A significant step in the space food system history was the establishment of the ISS in 1998, with a range of meals and a supply chain from Earth to the ISS, designed to feed around six astronauts at a time [23]. In the following years the provision of food has been further evolved to provide nutritious, tasty food in different formats, and more than 200 different menu items are now available on the ISS [23].

The current plan of the U.S. National Aeronautics and Space Administration (NASA) is for a crewed mission to Mars and back lasting around 2.5 to 3.0 years that will be a rather challenging task for the food supply chain taking into account the impossibility of a resupply chain, the amount of food required to nourish the space crew *vs.* the limited space to store it as well as the still unknown effects of the environmental conditions (e.g. high level of radiation) that could compromise the nutritional and sensory quality of the food itself and impair its nutritional value [26, 27].

13.4 Space Food Types: Processing Aspects and Storability

In Table 13.1, the full list of the main space food categories, their packaging, preparation requirements, and storability (or "shelf-life") are summarized.

Table 13.1 Space food classification and the related food process to manufacture them, packaging conditions, and related shelf-life.

Food type	Technology/ies	Packaging	Storage conditions	Preparation requirements	Average storability/ shelf-life
Rehydratable foods	Drying, Freeze-drying	Flexible plastic pouches	Ambient	Re-hydration (ambient, hot water)	Up to 36 months
Intermediate-Moisture Foods (IMF)	Drying, baking, mixing with humectants (e.g. sugars, salt, glycerol), osmotic dehydration, alone or in combination	Plastic pouches Aluminum cans and tubes	Ambient	no	Up to 12 months
Thermostabilized	Sterilization/canning	Metal (aluminum, tin) can Retort pouches	Ambient	Re-heating	Up to 36 months
Irradiated	High-energy radiations	Plastic pouches	Ambient	Re-heating	Up to 1 year
Natural form foods	Natural drying; use of humectants, baking/cooking (for cookies and granola bar)	Plastic pouches	Ambient	No	Up to 10–12 months
Chilled	Refrigeration	Plastic pouches	$T < 4\,°C$	No/re-heating	Up to 30 days
Fresh foods	Not applicable	Plastic pouches	Ambient	No	Up to 1–2 weeks

T: Temperature.

The key requirements for space food that need to be satisfied before the start of the space flight and kept for the entire duration of the mission include:

- *Safety*: foods aimed at crews in space missions have to guarantee the absence of microbiological, chemical, or physical risks. Thus, since the earlier missions, space foods have been obtained by process technologies and preparation conditions able to achieve the highest safety and lowest health risk, mainly as referred to the presence of pathogenic microorganisms and potential growth during storage and/or related illnesses.
- *Stability*: foods must be stable for the entire duration of the space mission, i.e. no main or significant degradative reactions (chemical, biochemical) and no microbial growth has to occur affecting the food safety and quality. Along the space history the duration of the missions has significantly prolonged and the new scenarios forecast three- up to five-years space flights. The conventional and more intense processes applied to obtain some space food types (e.g. thermostabilized, freeze-dried) allow to keep a good quality and nutritional value up to two to three years. Longer times are a challenging task based on the current knowledge and availability of food technologies and, thus, modern approaches are required. For the European Space Agency (ESA), the current minimum shelf-life on launch day is at least nine months without preservatives [28].
- *Nutritional value*: the highest one compatible with the need to guarantee safety and stability.
- *Sensory properties*: in general flavor, taste, and texture may affect palatability and acceptability of food and, indirectly, also the appetite.

The history of space foods and related manufacturing process has developed based on these fundamental aspects along with the availability of packaging able to meet the specific needs and the main constraints related to the volume occupancy, weight, availability of water, and/or equipment to store foods. As regards packaging, it is known its role in contributing to some of the requirements of space food, e.g. the protection from environmental contamination (especially after sanitization), sizing/portioning for the individual crew members, and the lowering the risk of food particles and crumbs spilling into machinery during eating. Metal (e.g. cans, tubes) and high-barrier, flexible plastic materials were mainly used and their choice depends also on the nature of the space food system, the technology applied to produce it, and the expected stability.

The first foods launched to space were mainly freeze-dried or dried products and intermediate-moisture foods. In the Mercury project three foods belonging to these categories were used: i.e. a concentrated, low-moisture applesauce packed in a tube [29], partly dried compressed food cubes [2], and freeze-dried powders, which can be eaten after rehydration [30].

From the Apollo missions (1968–1972) the study of physiological processes in microgravity environment, the advancements in processing technologies and availability of new packaging materials, and the need of sensory satisfaction of the crew contributed to a progressive improvement in space food types availability and variety. Thermostabilized foods, stable at ambient temperature, were introduced in the space food list thanks to the use of retort pouches and cans, and the availability of hot water made it easier to rehydrate the dehydrated products [31]. Refrigerators, freezers, and heaters were set in Sky Lab cabinets and, thanks to this, crews could enjoy fresh and frozen foods (e.g. ice cream, filet steak, lobster, chilled drinks, and desserts). However, the high-energy demand of these apparatus along with the short shelf-life of chilled foods represent a main limit in their usage, especially in a long-term space flights perspective.

The main space food type characteristics and processing aspects (see Table 13.1) are here below briefly described.

Rehydratable foods include foods and dishes that are preliminarily dried or freeze-dried to remove water and then rehydrated during the mission using the potable water supply. Thanks to the development of the modern freeze-drying technologies, rehydratable foods include a large variety of staple foods and beverages (e.g. tea) as well as side and mixed dishes (chicken and vegetables), desserts, ice-cream, and cornbread dressing [32]. The advantages of these products are their stability at ambient temperature thanks to the very low water activity (<0.2) and moisture content that limits degradation reactions and microbial growth, low weight and ease in their preparation as they are packed in trait packs with a one-way intake valve bags for rehydration by ambient or hot water. Contrary to conventional drying technologies, freeze-drying causes desiccation by water sublimation of a food preliminarily frozen under low pressure and temperature and maintains some of the structure of the product. The process conditions applied guarantee a long shelf-life of the products if packed under high barrier properties against moisture uptake, a minimum nutrients loss, the retention of the flavor compounds and pigments that lead to high-quality and nutrient products.

Natural form foods comprise fresh products, like fruits and vegetables, dry fruits (e.g. nuts) as well as some processed "natural" products (e.g. cookies or granola bars). Fresh vegetables and fruits need refrigeration and have a short shelf life, but they are important for the psychological status of the astronauts [33]. On the contrary, cookies and granola/fruit bars are processed and/or compressed to obtain products stable at ambient conditions, easy to be stored (low volume) while from the nutritional point of view they are in general formulated as mix of high protein ingredients, fat, sugars, fruit and nuts. Different from the fresh foods, these products lose the naturalness during processing and may have a relative low sensory acceptability; moreover, their high-energy value needs to be considered in the daily food intake.

Intermediate-Moisture Foods are products either processed by dehydration and/or formulated with humectants (i.e. ingredients able to "bind" free water like sugars, glycerol, salt). Their relatively long stability is based on the relatively low moisture content and water activity (<0.85) and/or by a combination of various "hurdle" effects obtained by processing or formulation (e.g. low pH, natural presence of antioxidants) to inhibit microbial growth and biochemical degradation reactions without the need of refrigeration. Examples are dried fruits (e.g. dried dates and figs), jams and jellies, marshmallows, dry meat products, fruit cakes. Traditional or conventional food processes (drying, evaporation, mild heating) are generally used to obtain these products. From the nutritional point of view, the amount of salt or sugar required to achieve the low water activity significantly modifies the sensory (flavor, texture) properties and acceptability along with their nutritional value in respect to the corresponding fresh products.

Thermostabilized foods represent a large part of the space menu. NASA-thermostabilized products include pouched soups, sides, desserts, puddings, and entrees [23], and their variety has progressively increased with the inclusion also of gourmet recipes. They are products obtained by intense heat treatments of sterilization ($T > 110\,°C$), and the process conditions determine the complete thermal death of all pathogenic and spoilage microorganisms as well as of spores and the inactivation of degradative enzymes to achieve shelf-stable products. Packaging, in metallic cans or thermostable flexible pouches, is made either before or after the stabilization process. In terms of moisture and appearance they are the products that are closer to the fresh ones even if their formulation includes thickening agents to reduce the influence of weightlessness and the intense heat treatment causes significant changes in the qualitative, nutritional, and sensory properties. The lower content of thermolabile vitamins, excessive softening (due to a side effect of over-cooking), and changes in aroma and flavor are the main drawbacks of these products, aspects that could reduce their nutritional value and overall acceptability.

Irradiated products: they are products subjected to high-energy rays produced by ionizing radiation. This technology is applied and allowed to process foods only to a limited extent and for stabilization purposes thanks to the significant microbial lethality effects [23]. The Food and Drug Administration (FDA) in 2009 allowed NASA for the preparation of nine irradiated items to commercial sterility including a cooked steak, packed under vacuum in flexible pouch, subjected to irradiation at ambient temperature, currently used on the shuttle [34]. It is recognized that this technology, which can be applied also on packed products, in general causes only minor effects on food quality properties. However, for sterilization purposes the effective dose of irradiation, which depends on the product and the radiation source, is rather high and this can trigger some reactions (e.g. oxidation of lipids) that could continue to develop during storage and affect the sensory acceptability, the health and nutritional value of the products.

Refrigerated and frozen foods: these are products whose nutritional value is generally high, but they have been made available only in the Skylab missions as the spacecraft are not generally equipped with refrigerator or freezer apparatus. Technology advancements are providing innovative refrigerators and freezers intended for use in space missions that allow the storage of foods that require, for their safety or stability, chilling or frozen conditions [35]. This is enhancing the variety of dishes of the daily diet with freshly made foods and frozen foods like ice-creams.

Eventually, fresh foods (fruits and vegetables) have been included in space missions but only in limited amount due to their intrinsic short shelf-life, mainly for a psychological support in the initial time of the space missions and not for meeting nutritional requirements [23].

13.5 Safety Management, Quality, and Stability of Space Food Systems

Space foods have to be intrinsically safe, nutritious, and acceptable to avoid any negative effect on crew health status and performances during the overall spaceflight [36]. To tackle this challenge safety and quality management procedures have to be coupled with the selection and adoption of process technologies applied during food manufacturing and/or dish preparation that could guarantee the required level of safety, nutritional value, and acceptability.

Hazard Analysis and Critical Control Point (HACCP) was developed in the late 1960s as a systematic and preventive approach to the identification, assessment, and control of hazards, in particular those related to microbial contamination and growth, and to guarantee "zero defects" of space food [23]. Since then the HACCP principles are used and applied for food safety control in many countries all over the world. The use of HACCP principles accompanied by strict good manufacturing practices as well as Quality Control and Assurance approaches is associated with the prevention of foodborne diseases during space missions.

Food processing, while it determines the required food safety and desired stability, is also negatively affecting the final quality of the foods and their acceptability. Actually, it is recognized that some process conditions, like those required to achieve the desired safety (e.g. high temperatures, high-energy radiations), could degrade labile biomolecules and nutrients that could compromise, to a certain extent, the nutritional and healthy value of the products.

The acceptability of foods is a complex concept involving external (e.g. diet and cultural habits, environmental, timing, social context), individual, and food-related aspects. Currently the sensory (or 'organoleptic') properties (e.g. taste, texture, aroma) of space foods are evaluated on the ground by a panel of at least 30 consumers [23]. However, the space environmental conditions (e.g. competing odors in the cabin) and the psychological and physiological (sinus congestion due to fluid shifts) conditions of the crews could impair their flavor perception. Moreover, the evolution of the degradation processes during the mission duration could develop off-flavors or modify quality properties that significantly reduce food acceptability and food intake and, thus, impair the nutritional and health status of astronauts [1].

In Table 13.2, the main processes applied to space foods are listed along with the effects on qualitative and nutritional attributes that could negatively affect their acceptability both just after their production and during storage.

Processed space foods are characterized by rather different storability or shelf-life (see also Table 13.1), which in some cases, depend on environmental conditions (e.g. refrigerated) and packaging characteristics.

In general, intense processes such as canning and drying are proven technologies used for long shelf-life microbial stability that guarantee a high safety level of the final product. However, they can induce degradation reactions during processing that reduce both their nutritional value and qualitative properties (e.g. flavor, taste, and color). Moreover, degradation reactions and phenomena could occur and evolve during the spaceflight mission time and lead to the formation of undesired

Table 13.2 Effect of processing on quality and nutritional attributes of space food systems, changes induced by processing during storage, and alternative technological approaches for future high-quality space food products.

Process	Effect on flavor and sensory properties	Effects on nutritional and health value	Qualitative changes during storage	Alternative and innovative technological solutions
Thermo-stabilization (sterilization/ canning)	• Main degradation natural aroma and pigments. • Formation of off-flavors (e.g. WOF in meat), undesired colors • Texture (hardening/softening) • Water loss	• Main reduction of thermolabile macro- and micro-nutrients (e.g. vitamins) • Formation of oxidation compounds	• Lipid oxidation • Vitamins loss • Color and flavor changes	• Innovative stabilization systems (Pressure assisted-microwave assisted-sterilization; -irradiation) • Use of natural antioxidant compounds/ingredients
Drying	• Degradation and loss of natural aroma compounds • Formation of undesired volatile compounds (e.g. Maillard/non-enzymatic browning reactions) • Hardening • Structural collapse	• Main reduction of thermolabile macro- and micro-nutrients (e.g. vitamins) • Formation of oxidation compounds • Loss of bioavailability and bioaccessibility of nutrients	• Loss of rehydration ability • Lipid oxidation • Vitamins loss • Color and flavor changes	• Alternative mild drying technologies (e.g. micro-wave drying)
Freeze-drying	• Loss of aroma compounds (to a limited extent)	No/limited loss of vitamins	High tendency to water uptake and to oxidize during storage (proper packaging need)	Microwave-heating
Irradiation	• Limited formation of undesired volatile compounds (e.g. from oxidative reactions)	• Limited degradation of micro-nutrients	• Lipid oxidation • Vitamins loss • Color and flavor changes • Loss of nutritional value	Thermo-irradiation
Intermediate-Moisture Foods	• Change of natural aroma and pigments due to degradation/ formation of new volatile compounds • Hardening and texture changes (if partially dried)	• Change in nutritional value (higher content of salt and sugar) • Loss of bioactives	• Loss of nutritional value • Lipid oxidation • Vitamins loss • Color and flavor changes • Loss of nutritional value	Hurdle technology

off-flavors, or change the physical, structural, and colloidal properties of the product (e.g. starch retrogradation and crystallization with hardening, coalescence of lipid separation of emulsified products) that overall may reduce the effective duration of the product based on the sensory acceptability.

Accelerated shelf-life test carried out on 13 thermostabilized space foods items (meat, fruit, vegetable sides and starch dishes, egg products), with long shelf-life stored at different temperatures, highlighted a different stability and acceptability over time, and actually only meat, starchy and vegetable dishes maintained their quality over three years without refrigeration with the appearance of off-flavors being the main cause of unacceptability [37]. If five year-time space missions are expected to occur, food space shelf-life is still a challenge to be tackled and alternative process technologies and/or food production need to be developed.

Despite the main impact on mass, volume, and waste allocations in space missions, packaging plays a main role in maintaining safety, nutritional value, and sensory acceptability of space foods. It acts as physical barrier against external (e.g. microbial, foreign materials) contamination, as well as environmental factors (e.g. oxygen, light, moisture) that trigger food degradation or qualitative changes that impair storability and/or usage. Moisture uptake in dried products could limit their rehydratability, while the entrance of oxygen in the package contributes to fats oxidation and rancidity of the product. Metallic, impermeable packaging is used for some space foods (e.g. thermostabilized cans) and as alternative plastic films with a high barrier properties are utilized for making flexible pouches or bags. Oxygen and water vapor permeability limits have been imposed by NASA. However, for storage purposes and to achieve a higher protection against moisture and gas permeation/diffusion, two packaging are also used [23].

New packaging materials with low mass, high barrier (e.g. metallized, aluminum laminated) improved protection abilities and fitting waste management in the space (incineration) are under development, while innovative technologies and active packaging could be a possible alternative or an additional action to be developed.

13.6 Future Trends in Food Supply to Meet Nutritional Challenges

Traveling to Moon and beyond as planned for twentieth and thirtieth centuries by several nations provides a real challenge for food supply and thereby for an adequate provision of energy, nutrients, and bioactive compounds. Today, astronauts on the ISS are provided with a standard meal (about 80% of their food) plus a personal container (one per month) to meet the respective individual taste of each astronaut (about 20% of the food amount) [38]. A food warmer is provided which makes food, which is intended to be eaten warm, more palatable. Whether a food warmer will be available for future exploration missions is not yet clear. To ensure the somewhat tailored menu for the astronauts, ISS is regularly resupplied with food and even some fresh fruits and vegetables every three to four months, but for longer missions (i.e. Moon, Mars) it will be impossible because of the distance. Carrying all food for three to five astronauts for up to three years in the rocket corresponds to tons of food that cannot fit the weight and volume availability.

However, for mission success it is essential to supply adequate nutrients. Growing food in a greenhouse, such as vegetables or fruit, could be one way of providing nutrition for exploration missions. Additional alternatives under study are insects as a new valuable protein food source, which would be reared and used for different products as an ingredient.

13.7 Food Processing and Design for Future Space Food

A long-term crewed mission to Mars and back is a demanding target also in terms of the availability of processed space foods able to guarantee a sufficient high quality and palatability for a longer shelf-life than the actual and to counteract the negative effects on the nutritional status and metabolisms of the crew. Improvements in conventional technologies and alternative and innovative processing solutions for space foods are, thus, needed and currently under development with promising results.

Freeze-drying has been used for food production since the early space missions, but in recent times advancements have been made allowing to stabilize high-quality and nutritionally valuable dried foods and dishes like entrees, ethnic foods, ready-to-eat rice, ice-cream or gourmet desserts [30, 39]. To decrease the energy consumption for water sublimation and increase the sustainability of this process, microwave heating is under study and/or the use, as initial materials to be processed, of foods partly dehydrated by other technologies (e.g. osmosis).

Irradiation is commonly used by NASA to sterilize and extend the shelf-life of space food without the need of heat treatments, with minor effects on the nutritional properties but a potential negative impact on their sensory properties during the storage time in the space flight. To avoid this undesired side effect, under investigation is the reduction of the irradiation dose and the combination of the treatment with other technological actions (heat treatments, addition of antioxidants) [30].

To achieve a longer shelf-life than that obtained by the current processes, the coupling of conventional and/or innovative technologies represents a main options for the future space foods. Thermostabilized foods could enhance their sensory acceptability and nutritional value by improving the efficiency of the sterilization process by microwave assisted-technology as investigated by NASA [40]. This is a new in-container processing that reduces the rate of heating allowing a higher nutrient retention than conventional technology.

High hydrostatic pressure is one of the most successful innovative, non-thermal, in-container processing developed for various foods (e.g. fruit and vegetable derivatives, fish and meat products and dishes) on Earth. By applying pressures up to 600 MPa, this technology determines microbial lethality and enzymatic inactivation with stabilization effects similar to a pasteurization process [41, 42] while keeping the nutritional, qualitative, and higher fresh-like properties. Due to equipment constraints in terms of cost and plants as well as the higher process pressures (>600 MPa) required to obtain the safety requirements for space foods, the HHP technology is often used in combination with relative high temperatures, lower than those used in conventional sterilization processes, in the so-called pressure-assisted thermal sterilization (PATS). This process greatly shortens the sterilization time and improves the quality of low-acid food [43]. In particular, NASA studies highlighted that better color and texture and vitamin content of PATS processed fruits could be obtained with a three-year shelf-life, while the further combination of these processed products with refrigeration could lead to a five-year storability [44].

Additional requirements that are relevant for the manufacturing of future space food are related to the sustainability and reduction of packaging waste, one of the most critical issues in space mission and the development of innovative high-barrier plastic materials and edible coatings need to be considered [30].

The scientific advancements and knowledge about the challenges to keep the nutritional and health status of astronauts highlight the need of improving the nutritional value of the space foods and the fortification and enhancement of the content of vitamins, bioactives, and the dietary fiber intake. The supplementation with pre-, or pro-, or synbiotics might maintain the health of space explorers during their missions [45].

Alternative actions for future space foods meeting also the food supply needs include the development of more energy-dense food might need to be developed to meet energy expenditure with a smaller food volume and weight probably as a high-energy snack which could replace a meal, while the risk is to diminished food variability. NASA has recently tested - exchanging one meal (breakfast) by an energy-dense bar in a closed-chamber study resulting in a reduced food intake, lost weight of the subjects that showed an attenuation in mood and neurobehavioral functioning [46].

Moreover, in the perspective of an "in-house" space food production, 3D food printing represents an innovative promising technology based on "layer by layer" printing and stack forming approach [30]. Optimized mix of dry and liquid ingredients is used to mold products with different shapes and, by changing formulation, foods for personalized nutrition purposes could be made. Some projects carried out by NASA and other space research centers have already designed pilot 3D printers (e.g. Zero-G-Printer, launched into orbit on 21 September 2014) and produced the first 3D-printed products in the space [30]. However, further research on safety, stability, palatability of such kind of recipes as well as bioavailability of nutrients need to be answered before this kind of food can be offered to space explorers [41].

13.8 Conclusion

If we are to achieve a nutritionally complete diet that supports astronauts throughout the 2.5- to 3.0 year crewed mission to mars and back that is currently being planned by the U.S. NASA, new solutions to provide safe, high-quality food with high nutritional value with exceptionally long shelf life need to be developed. Innovative technologies, hurdle technology approaches, and new packaging materials could offer solutions for this and other future space food systems while recognizing that nutritional bioaccessibility and bioavailability also need to be considered in parallel with enrichment of foods with micro- and macro-nutrients as well as of bioactive components.

References

1 Taylor, A.J., Beauchamp, J.D., Briand, L. et al. (2020). Factors affecting flavor perception in space: does the spacecraft environment influence food intake by astronauts? *Comprehensive Reviews in Food Science and Food Safety* 19 (6): 1–37.

2 Smith, S.M., Zwart, S.R., Douglas, G.L., and Heer, M. (2021). Human adaptation to spaceflight: the role of food and nutrition (NP-2021-03-003-JSC). In: *National Aeronautics and Space*, 2nde. Houston: Lyndon B Administration Johnson Space Center.

3 Ferrando, A.A., Paddon-Jones, D., and Wolfe, R.R. (2002). Alterations in protein metabolism during space flight and inactivity. *Nutrition* 18: 837–841.

4 Baek, K., Barlow, A.A., Allen, M.R., and Bloomfield, S.A. (2008). Food restriction and simulated microgravity: effects on bone and serum leptin. *Journal of Applied Physiology* 104: 1086–1093.

5 Florian, J.P., Baisch, F.J., Heer, M., and Pawelczyk, J.A. (2015). Caloric restriction decreases orthostatic tolerance independently from 6 degrees head-down bedrest. *PLoS One* 10: e0118812.

6 Adams, G.R., Caiozzo, V.J., and Baldwin, K.M. (2003). Skeletal muscle unweighting: spaceflight and ground-based models. *Journal of Applied Physiology* 95: 2185–2201.

7 Carter, J.R. and West, J.B. (2013). Space physiology within an exercise physiology curriculum. *Advances in Physiology Education* 37: 220–226.

8 Laurens, C., Simon, C., Vernikos, J. et al. (2019). Revisiting the role of exercise countermeasure on the regulation of energy balance during space flight. *Frontiers in Physiology* 10: 321.

9 Ihle, R. and Loucks, A.B. (2004). Dose-response relationships between energy availability and bone turnover in young exercising women. *Journal of Bone and Mineral Research* 19: 1231–1240.

10 Houston, D.K., Nicklas, B.J., Ding, J. et al. (2008). Dietary protein intake is associated with lean mass change in older, community-dwelling adults: the health, aging, and body composition (health ABC) study. *The American Journal of Clinical Nutrition* 87: 150–155.

11 Pasiakos, S.M., McLellan, T.M., and Lieberman, H.R. (2015). The effects of protein supplements on muscle mass, strength, and aerobic and anaerobic power in healthy adults: a systematic review. *Sports Medicine* 45: 111–131.

12 Stein, T.P. and Blanc, S. (2011). Does protein supplementation prevent muscle disuse atrophy and loss of strength? *Critical Reviews in Food Science and Nutrition* 51: 828–834.

13 Heer, M., Baecker, N., Frings-Meuthen, P. et al. (2016). Effect of high protein intake on bone turnover in long-term bed rest in women. *Applied Physiology, Nutrition, and Metabolism* 42 (5): 537–546.

14 Remer, T., Krupp, D., and Shi, L. (2014). Dietary protein's and dietary acid load's influence on bone health. *Critical Reviews in Food Science and Nutrition* 54: 1140–1150.

15 Zwart, S.R., Davis-Street, J.E., Paddon-Jones, D. et al. (2005). Amino acid supplementation alters bone metabolism during simulated weightlessness. *Journal of Applied Physiology* 99: 134–140.

16 Stein, T.P., Leskiw, M.J., Schluter, M.D. et al. (1999). Protein kinetics during and after long-duration spaceflight on MIR. *The American Journal of Physiology* 276: E1014–E1021.

17 Stein, T.P. (2002). Space flight and oxidative stress. *Nutrition* 18: 867–871.

18 Rizzo, A.M., Corsetto, P.A., Montorfano, G. et al. (2012). Effects of long-term space flight on erythrocytes and oxidative stress of rodents. *PLoS One* 7: e32361.

19 Hosseini, B., Berthon, B.S., Wark, P., and Wood, L.G. (2017). Effects of fruit and vegetable consumption on risk of asthma, wheezing and immune responses: a systematic review and meta-analysis. *Nutrients* 9: 341.

20 Gilbert, J.A., Blaser, M.J., Caporaso, J.G. et al. (2018). Current understanding of the human microbiome. *Nature Medicine* 24: 392–400.

21 Haase, S., Haghikia, A., Wilck, N. et al. (2018). Impacts of microbiome metabolites on immune regulation and autoimmunity. *Immunology* 154: 230–238.

22 Voorhies, A.A., Mark, O.C., Mehta, S. et al. (2019). Study of the impact of long-duration space missions at the International Space Station on the astronaut microbiome. *Scientific Reports* 9: 9911.

23 Cooper, M., Douglas, G., and Perchonok, M. (2011). Developing the NASA food system for long-duration missions. *Journal of Food Science* 76: R40–R48.

24 Desai, M.S., Seekatz, A.M., Koropatkin, N.M. et al. (2016). A dietary fiber-deprived gut microbiota degrades the colonic mucus barrier and enhances pathogen susceptibility. *Cell* 167: 1339–1353.

25 Aune, D., Keum, N., Giovannucci, E. et al. (2016). Whole grain consumption and risk of cardiovascular disease, cancer, and all cause and cause specific mortality: systematic review and dose-response meta-analysis of prospective studies. *BMJ* 353: i2716.

26 Durante, M. (2014). Space radiation protection: destination Mars. *Life Sciences and Space Research* 1: 2–9.

27 Catauro, P.M. and Perchonok, M.H. (2012). Assessment of the longterm stability of retort pouch foods to support extended duration spaceflight. *Journal of Food Science* 71 (1): S29–S39.

28 ESA (2017). Space food. In: *Astronaut Nutrition*, 1–28. European Space Agency.

29 Lachance, P.A., Michel, E.L., and Nanz, R.A. (1967). Evolution of space feeding concepts during the Mercury and Gemini space programs. *Food Technology* 21: 52–58.

30 Jiang, J., Zhang, M., Bhandari, B., and Cao, P. (2020). Current processing and packing technology for space foods: a review. *Critical Reviews in Food Science and Nutrition* 60: 3573–3588.

31 Perchonok, M. and Bourland, C. (2002). NASA food systems: past, present, and future. *Nutrition* 18 (10): 913–920.

32 Bannister, J.R. (1968). Food for Spaceflight. *Spaceflight* 10: 118–121.

33 Cooper, M., Perchonok, M., and Douglas, G.L. (2017). Initial assessment of the nutritional quality of the space food system over three years of ambient storage. *NPJ Microgravity* 3: 17.

34 Song, B.S., Park, J.G., Park, J.N. et al. (2009). High dose processing and application to Korean space foods. *Radiation Physics and Chemistry* 78 (7–8): 671–674.

35 Bustos, M. (2021). Microgravity refrigerator could store food for astronauts in the future. *The Science Times* https://www.sciencetimes.com/articles/31426/20210528/microgravity-refrigerator-store-food-astronauts-future.htm.

36 Douglas, G., Cooper, M., Bermudez-Aguiree, D., and Sirmons, T. (2016). Risk of performance decrement and crew illness due to an inadequate food system. https://humanresearchroadmap.nasa.gov/Evidence/reports/AFT.pdf.

37 Perchonok, M. and Antonini, D. (2008). Thermostabilized shelf life study. In: *Proceedings from the NASA Human Research Program Investigators Workshop, 4–6 February 2008*. League City, Texas: NASA http://hefd.jsc.nasa.gov/projects.htm after clicking on Thermostabilized Shelf Life Study (Food) link.

38 Douglas, G.L., Zwart, S.R., and Smith, S.M. (2020). Space food for thought: challenges and considerations for food and nutrition on exploration missions. *The Journal of Nutrition* 150: 2242–2244.

39 Palmer, R. (2013). A history of ice cream innovations, from ancient China to NASA astronauts and Dippin' Dots. *International Business Times* https://www.ibtimes.com/history-ice-cream-innovations-ancient-china-nasa-astronauts-dippin-dots-1353825.

40 Perchonok, M. H. and Catauro, P.M. (2009). Thermostabilized shelf life study. In edited by NASA. https://ntrs.nasa.gov/api/citations/20090006887/downloads/20090006887.pdf.

41 Huang, H.W., Wu, S.J., Lu, J.K. et al. (2017). Current status and future trends of high-pressure processing in food industry. *Food Control* 72 (12): 1–8.

42 Rastogi, N.K., Raghavarao, K.S.M.S., Balasubramaniam, V.M. et al. (2007). Opportunities and challenges in high pressure processing of foods. *Critical Reviews in Food Science and Nutrition* 47 (1): 69–112.

43 Balasubramaniam, V.M., Barbosa-Canovas, G.V., and Lelieveld, H.L.M. (2016). High-pressure processing equipment for the food industry. In: *High Pressure Processing of Food*, 39–65. New York, US: Springer.

44 Cooper, M.R. and Douglas, G.L. (2014). Integration of product, package, process, and environment: a food system optimization. In edited by NASA.

45 Turroni, S., Magnani, M., Kc, P. et al. (2020). Gut microbiome and space travelers' health: state of the art and possible pro/prebiotic strategies for long-term space mMissions. *Frontiers in Physiology* 11: 553929.

46 Sirmons, T.A., Roma, P.G., Whitmire, A.M. et al. (2020). Meal replacement in isolated and confined mission environments: consumption, acceptability, and implications for physical and behavioral health. *Physiology & Behavior* 219: 112829.

Part VI

Advanced Materials

14

Metal Alloy Synthesis in Microgravity

Jessica J. Frick and Debbie G. Senesky

Stanford University, Department of Aeronautics and Astronautics, Durand Building, 496 Lomita Mall, Stanford, CA 94305, USA

14.1 Historical Context

Metallurgy is one of the greatest technological advancements in human history, and is, above all, the study of metallic alloying. Metal alloys consist of a primary metal matrix diluted with one or more other element for the purpose of advancing desired properties in the resulting material. For example, to achieve the signature pink hue in rose gold jewelry, a Au matrix is diluted with variable concentrations of Ag and Cu metals. Since the Bronze Age, metal alloys have been a key part of our species' world-building – the term "Bronze Age" itself comes from the alloy bronze (copper (Cu) diluted with tin (Sn)) which was the hardest metal used throughout humanity during the archeological period spanning approximately 3300 to 1200 BCE. In today's world, the use of metal alloys is so prevalent, it is impossible for one to go throughout their day without directly interacting with such a material. Figure 14.1 provides a general overview of the seemingly endless applications of society's use of metal alloys.

Nearly all metal alloys are produced by heating their elemental starting components into a liquid phase and subsequently cooling the liquid mixture until it freezes into a solid. Variables during this synthetic process (e.g. composition, temperature, atmosphere, pressure, reaction time) can be tuned to achieve a desired outcome. This concept can be easily analogized to how temperature, ingredients, and recipe preparation effect chocolate chip cookies: higher temperatures give a crisp exterior and a slightly doughy center; using butter instead of vegetable oil creates a fudgier texture; and letting the dough chill rather than baking immediately allows the flour to absorb more moisture resulting in a chewier cookie.

One variable that cannot so easily be manipulated during metal alloy synthesis is Earth's gravity. This has always posed a challenge in materials science, because acceleration-dependent processes, like convective flows, are highly influential conditions in an alloy recipe. Convective processes occur when temperature and/or compositional gradients (if a fluid is not homogeneous) induce density gradients.

In-Space Manufacturing and Resources: Earth and Planetary Exploration Applications, First Edition.
Edited by Volker Hessel, Jana Stoudemire, Hideaki Miyamoto, and Ian D. Fisk.

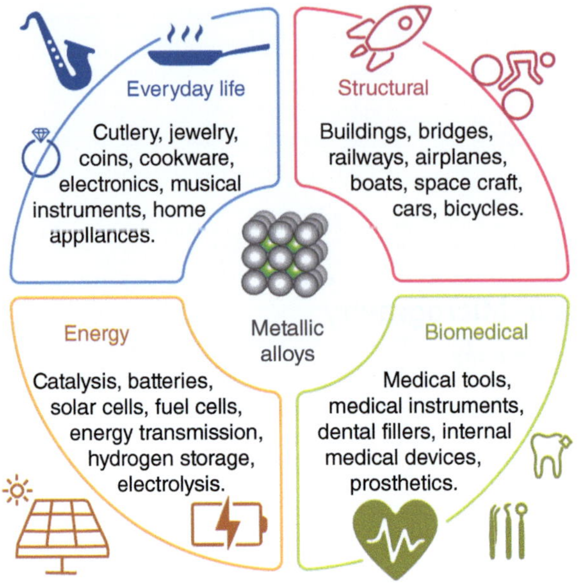

Figure 14.1 Everyday life, structural, biomedical, and energy applications of metal alloys.

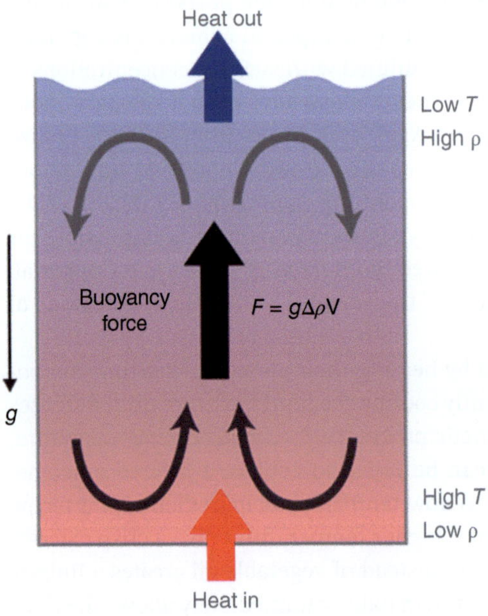

Figure 14.2 Thermal-gradient-driven buoyancy convection force under Earth's gravity.

This is commonly referred to as buoyancy-driven convection. Figure 14.2 gives a pictorial representation of buoyancy force due to a thermal gradient under gravity conditions. For an exhaustive review of convection phenomena, both independent and dependent of acceleration, we refer the reader to Chapter I Phenomenology in "Materials Science in Space with Application to Space Processing" [1].

In materials science processing, alloys are generally prepared for a specific application in mind – therefore, it is crucial to produce the desired properties by exercising some degree of control over the shape, size, composition, and defect structure during their growth. Compositional control is by far the most important for metal alloys and is commonly dictated by the aforementioned convective flows. The microgravity environment afforded by low-Earth orbit (LEO) laboratories (i.e. the constant state of free fall) enables scientists to finally adjust the knob of acceleration in materials synthesis for extended periods of time. In addition to the reduction of density-gradient driven flows, microgravity alloy solidification offers other unique features like containerless handling of melts, the reduction of sedimentation (and Stokes flow), and the elimination of hydrostatic pressure gradients. The ability to "turn off" (or significantly reduce) these natural Earth phenomena also enables scientists to study phenomena typically hidden by gravitational effects, e.g. acceleration-independent convection driven from surface-tension, thermoacoustics, phase changes (the solid–liquid interface in alloy crystal growth), and electric and magnetic fields.

Overall, the twofold advantages of microgravity conditions enable alloy synthesis experiments to be conducted for both technological aspects, such as discovering solutions to manufacturing problems, and scientific aspects, such as discovering new materials or unknown physical and chemical phenomena that until now have remained hidden under Earth's gravity. This chapter will provide an overview of the various investigations in metal alloy synthesis in microgravity within the last decade. This incorporates experiments performed in the research laboratories on the International Space Station (ISS) and China's Tiangong-2 (TG-2) space laboratory.

14.2 Metal Alloy Synthesis on the International Space Station (ISS)

The ISS hosts two main facilities that can cater to alloy synthesis: the Materials Science Research Rack (MSRR), launched in 2009 [2], and the Microgravity Science Glovebox (MSG), launched in 2002 [3]. The MSRR houses the Materials Science Laboratory (MSL) developed through an international cooperative agreement between NASA and the European Space Agency (ESA). The MSL is designed to accommodate two furnace module inserts: the low gradient furnace (LGF), for vapor transport (VT) or Bridgman processing [4, 5], and the solidification and quenching furnace (SQF), for rapid-displacement (50–100 mm within 1 second) Bridgman processing [6]. Both furnaces have a maximum process temperature of 1400 °C, a maximum furnace displacement of 150 mm, and the ability to use a rotating magnetic field to influence the movement of the internal liquid melt [5].

The MSG is a multi-purpose, sealed environment facility designed to contain experiments with hazardous, liquid, or small-particle materials. Among this facility's many purposes, it operates the solidification using a baffle in sealed ampoules (SUBSA) furnace [7]. Its name originates from the title of one of the first materials science investigations sponsored by the United States to operate within the MSG [3]. This furnace is used for gradient freeze processing [4] and can reach a

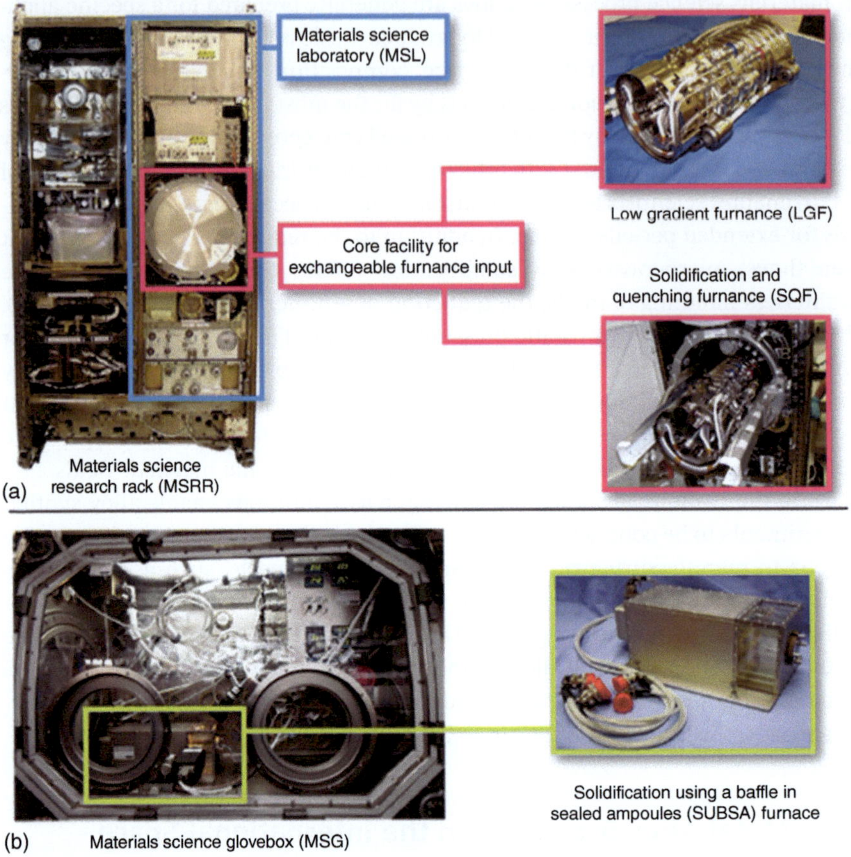

Figure 14.3 Overview of high process temperature furnaces on the ISS. (a) The Materials Science Research Rack (MSRR) is home to the Materials Science Laboratory (MSL) which can host two exchangeable furnace inputs, the low gradient furnace (LGF) and the solidification and quenching furnace (SQF). (b) The Microgravity Science Glovebox (MSG) is home to the solidification using a baffle in sealed ampoules (SUBSA) furnace. Source: NASA.

maximum temperature of 850 °C. Figure 14.3 shows pictures (Courtesy: NASA) of the MSRR, MSG, and their furnace inserts. For full furnace capabilities and critical performance parameters, please see Ref. [7] for the SUBSA furnace, and Ref. [5] for the LGF and SQF.

Since the MSL entered orbit in 2009, ESA has led four major, on-going projects using this facility: columnar-to-equiaxed transition in solidification processing (CETSOL), microstructure formation in casting of technical alloys under diffusive and magnetically controlled connection conditions (MICAST), solidification along a eutectic path in ternary alloys (SETA), and metastable solidification of composites: novel peritectic structures and *in situ* composites (METCOMP). Table 14.1 lists the material of interest for each investigation (W. Sillekens, personal communication).

The CETSOL, MICAST, and SETA investigations all have backgrounds rooted in aluminum casting alloys (i.e. a liquid alloy is poured in a mold of the desired shape and is allowed to solidify). These materials make up a huge portion of

Table 14.1 Material overview of the on-going investigations on the ISS utilizing both the LGF and SQF furnace inserts of the MSL facility.

Investigation	Batch	Material
CETSOL	1	Al–7Si; Al–7Si--0.5AT5B[a)]
	2a	Al–7Si; Al–7Si–0.5AT5B[a)]
	3a	Al–xCu; Al–xCu–0.1AT3B[a)](x = 4, 10, 20)
MICAST	1	Al–7Si; Al–7Si–1Fe
	2a	Al–7Si; Al–7Si–1Fe; Al–7Si–1Mn
	3a	Al–12.6Si; Al–18Si; Al–18Si–10Cu; Al–6Si–4Cu–1Fe; Al–6Si–4Cu–2Fe
SETA	2a	Al–3Si–1Mn–0.3TiB_2
	2b	Al–17Cu–1Ag; Al–13Cu–18Ag
METCOMP	2b	Cu–xSn (x = 10, 19, 20, 21, 23.5, 29)

For each batch, several cartridges for one or more investigation were sent to the ISS for processing. Batches 1, 2a, and 2b have been processed. Batch 3a is currently being processed at the time of this publication. The number in front of an elemental or alloy component denotes its weight percent.
a) AT5B and AT3B are shorthand for Al-Ti-B grain refiners used in aluminum alloy casting to reduce the grain size.

alloy structural parts in automobiles, aircrafts, bicycles, and outdoor equipment with exposure to corrosion like road signs and lighting fixtures [8]. As with any synthetic technique with a liquid-to-solid phase transition, the influence of convective flows is one of the most challenging aspects in aluminum casting. Physical properties such as the alloys' yield, fracture, and fatigue strengths can vary widely when varying convective flows influence the solidifying microstructure [9, 10]. These aluminum alloy–focused investigations seek to better understand the extent of influence gravitational variation has on the solidification process, in hopes to lead to better experimental manufacturing techniques and to develop better theoretical models to predict the effect of gravitational variations on alloy microstructure.

Although rooted in the same overarching goal, each investigation supports a complementary branch of examination into the complex physical principles that govern alloy crystallization. The main objective of the CETSOL investigation is to better understand the phenomenon known as "columnar-to-equiaxed transition" (CET) during solidification of metal alloys – a process which determines grain structure of metal alloys via growth competition of elongated, columnar crystals and rounder, equiaxed crystals [11–13]. Al-7 wt% Si alloys, with and without grain refiners, were directionally solidified on the ISS to study this phenomenon. Results so far indicate CET occurs only for refined alloys, with columnar growth occurring in nonrefined samples [14]. Additionally, the influence of experimental variables like growth velocity and temperature gradient was studied. Detailed analysis showed that an increase in growth velocity leads to a size decrease of the dendrite microstructure and a more homogeneous eutectic distribution, and an increase in temperature gradient leads to a finer dendrite structure [15, 16].

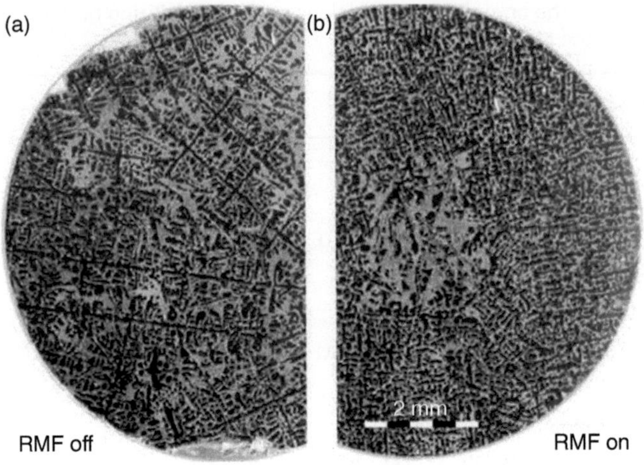

Figure 14.4 Macrographs of Al–7Si–1Fe alloy directionally solidified within MSL SQF without RMF (a) and with RMF (b). Under pure diffusive conditions, a homogeneous distribution of dendrite structures occurs. With RMF on (i.e. under the presence of forced fluid flow), Si strongly segregates to the center of the sample. Source: Enz et al. [27]. Adapted with permission from Springer Nature.

The MICAST investigation uses the magnetic field option in the MSL to manipulate magnetically controlled convective flows (which are independent of acceleration) during solidification of aluminum alloys. Synthesizing aluminum alloys in the presence of a magnetic field itself is not a new idea [17–20]; however, the prolonged microgravity environment effectively eliminates any influence of Earth's acceleration during this process, enabling scientists to truly isolate the system and glean understanding of aluminum alloy solidification under the presence of artificial fluid flow. Under the influence of Earth's gravity, a "steepling" type of thermosolutal convection is known to develop during directional solidification of metal alloys. This is due to radial thermal gradients in the melt which result in a convex-shaped liquid–solid interface, leading to severe radial macrosegregation [21–23]. This behavior can be eliminated in microgravity (purely diffusive growth environment) and can be turned back on using a rotating magnetic field (RMF) (0–3.5 mT) to create forced melt flow [24–26]. Macrographs in Figure 14.4 show this clear "off-and-on" behavior in microgravity-synthesized Al–7Si–1Fe with RMF off (left) and RMF on (right) [27]. The forced fluid flow induces macrosegregation of Si to the alloy center, similar to radial macrosegregation from convection under Earth's gravity. Slow directional solidification velocities were shown to be more susceptible to macrosegregation under induced melt flow [28]. Additionally, two studies published in 2019 [29] and 2021 [30] focused on primary dendrite trunk diameter and array morphology, respectively, in Al-7 wt% Si alloys and concluded that thermosolutal convection during directional solidification both reduces the spacing between primary dendrites and increases trunk diameter. These conclusions were formed after *not* observing these phenomena in alloys synthesized in microgravity, under diffusive transport conditions.

Table 14.2 Overview of in-progress investigations on the ISS studying metal alloy synthesis in microgravity.

Investigation	Material	Scientific goals/findings
SUBSA-CETSOL	Al-x wt% Cu ($x = 4, 10, 18$)	Investigate whether a CET in the grain structure of metal alloys is present in the absence of convection and flow of crystal fragments in the melt
	Al-7 wt% Si	
SUBSA-DFM	Pb–Sn	Experimentally isolate the influence of heating and cooling rate effects on metal alloy dendrite shape to produce benchmark data for accurate process modeling design
GEDS	W–Ni–Cu–Mn	Study the degree of incomplete densification in commercially used alloys when solidified in a reduced gravity environment to better predict and calculate distortion
FAMIS	$Zr_{57}Nb_5Cu_{15.4}Ni_{12.6}Al_{10} + W$	Synthesize bulk, homogeneous samples of BMGs and metal matrix composites (BMGMCs) in microgravity to circumvent issues of crystalline phase sedimentation
	$Zr_{48}Cu_{47.5}Al_4Co_{0.5} + W$	

This information was gathered from a combination of personal communication with the PI of each investigation and the Space Station Research Explorer on NASA.gov.
University in the Department of Materials Science and Engineering.

The SETA investigation studies the formation of eutectics which is governed by three different growth mechanisms: coupled, divorced, and symbiotic [31]. SETA focuses primarily on the symbiotic growth in Al–Mn–Si alloys and coupled growth in Al–Cu–Ag alloys. The METCOMP investigation is an extension of SETA into systems outside aluminum alloys. This investigation looks at copper alloy systems, specifically binary Cu–Sn peritectic alloys and their growth mechanism of competing phases. For detailed descriptions of these four on-going projects, we direct the readers to ESA-funded experiment database, the erasmus experiment archive (EEA).

In addition to the on-going investigations mentioned above, there are four other metal alloy synthesis investigations currently in progress: (i) the effect of convection on the columnar-to-equiaxed transition in alloy solidification (SUBSA-CETSOL) investigation, (ii) the dendrite fragmentation and morphology during melting and solidification (SUBSA-DFM) investigation, (iii) the gravitational effects on distortion in sintering (GEDS) project, and (iv) the formulation of amorphous metals in space (FAMIS) experiment. Table 14.2 presents an overview of the metal alloys of interest and scientific goals of each investigation.

The SUBSA-CETSOL investigation is led by Prof. Christoph Beckermann and his team of researchers at the University of Iowa in the Department of Mechanical

Engineering. This investigation is an extension of the ESA-funded CETSOL project discussed above. A series of Al-x wt% Cu alloys ($x = 4, 10, 18$) will be processed, each representing a different solid-to-melt density ratio [32]. For high solute concentrations processed under Earth's gravity, convection of the melt and floatation of any solid crystal fragments is anticipated to form a solidified metal alloy with a CET in the grain structure. However, it is unclear whether a CET will occur under microgravity, where movement of the melt and any unattached solid fragments is prevented. A fourth sample, Al-7 wt% Si, will be processed as a control for facile comparison to the ESA-sponsored project of the same name [32].

Dendritic solidification, characterized by the widely recognized tree-like structure, has large consequences regarding the materials' resulting physical properties and is one of the most common forms of solidifying metals and alloys. Prof. Voorhees' research team is part of the scientific community aiming to accurately model this critical and complex process through reconstructions of experimental samples. Their particular area of interest is understanding how the rate of heating and cooling affects the shape of these dendrites. To isolate these effects for study, the researchers are designing a synthetic experiment to take place on the space station, removing the overwhelming effects of sedimentation and convection from the equation. The well-known and studied Pb-Sn alloy system will be used in this experiment as a reliable alloy platform [33]. Both the SUBSA-CETSOL and SUBSA-DFM investigations have an anticipated payload return in July 2021 [P. Voorhees, personal communication].

The GEDS investigation, which began in 2010, is led by Prof. Randall M. German of San Diego State University. This program seeks to answer how a reduced gravity environment influences densification in metal alloys – a process which critically influences the degree of distortion of these alloys used in structural applications. Tungsten heavy alloys (WHA) were chosen as the alloy of study for this program, as they represent a large segment of commercial high-density alloy used in the aerospace and automotive industries. Microgravity processed samples were returned to the PI in June 2021 – results of this investigation are pending [R. German, personal communication].

FAMIS, the more recent metal alloy investigation using the LGF, is led by Dr. Douglas Hofmann of the Materials Development and Manufacturing Technology Group of NASA's Jet Propulsion Laboratory (JPL). This program focuses on microgravity processing of bulk metal glasses (BMGs) – a unique class of non-crystalline metal alloy with attractive attributes for numerous engineering applications [34]. The BMG class itself has been further dissected into subdivisions of monolithic BMGs and BMGs with metal matrix composites (BMGMCs). BMGMCs integrate a soft, ductile crystalline phase into the brittle BMG matrix to improve notable detractions of BMGs such as ductility, toughness, and fatigue [35]. The crystalline phase, which can have up to three times high density than the BMG melt, is subject to sedimentation when synthesized under Earth's gravity – leading to nonhomogeneous samples (Figure 14.5). The FAMIS investigation plans to synthesize BMGMCs with dense inclusions of W crystalline phases in microgravity to circumvent sedimentation,

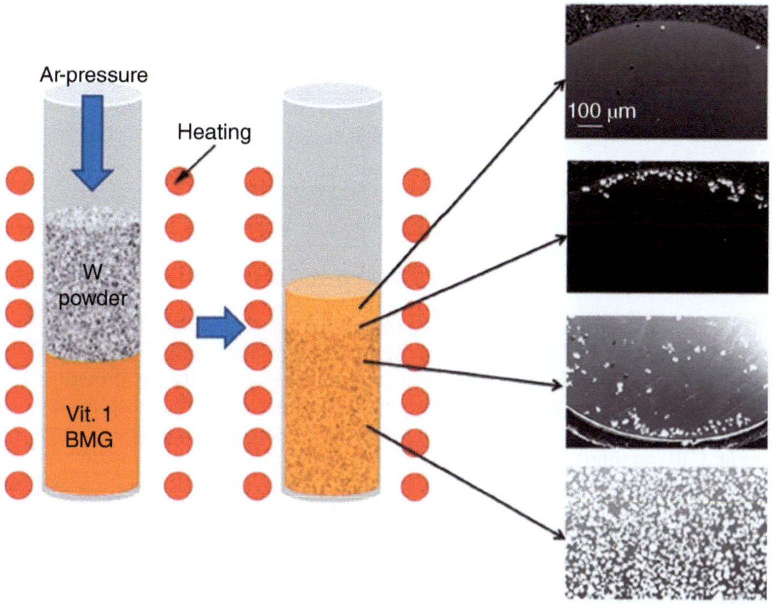

Figure 14.5 Schematic showing the influence of sedimentation when synthesizing BMGMCs on Earth. When processed in a standard vertical Bridgman furnace, the higher-density W diluent sinks into the BMG melt (Vit. 1), producing a nonhomogeneous distribution. The micrographs at right (scale bar, 100 μm) highlight the inhomogeneity of W concentration (bright sports) at different vertical samplings of the bulk. Source: Hofmann and Roberts [36]. Adapted with permission from npj Microgravity.

resulting in bulk, homogenous BMGMC samples [36]. The FAMIS payload is anticipated to launch in 2022 [D. Hofmann, personal communication].

14.3 Metal Alloy Synthesis on the Tiangong-2 (TG-2) Space Laboratory

From 2016 to 2019, the Chinese Tiangong-2 (TG-2) space laboratory conducted 18 Materials Science experiments. These experiments were performed using a multiple materials processing furnace (MMPF) mounted in the TG-2 orbital module [37]. In 2019, two investigative findings on metal alloy synthesis were published – these works, reported by Hongxiang Jiang et al., focused on the solidification of immiscible Al–Bi–Sn metal alloys: 3.6 wt% Al-1 wt% Bi–Sn [38] and Al-3.6 wt% Bi-1 wt% Sn [39]. Immiscible alloys, characterized by the presence of a miscibility gap in their phase diagram, have strong structural applications due to their unique ability to form load-bearing composite micro- and macro-structures that can be an inherent phenomenon during their solidification [40]. Both studies gave similar, positive results with respect to the distribution of the minority phases within the majority metal matrix when synthesized in microgravity – due in part to the weak convective flow

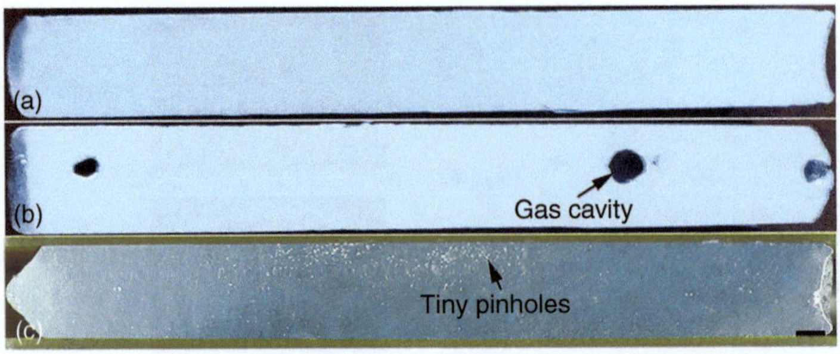

Figure 14.6 Longitudinal section photographs of the as-synthesized crystal boules of 3.6 wt% Al-1 wt% Bi–Sn solidified (a) in microgravity, (b) anti-gravitationally (i.e. solidification direction opposite of gravity) on Earth, and (c) gravitationally (i.e. growth direction in line with gravity) on Earth. The microgravity-synthesized sample displays a beautiful, bubble-free ingot compared to the two Earth-grown samples. Scale bar, 2 mm. Source: Jiang et al. [38]. Adapted with permission from npj Microgravity.

of the matrix melt and the diminished sedimentation (Stokes flow) of the minority phase. These experimental results are helpful for the clarification of the kinetic details of the solidification microstructure evolution of immiscible alloys – critical information required to create accurate models of crystal growth in reduced gravity environments. The Sn-majority study also revealed that these same conditions helped reduce/diminish the undesired porosity of the sample typically incited by the movement of bubbles (Figure 14.6).

Two additional investigations on metal alloy synthesis were mentioned in a review document titled, "Materials Experiment on Tigangong-2 Space Laboratory," which offers an overview of the investigations processed on TG-2 using the MMPF [37]. The *Solidification of Multicomponent and Multiphase Alloys in Space* investigation studied the solidification of Ag–Cu–Ge and Ag–Cu–Sb metal alloys. These ternary alloys were selected in hopes of gaining a deep understanding of competitive nucleation and cooperative growth of a mixture with multiple phases. Interestingly, the microgravity grown samples result in the primary metal phases, Ge and Sb, distributed in the middle of or inside of the ternary alloy phase, respectively. Earth-synthesized samples result in Ge on top of and Sb on the surface of the ternary phase.

In the *Solidification of Al-based Single Crystal Alloy in Space* investigation, the AlZnMgCu alloy was chosen for study due to its high similarity in solidification and microstructure with a Ni-based superalloy. Even in a reduced gravity environment, the complicated quaternary metal alloy still resulted in micro- and macroscopic segregation of the elements; however, the extent of segregation was found to be slighter, especially in the radial direction, and the element distributions were more uniform in the microgravity-synthesized samples. What was most interesting, though, was that among the diluent elements, the segregation of Cu was most influenced by gravity, and Mg segregation was the least influenced. The current report of this study is brief, so an in-depth discussion on this finding is not included. We are looking forward to future publications on this work.

14.4 Future Outlook

Within the last decade, materials scientists have maintained a conservative approach with their newly expanded synthetic tool kit. The majority of metal alloy synthesis in microgravity has been performed on systems that are well-known and commonly used on Earth (e.g. aluminum alloys). These investigations have laid the crucial experimental groundwork required for scientists to develop theoretical models to more accurately predict behavior of metal alloy solidification in a variety of synthetic environments. With this information in hand, we are now approaching an important branching point in microgravity materials science, which, coincidentally, coincides with the shift from government-run to privately operated microgravity research platforms:

(1) Scientists and engineers must move forward with experimentally based growth models to improve terrestrial manufacturing processes and construct informed cost–benefit analyses to determine what metal alloy families can be most adventitiously manufactured in space.
(2) Scientists must begin more exploratory work to fully utilize the unique synthetic platform of microgravity. Higher-risk, progressive science needs to happen to discover new materials or create those which on Earth have been impossible to create. Additionally, and to truly utilize a reduced gravity environment, researchers need to re-think experimental design through a different lens – instead of adapting Earth hardware to microgravity (like the design of the LGF, SQF, and SUBSA furnaces), we must design hardware with operational mechanics specific to a laboratory suspended in 3D space. Only then can we achieve scientific insight required to generate new and innovative processing techniques.

References

1 Steg, L. (ed.) (1977). *Materials Sciences in Space with Application to Space Processing*. New York: American Institute of Aeronautics and Astronautics.
2 Pettigrew, P.J., Kitchen, L., Darby, C. et al. (2003). Design features and capabilities of the First Materials Science Research Rack (MSRR-1). In: *2003 IEEE Aerospace Conference Proceedings*, (Cat. No.03TH8652), 2003, pp. I–55, doi: 10.1109/ AERO.2003.1235040.
3 Spivey, R. and Flores, G. (2008). An overview of the microgravity science glovebox (MSG) facility, and the gravity-dependent phenomena research performed in the MSG on the International Space Station (ISS). In: *46th AIAA Aerospace Sciences Meeting and Exhibit*. Reno, NV, USA (7–10 January 2008).
4 Isshiki, M. and Wang, J. (2017). Wide-Bandgap II-VI Semiconductors: Growth and Properties. In: *Springer Handbook of Electronic and Photonic Materials* (ed. S. Kasap and P. Capper), 1–1. Cham: Springer International Publishing.

5 Marshall Space Flight Center. (2012). MSFC MSRR-1 Sample Cartridge Assembly (SCA) Capabilities Document. MSFC-RQMT-3633.

6 Carswell, W., Kroeger, F., and Hammond, M. (2003). QMI: a furnace for metals and alloys processing on the International Space Station. In: *2003 IEEE Aerospace Conference Proceedings*, vol. 1, 1–74. MT, USA: Big Sky (8–15 March 2003).

7 Spivey, R., Gilley, S., Ostrogorsky, A. et al. (2003). SUBSA and PFMI Transparent Furnace Systems Currently in use in the International Space Station Microgravity Science Glovebox. In: *41st AIAA Aerospace Sciences Meeting and Exhibit*. Reno, NV, USA (6–9 January 2003).

8 Vargel, C. (2020). Aluminium casting alloys. In: *Corrosion of Aluminium*, 525–537. Elsevier Ltd.

9 Fang, X., Shao, G., Liu, Y.Q., and Fan, Z. (2007). Effects of intensive forced melt convection on the mechanical properties of Fe containing Al–Si based alloys. *Materials Science and Engineering: A* 445–446: 65–72.

10 Kok, Y., Tan, X.P., Wang, P. et al. (2018). Anisotropy and heterogeneity of microstructure and mechanical properties in metal additive manufacturing: a critical review. *Materials & Design* 139: 565–586.

11 Zimmermann, G., Sturz, L., Billia, B. et al. (2011). Investigation of columnar-to-equiaxed transition in solidification processing of AlSi alloys in microgravity – The CETSOL project. In: *International Symposium on Physical Sciences in Space, Journal of Physics: Conference Series*, 327, 012003. Bonn, Germany (11–15 July 2011).

12 Billia, B., Gandin, C.-A., Zimmermann, G. et al. (2005). Columnar-equiaxed transition in solidification processing: the ESA–MAP CETSOL project. *Microgravity - Science and Technology* 16: 20–25.

13 Kurz, W., Bezençon, C., and Gäumann, M. (2001). Columnar to equiaxed transition in solidification processing. *Science and Technology of Advanced Materials* 2 (1): 185.

14 Zimmermann, G., Sturz, L., Nguyen-Thi, H. et al. (2017). Columnar and equiaxed solidification of Al-7 wt.% Si alloys in reduced gravity in the framework of the CETSOL project. *The Journal of The Minerals, Metals & Materials Society* 69 (8): 1269–1279.

15 Li, Y.Z., Mangelinck-Noël, N., Zimmermann, G. et al. (2019). Comparative study of directional solidification of Al-7 wt% Si alloys in Space and on Earth: effects of gravity on dendrite growth and columnar-to-equiaxed transition. *Journal of Crystal Growth* 513: 20–29.

16 Li, Y.Z., Mangelinck-Noël, N., Zimmermann, G. et al. (2018). Effect of solidification conditions and surface pores on the microstructure and columnar-to-equiaxed transition in solidification under microgravity. *Journal of Alloys and Compounds* 749: 344–354.

17 Vivès, C. (1996). Effects of forced electromagnetic vibrations during the solidification of aluminum alloys: Part I. solidification in the presence of crossed alternating electric fields and stationary magnetic fields. *Metallurgical and Materials Transactions B* 27 (3): 445–455.

18 Mizutani, Y., Ohura, Y., Miwa, K. et al. (2004). Effect of the electromagnetic vibration intensity on microstructural refinement of Al-7% Si alloy. *Materials Transactions* 45 (6): 1944–1948.

19 Dong, J., Zhao, Z., Cui, J. et al. (2004). Effect of low-frequency electromagnetic casting on the castability, microstructure, and tensile properties of direct-chill cast Al–Zn–Mg–Cu alloy. *Metallurgical and Materials Transactions A* 35 (8): 2487–2494.

20 Yubo, Z., Jianzhong, C., Zhihao, Z. et al. (2005). Effect of low frequency electromagnetic field on casting crack during DC casting superhigh strength aluminum alloy ingots. *Materials Science and Engineering: A* 406 (1): 286–292.

21 Dupouy, M.D., Camel, D., and Favier, J.J. (1989). Natural convection in directional dendritic solidification of metallic alloys. I. Macroscopic effects. *Acta Metallurgica* 37 (4): 1143–1157.

22 Dupouy, M.D., Camel, D., and Favier, J.J. (1992). Natural convective effects in directional dendritic solidification of binary metallic alloys: dendritic array primary spacing. *Acta Metallurgica et Materialia* 40 (7): 1791–1801.

23 Dupouy, M.D. and Camel, D. (1998). Effects of gravity on columnar dendritic growth of metallic alloys: flow pattern and mass transfer. *Journal of Crystal Growth* 183 (3): 469–489.

24 Zhang, H., Wu, M., Zheng, Y. et al. (2018). Macrosegregation formation in an Al–Si casting sample with cross-sectional change during directional solidification. *Transactions of the Indian Institute of Metals* 71 (11): 2639–2643.

25 Lauer, M., Ghods, M., Angart, S.G. et al. (2017). Macrosegregation during Re-melting and holding of directionally solidified Al-7 wt.% Si alloy in microgravity. *The Journal of The Minerals, Metals & Materials Society* 69 (8): 1289–1297.

26 Li, Y.Z., Mangelinck-Noël, N., Zimmermann, G. et al. (2020). Modification of the microstructure by rotating magnetic field during the solidification of Al-7 wt.% Si alloy under microgravity. *Journal of Alloys and Compounds* 836: 155458.

27 Enz, T., Steinbach, S., Simicic, D. et al. (2011). First experiments using the materials science laboratory on board the International Space Station. *Microgravity Science and Technology* 23 (3): 345–353.

28 Steinbach, S., Ratke, L., Zimmermann, G., and Budenkova, O. (2016). Formation of intermetallic phases in AlSi7Fe1 alloy processed under microgravity and forced fluid flow conditions and their influence on the permeability. *4th International Conference on Advances in Solidification Processes, IOP Conference Series: Materials Science and Engineering*, 117, 012019. Windsor, UK (8–11 July 2014).

29 Upadhyay, S., Tewari, S., Ghodes, M., Grugel, R., Poirier, D., and Lauer, M. (2019). Primary Dendrite Trunk Diameter in Al-7wt% Si Alloy Directionally Solidified Aboard the International Space Station. Joint 5th International Conference on Advances in Solidification Processes & 5th International Symposium on Cutting Edges of Computer Simulation of Solidification, Casting and Refining, *IOP Conference Series: Materials Science and Engineering*, 529, 012022. Salzburg, Austria (17–21 June 2019).

30 Ghodes, M., Upadhyay, S.R., Rajamure, R.S. et al. (2021). Primary dendrite array morphology in Al-7 wt% Si alloy samples directionally solidified aboard the International Space Station. *Journal of Crystal Growth* 562: 126077.

31 Stefanescu, D.M. (2009). Eutectic solidification. In: *Science and Engineering of Casting Solidification*, 2e, 1–52. Boston, MA: Springer.

32 NASA (2020). Investigation Summary Form (ISF) on SUBSA-CETSOL. http://www.nasa.gov/iss-science/.

33 NASA (2020). Investigation Summary Form (ISF) on SUBSA-DFM. http://www.nasa.gov/iss-science/.

34 Ashby, M.F. and Greer, A.L. (2006). Metallic glasses as structural materials. *Scripta Materialia* 54 (3): 321–326.

35 Hofmann, D.C., Suh, J.-Y., Wiest, A. et al. (2008). Designing metallic glass matrix composites with high toughness and tensile ductility. *Nature* 451 (7182): 1085–1089.

36 Hofmann, D.C. and Roberts, S.N. (2015). Microgravity metal processing: from undercooled liquids to bulk metallic glasses. *npj Microgravity* 1: 15003.

37 Xiangyang, L.I. and L.Y., and LI Xiangyang, L.Y. (2018). Materials experiment on Tiangong-2 space laboratory. *Chinese Journal of Space Science* 38 (5): 829–835.

38 Jiang, H., Li, S., Zhang, L. et al. (2019). Effect of microgravity on the solidification of aluminum–bismuth–tin immiscible alloys. *npj Microgravity* 5 (1): 26.

39 Li, W., Jiang, H., Zhang, L. et al. (2019). Solidification of Al–Bi–Sn immiscible alloy under microgravity conditions of space. *Scripta Materialia* 162: 426–431.

40 Zhao, J.-Z., Ahmed, T., Jiang, H. et al. (2017). Solidification of immiscible alloys: a review. *Acta Metallurgica Sinica* 30: 1–28.

15

Layer-by-Layer Deposition in Microgravity for Enhanced Thin-Film Production

Daniel B. Lawrence[1], Jordan A. Greco[1,2], Robert R. Birge[1,2], and Nicole L. Wagner[1,2]

[1] *LambdaVision Incorporated, 400 Farmington Avenue, Farmington, CT 06032, USA*
[2] *University of Connecticut, Department of Chemistry, 55 North Eagleville Road, Storrs, CT 06269, USA*

15.1 Background

Thin-film production via layer-by-layer (LBL) assembly has been implemented to manufacture technologies with diverse applications, including those involving advanced materials, bioengineering, drug delivery, and regenerative medicine (Richardson et al. 2016; Birge et al. 1999). The LBL manufacturing technique, originally proposed by Decher and Hong (1991), is broadly defined by the alternating deposition of discrete building blocks that coat a surface or interface. The methodology conventionally utilizes the sequential exposure of positively charged and negatively charged substances to assemble a multilayered construct (Gentile et al. 2015); however, other molecular interactions (e.g. hydrogen bonding, covalent bonding) have since been utilized to create stable and uniform thin films with unique activities (Borges and Mano 2014). Manufacturing using LBL assembly allows for precise control of coating properties, mild conditions and low-cost processes, and versatility for coating diverse surfaces.

The sequential deposition of layers is achieved using a number of different methods, including immersion (He et al. 1998), spray coating (Schaaf et al. 2012), spin coating (Lee et al. 2001), and fluidic channels (Wang et al. 2011) (Figure 15.1). Though there have been many advancements that have expanded the definition of LBL to include additional techniques, e.g. three-dimensional printing (Andres and Kotov 2010) and bioprinting (Wilson and Boland 2003), we focus primarily on the conventional immersion and fluidic techniques to discuss our application. These traditional LBL methods are often facilitated by automated dipping machines or fluidic chamber assemblies. Substrates are exposed to solutions containing the polyions of interest while rinsing steps are used to remove unbound material. The deposition process relies on diffusion-driven kinetics to transport and promote adsorption of layering materials onto the substrate. Improvements on this technique have been hardware-focused and related to modifying the timing and sequence

In-Space Manufacturing and Resources: Earth and Planetary Exploration Applications, First Edition. Edited by Volker Hessel, Jana Stoudemire, Hideaki Miyamoto, and Ian D. Fisk.

Figure 15.1 Examples of conventional LBL assembly techniques, including the use of dip coating, spraying, spin coating, and fluidic chamber designs.

of deposition periods. However, little has been done to enhance the efficiency of diffusion-mediated deposition processes and to moderate the inherent physical processes (e.g. convective motion, solution gradients) that can negatively impact deposition of layers. In this chapter, we discuss the use of a microgravity environment to improve the diffusive transport of polyions to film surfaces, thus improving the homogeneity, orientation of layers, and stability of the LBL-assembled thin films.

LambdaVision, a University of Connecticut spinoff company, is developing a protein-based artificial retina technology to treat blind patients with advanced retinal degenerative diseases. The company is one of the first groups to investigate the effectiveness of LBL assembly in microgravity, and has utilized principles and findings from other microgravity-based investigations involving the self-assembly and deposition of layered materials via weak interlayer interactions (e.g. electrostatic adsorption, dipole–dipole interaction, van der Waals forces) (Ahari et al. 1997; Dag et al. 1997) to guide our manufacturing efforts. A microgravity environment is known to eliminate buoyancy-driven convective turbulence and Stokes sedimentation effects, which leads to a decrease in solute aggregation, a reduction of intralayer defects, efficient interlayer adsorption, and an increase in the optical clarity of the multilayered film (Dag et al. 1997). The influence of microgravity on self-assembly has been demonstrated with protein crystallization (DeLucas et al. 1989), including bacteriorhodopsin (BR) (Wagner and Linhardt 1991), as well as microporous materials (Ahari et al. 1997; Dag et al. 1997), and we expect that the three-dimensional assembly of the protein-based artificial retina technology will similarly benefit from a low-Earth orbit (LEO) environment throughout the protein/polymer deposition cycles in our manufacturing approach.

Utilization of the capabilities of the International Space Station (ISS) and private partners to stimulate demand and catalyze markets leading to broad LEO-based commercial opportunities is of high importance and is driving research and development of terrestrial-based applications that can benefit from a microgravity environment. The small size of the artificial retina, low volume and mass of the solutions and manufacturing apparatus, limited voltage requirements, and minimal need for human intervention allow for LEO commercialization to be an economically feasible opportunity. Furthermore, because the LBL method can be used to produce thin films with controlled micro-scale properties, it can be used in many other high-value applications, all while being inherently small in scale and power consumption. In the following sections, we discuss potential applications of LBL-assembled films, including LambdaVision's artificial retina technology, and we will provide an overview of the impact that a microgravity environment can have across many diverse fields.

15.2 Layer-by-Layer Applications

15.2.1 Terrestrial Applications of Significance

LBL assembly is a manufacturing method with diverse applications. Areas actively under research include medicine, nano-electronics, energy production and storage, gas/solute separation, and specialty coatings. In the field of medicine, LBL has been used to enhance scaffolds for tissue engineering (Shah et al. 2011) and for drug-delivery systems (Scheffler et al. 2020). LBL-assembled films can be loaded with higher drug concentrations than traditional encapsulation methods (Hammond 2012), and polyion selection allows for custom release profiles. For example, bone morphogenic protein-2 (BMP-2) is a potent inducer of osteoblast differentiation and is used to enhance bone graft therapies. However, the molecule has a short half-life and high doses are associated with serious adverse events. LBL coating of orthopedic implants with BMP-2 can increase the half-life nearly threefold (Martin et al. 2021).

In addition to improved drug loading and tailored release profiles, LBL encapsulation of pharmaceuticals benefits from the mild manufacturing conditions involved. Denaturing of biological therapeutics like proteins, deoxyribonucleic acid (DNA), and ribonucleic acid (RNA) is reduced by LBL processing due to the elimination of solvents used in standard encapsulation (Becker, Johnston and Caruso 2010). These features have been leveraged for development of shelf-stable microneedle-delivered vaccines (Turvey et al. 2019). For traditional delivery of vaccines, adjuvants can be incorporated to finely tune the resulting immune response (Tang et al. 2019).

LBL manufacturing techniques have been used in the development of novel diagnostic devices, as well. Amperometric glucose sensors based on this approach show excellent sensitivity and linear range (Wu, Lu and Ye 2019). The LBL construction of the sensors is readily amenable to miniaturization, so the sensors can be incorporated into *in vivo* and *ex vivo* diagnostic devices. LBL approaches are being used to detect a variety of analytes, positioning the technology to drive the next generation of biosensors.

Additionally, the ability of LBL coatings to produce functional membranes has applications in energy storage and production. For example, polyelectrolyte coatings that facilitate proton conduction while maintaining anion impermeability may replace Nafion™ in fuel cell proton exchange membranes. In addition to fuel cell membranes, LBL coatings can functionalize the catalytic layers for biofuel cells (Kwon et al. 2018), or improve the working life of battery electrodes (Pomerantseva and Gogotsi 2017).

15.2.2 Potential Microgravity Applications

The cost of manufacturing items in LEO at present is heavily influenced by the mass and volume of material transported to and from orbit. Therefore, products must have a combination of the following characteristics: high-value, minimal volume, and improved quality over terrestrially manufactured items. LBL is a method used to produce films with controlled micro-scale properties; thus, many applications are inherently low volume. Among technologies that benefit from a LBL manufacturing approach, many of the biotechnology and biopharmaceutical applications discussed above show particular potential for production of low-volume, high-value products in LEO.

While biotechnology shows unique potential for LBL manufacturing on-orbit for terrestrial use, the broader array of LBL applications can be leveraged for more efficient manufacturing of materials for use in space. Once a LBL manufacturing system is in place, only the cost of launching feedstocks needs to be considered. Saving the expense of payload volume is already being explored with other missions, such as the planned launch of the Made in Space, Inc. On-Orbit Servicing, Assembly, and Manufacturing (OSAM-2) in-space manufacturing platform (Harbaugh 2020). As large-scale manufacturing in space for space continues to develop, the needs to treat the various generated components to handle the harsh conditions of space will become increasingly important. LBL manufacturing can be used to apply coatings that can help prevent erosion and degradation.

15.2.3 Benefits of Microgravity

LBL manufacturing conducted on Earth can face issues with aggregation of solutes (Richardson, Bjornmalm and Caruso 2015), and gravity has been shown to influence self-assembly processes that are dependent upon weak intermolecular interactions (Dag et al. 1997; DeLucas et al. 1989; Wagner and Linhardt 1991; Ahari et al. 1997). A microgravity environment reduces buoyancy-driven convective turbulence and Stokes sedimentation, leaving Brownian motion as the primary driver of solute transportation. By reducing the influence of these additional forces, a reduction in perturbation-induced imperfections can be achieved for sensitive self-assembly applications. This effect is highlighted by previous work that demonstrated improved layer registry in layered microporous tin(IV) sulfides, which are sensitive to disturbances during crystallization on Earth (Dag et al. 1997).

LBL-assembled films that require the uniform orientation of layers (e.g. LambdaVision's artificial retina technology) or precise alignment between layers are most likely to benefit from manufacturing in a microgravity environment. These could take the form of active films that transport material across a substrate, passive films such as membranes or filters where small defects in layers could allow undesired flow of material, or films with multiple active components requiring precise interaction between multilayers. LBL products that are tolerant of layer inhomogeneity may not directly benefit from a microgravity environment, but may be integrated into a LEO manufacturing workflow.

15.2.4 Considerations for Microgravity LBL Manufacturing

Of the common methods used for LBL manufacturing (Figure 15.1), a fluidic approach is most easily adapted to operation in a microgravity environment. Fluidic

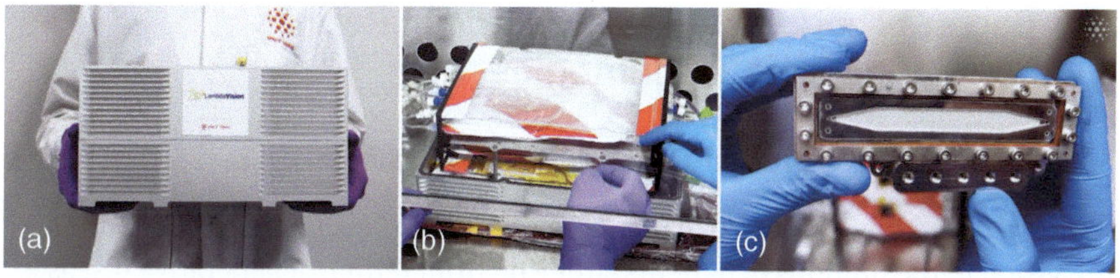

Figure 15.2 (a) The CubeLab™ developed by Space Tango and LambdaVision that was used on Northrop Grumman's 15th Commercial Resupply Services Mission for NASA (NG-15). (b) Media bags are suspended within the CubeLab™ framework and are used to store stock solutions of polyions and buffers for use in LBL manufacturing cycles. (c) The chamber that contains the scaffold support used for LBL assembly is specifically designed for use in a microgravity environment. Source: Space Tango Inc.

layering devices utilize closed systems, which are essential for fluid containment when manufacturing in LEO. Efficient use of space is also currently a priority when designing payloads for spaceflight. A fluidic approach provides flexibility in the layout of individual components, allowing for the compact arrangement of the hardware. Minimal design adjustments are needed to adapt terrestrial fluidic devices to LEO manufacturing, but compatibility of the payloads with the ISS facilities must be considered including payload dimensions, input–output interfaces, and voltage requirements. Collaborating with an implementation partner can minimize the time spent navigating these requirements and securing flights.

Figure 15.2 demonstrates the fluidic chamber hardware currently under development in collaboration with LambdaVision's implementation partner, Space Tango, Inc. (www.spacetango.com). Space Tango's CubeLab™ modules are generally designed for microgravity research and development, and each module is inserted into the TangoLab™ platform aboard the ISS to execute automated experiments and pilot manufacturing trials. The CubeLab™ payload shown in Figure 15.2a contains an automated LBL fluidic chamber assembly that can facilitate the production of LambdaVision's artificial retina technology and other LBL-assembled products. This closed-loop layering apparatus has been optimized for use in a LEO environment, where stock solutions of polyions and wash buffers are contained in media bags (Figure 15.2b) and infused into a custom chamber (Figure 15.2c) that contains an immobilized scaffold support for the multilayers. The hardware and software required for performing the LBL manufacturing cycles are designed to be self-contained and run autonomously on the ISS, which is important for efficiency and sterility of the proposed biomedical applications.

The biggest challenge to fluidic LBL manufacturing in microgravity is the formation and dynamic behavior of bubbles. Bubbles are also a problem for terrestrial fluidic devices; however, many terrestrial methods for bubble mitigation are not practical in microgravity. Design considerations should work to eliminate air from the system, when possible, or maintain liquid–air interfaces to prevent trapped bubbles. Extensive ground testing and a firm understanding of how gravity is influencing bubble behavior in the fluidic system are necessary to ensure success during manufacturing in LEO.

15.3 Bacteriorhodopsin-Based Artificial Retinas and Other Biophotonic Technologies

15.3.1 Significance of Protein-Based Artificial Retinas

The protein-based, ion-mediated artificial retina under development is intended to restore vision to the millions of patients suffering from retinal degenerative diseases, particularly retinitis pigmentosa (RP) and age-related macular degeneration (AMD). These diseases involve damage to the photoreceptor cells of the eye, and lead to a loss of independence for the individual and, eventually, blindness (Winter, Stuart and Rizzo III 2007). To date, no cure exists for patients with RP or AMD and limited treatments are available (Tamaki 2007).

The artificial retina is intended to first treat patients with advanced RP, with potential for a follow-on indication to the larger market of dry AMD. RP is the most common form of heritable retinal degeneration, affecting approximately 1.5 M people globally and <200 K in the US (1 in ~4000 individuals) (Berson 1996). RP is designated as a rare disease and qualifies for orphan designation by the Food and Drug Administration (FDA) (Weleber 2005). Our expected follow-on indication of AMD is the leading cause of irreversible vision loss and legal blindness in patients over the age of 55 (Friedman et al. 2004). Globally, an estimated 196 M people have some form of AMD (Wong et al. 2014), with almost 11 M suffering from moderate to severe vision impairment or blindness (Flaxman et al. 2017).

In addition to the common forms of retinal degeneration experienced terrestrially, LambdaVision's artificial retina technology has the potential to address visual complications experienced by astronauts who undergo extended-duration spaceflight. A recent ISS-based mouse study demonstrated a reduction in retinal thickness and performance, degradation of cone photoreceptors, increased oxidative stress to the ocular tissue, and disease-causing changes in gene expression (Overbey et al. 2019). While more research must be conducted to understand the effects of long-term spaceflight on the human anatomy, progressive visual impairment is a concern for future long-duration missions to the ISS and beyond. Artificial retinas or other end-stage treatments for retinal degeneration will not prevent these issues, but the technologies may offer terminal treatment options to reverse permanent loss of visual acuity.

15.3.2 Bacteriorhodopsin-Based Artificial Retinas

The BR-based artificial retina under development uses the photoactive protein, BR, to convert light energy into an ion gradient that activates the remaining neuronal network of degenerated retinas (Chen and Birge 1993). The implantable treatment is small, flexible, and is activated by incident light to generate a visual response. The artificial retina is a global treatment for all forms of RP and does not rely on the genetic cause of the disease. Moreover, the product does not require external electronics, which simplifies surgical implantation procedures. The implant replaces

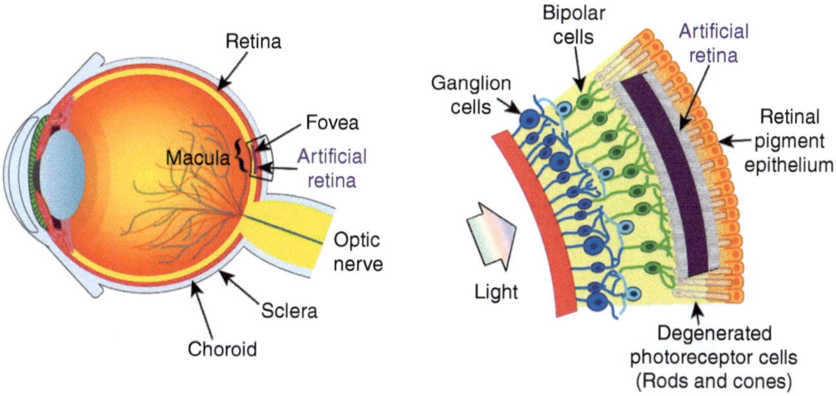

Figure 15.3 Application of the BR-based artificial retina technology as an implantable subretinal film. The subretinal BR-based artificial retina replaces the degenerated photoreceptor cells of the retina. Upon the absorption of light, the implant generates a proton gradient that stimulates the remaining retinal neurons.

the function of the photoreceptor cells and interfaces with the bipolar and ganglion cells to transmit visual information to the brain (Figure 15.3).

The artificial retina comprises multiple layers of oriented BR on a flexible, ion-permeable membrane. The multilayered architecture is enabled by a LBL electrostatic adsorption approach that is used to generate highly uniform films that can create a unidirectional proton gradient (He et al. 1998). To harness the ability of BR to generate a sufficient proton gradient for retinal stimulation, each protein layer is uniformly oriented in the LBL-assembled film, and the implant contains enough layers of BR to adequately absorb incident light. Using multilayers to achieve a high optical density mimics the native layering of the rhodopsin-containing disk membranes within the outer segments of rod and cone cells.

15.3.2.1 Bacteriorhodopsin

The protein-based artificial retina under development uses the light-activated protein, BR, as the photoactive biomaterial used to both absorb light and generate the proton gradient used to activate the remaining neurons of degenerated retinas (Chen and Birge 1993). BR (MW ≅ 26 kD) is a light-activated, transmembrane protein expressed by the salt marsh *archaeon*, *Halobacterium salinarum* (Figure 15.4a). The protein contains a covalently linked chromophore, all-*trans* retinal, which is responsible for absorbing light and undergoing an isomerization event to 13-*cis* retinal that initiates the proton pumping mechanism of the protein (Bogomolni et al. 1976). This process, conventionally known as the BR photocycle, involves a series of discrete photointermediates and is achieved in roughly 10–20 ms (Figure 15.4b). BR monomers are arranged in trimers within a two-dimensional crystalline lattice in its native lipid environment. This lattice of trimers is commonly referred to as the purple membrane (PM), fragments of which are isolated and purified before incorporation into the artificial retina.

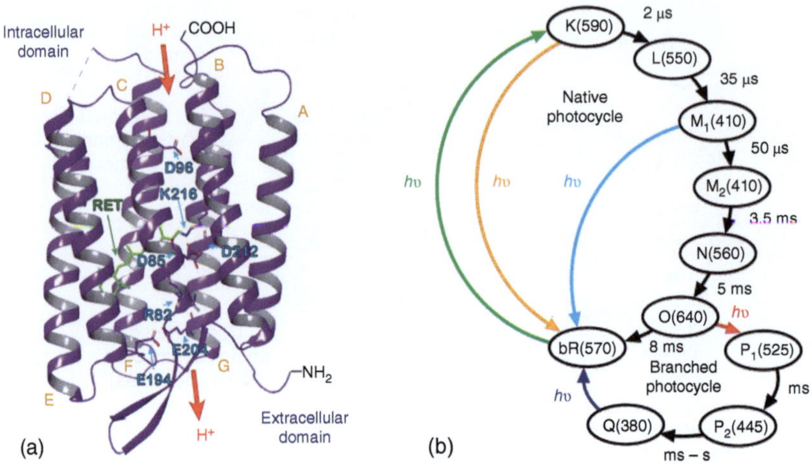

Figure 15.4 (a) Bacteriorhodopsin is a seven trans-membrane α-helical protein that is native to the archaeon, *H. salinarum*. All-trans retinal (RET) is covalently linked to K216 in helix G and is responsible for initiating a photoactivated proton-pumping mechanism. (b) The main and branched photocycles of BR are shown, including the photointermediate lifetimes and absorption maxima (in parentheses and in units of nm).

The inherent stability of the PM has led to the development of numerous BR-based bioelectronic applications (Birge et al. 1999; Hampp 2000). The proton-pumping function of the protein operates at a high quantum efficiency (65%) that is nearly identical in efficiency to the native photoreceptor pigment, rhodopsin (Oesterhelt, Brauchle and Hampp 1991). The protein is able to withstand temperatures above 80 °C (Jackson and Sturtevant 1978), and the cyclicity of the protein (i.e. the number of times that the protein can photochemically cycle before the sample degrades by 1/e) exceeds 10^6, which is considerably higher than most synthetic photochromic materials (Ranaghan et al. 2010). Moreover, the protein can maintain function following exposure to diverse chemical environments, including a broad pH range ($3 < pH < 10$) and encapsulation in non-native polymer matrices. Thus, the common perception that non-native biological materials are too fragile to be incorporated into implantable materials does not apply to BR.

15.3.2.2 Ion-Mediated Stimulation of the Retina

Retinal degenerative diseases are characterized by the loss of the light-sensitive rod and cone (photoreceptor) cells, and though some concomitant inner retinal degeneration and remodeling occurs (Marc et al. 2003), the bipolar and ganglion cell network remains largely viable during advanced stages of degeneration (Humayun et al. 1999). Retinal ganglion cells express acid-sensing ion channels, allowing them to generate action potentials in response to acidic stimuli (Lilley, LeTissier and Robbins 2004). The BR proton-pumping activity and the multilayered architecture allow the artificial retina to generate a local proton gradient that can briefly overcome interstitial pH levels ($7.2 < pH < 7.4$) and activate these channels. Due to the importance of generating a sufficient light-activated ion gradient, the quality

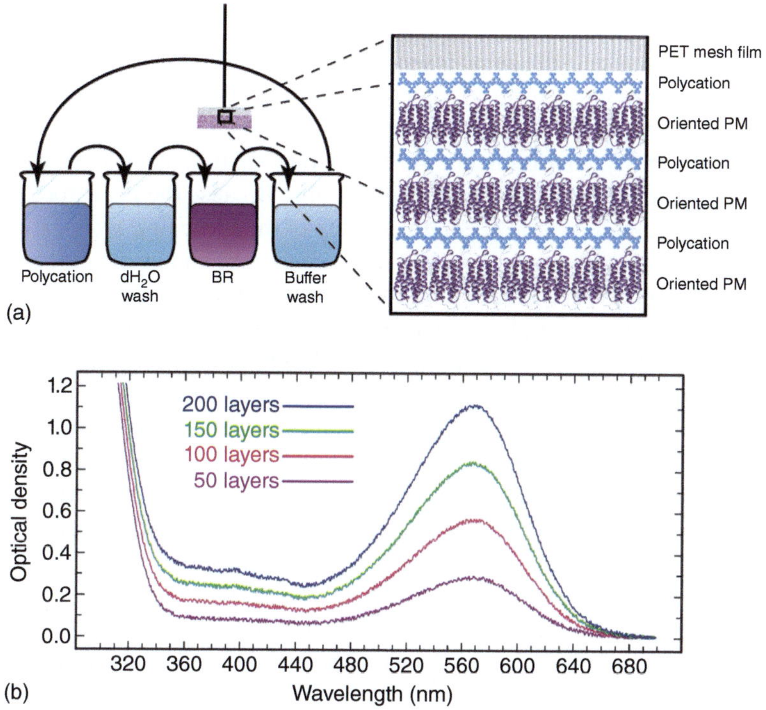

Figure 15.5 (a) Schematic of the dipping approach used to generate precision layers of protein on the PET mesh substrate. A series of solutions are implemented to accommodate the sequential LBL electrostatic adsorption of the polycation and BR. (b) Absorption spectra of the artificial retina films show that the optical density is directly proportional to the number of layers deposited following LBL assembly.

of manufacturing and the efficiency of the multilayered product are critical for successful patient outcomes.

15.3.3 Artificial Retina Manufacturing

The protein-based artificial retina comprises three main components: (i) a polyethylene terephthalate (PET) mesh substrate, (ii) BR contained within the PM, and (iii) a polycation binder. In a terrestrial setting, the multilayers are generated via sequential LBL electrostatic adsorption using a custom-made LBL dipping machine (Figure 15.5a) (He et al. 1998). The dipping approach allows for only one surface of the film to be coated with protein and polymer layers, which is important for the function of the technology. The process of depositing polycation/BR bilayers is repeated until the desired number of layers is obtained. A sufficient number of protein layers is required to adequately absorb incident light and generate a pH gradient for retinal stimulation. Figure 15.5b demonstrates that the optical density of the BR-based films is related to the number of layers deposited.

15.3.4 LBL-Assembled Artificial Retinas in Microgravity

The LEO environment of the ISS provides an excellent platform to remove the effects of gravity when manufacturing LBL-assembled thin films, and the advantages of microgravity could be essential in enhancing implant performance and stability. In a microgravity environment, we expect that each protein layer will adsorb more homogenously across the surface of the film. Because the LBL process is iterative, each subsequent layer relies on the quality of the preceding layers, and the improvement of homogeneity will lead to more consistent and stable electrostatic interactions. The protein will also orient uniformly so that all molecules are pointing in the same direction. Unidirectional orientation will strengthen the signal that is generated, which may allow us to use fewer layers of protein and reduce diffusion-related losses in resolution. Greater uniformity will lead to higher optical clarity and long-term structural stability due to enhanced electrostatic interaction between each of the layers. We are currently using microgravity to fully characterize these effects. Figure 15.2 depicts our current CubeLab™ LBL manufacturing prototype, which was built in collaboration with our implementation partner, Space Tango Inc. We continue to optimize hardware and software elements of the CubeLab™ apparatus for automated LBL assembly in microgravity on the ISS, while simultaneously developing an infrastructure to facilitate good manufacturing practices (GMP) in LEO for commercial products with application in clinical settings.

15.3.5 Other Bacteriorhodopsin-Based Biophotonic Applications

Shortly following the discovery of BR in 1971 (Oesterhelt and Stoeckenius 1971), Soviet scientists were the first to recognize the potential for developing biophotonic technologies based on the protein. Interest in creating biomolecular electronic devices was a byproduct of the Space Race, and the result was the first holographic processor based on BR thin films through the project, Biochrome (Bunkin et al. 1981). In subsequent decades of research, BR has emerged as a foundational optical material for use in biomolecular electronic applications (Birge et al. 1999; Hampp 2000). Recent advances in genetic engineering and directed evolution have further optimized BR for use in these promising applications (Wagner et al. 2013). Here, we briefly highlight select device applications of BR, with a focus on those that incorporate LBL for thin-film production. We also summarize our research on Space Shuttle missions in the 1990s, demonstrating the versatility of BR and the impact that microgravity can have on addressing solution gradients and optical clarity.

15.3.5.1 Thin-Film Applications

Beyond the artificial retina application, BR has been incorporated into LBL-assembled thin films for use in photovoltaics (Hong 1994), water-splitting devices (Johnson et al. 2018), and chemical sensors (Lanyi and Luecke 2001). Upon the absorption of a photon, BR transports a proton across the membrane, and this proton motive force results in a physical charge displacement that translates into an

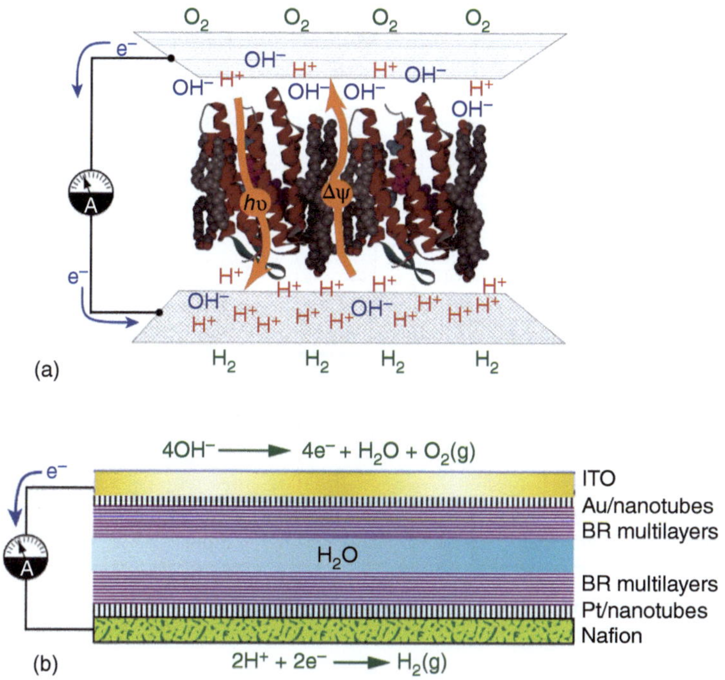

Figure 15.6 (a) A flow chart showing oriented BR that photochemically translocates protons toward the extracellular electrode creating a pH gradient that can be exploited for water splitting and photovoltaic energy. (b) A cell architecture with multilayers of BR that can be used to split water, including an indium tin oxide (ITO) anode with gold/nanotubes and Nafion with platinum on carbon nanotubes as the cathode.

electrical signal following light activation (Figure 15.6a). Magnification of the electrical signal occurs when oriented protein is condensed into a film on an electrically conductive surface (Váró and Keszthelyi 1983). BR-based heterostructures have demonstrated the ability to generate an electric current (Miyasaka and Koyama 1992), which has been implemented in a number of electrical circuit models (Birge and Rangarajan 2015), including those that can be used for water splitting (Figure 15.6b). The magnitude, duration, and polarity of the photovoltaic response is dependent on the quality of LBL thin-film assembly.

Multilayers of BR have also been examined as a biomaterial for use as a chemical sensor (Lanyi and Luecke 2001), in which deposited protein on a solid surface can be used as a medium to adsorb and detect foreign molecules (e.g. toxic agents in biological weapons). The BR photocycle (Figure 15.4b) can be monitored using time-resolved absorption spectroscopy to evaluate the concentration-dependent response following exposure to the target ligand.

15.3.5.2 Protein-Based Volumetric Memories and Associative Processors

Beyond the LBL-assembled thin-film applications described above, BR has been implemented into alternative architectural motifs, including holographic associative processors and three-dimensional (or volumetric) optical memory applications

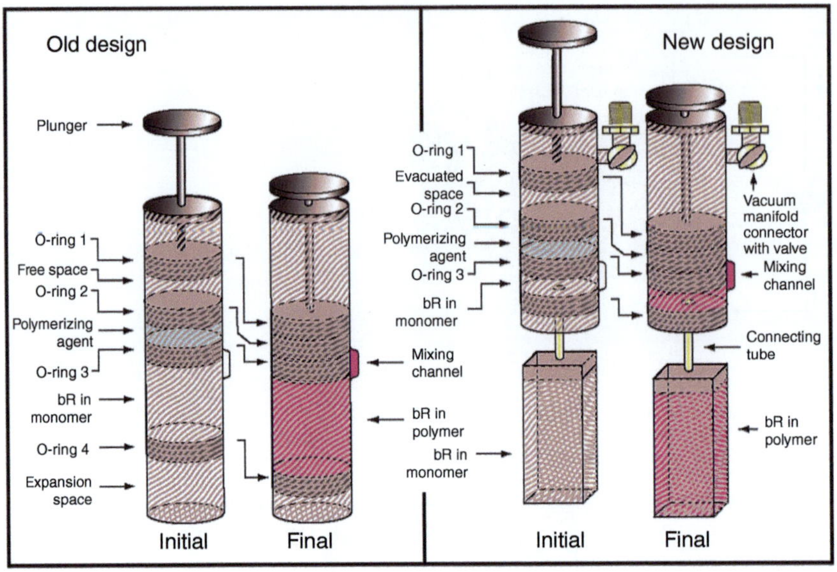

Figure 15.7 Microgravity polymerization test fixtures used within the Commercial Generic Bioprocessing Apparatus on Columbia STS-50 (1992), Endeavor STS-54 (1993), and Endeavor STS-69 (1995).

(Birge et al. 1999; Stuart et al. 1996). These technologies rely on the unique photochemistry of BR (Figure 15.4b) and the immobilization of BR in a polymer matrix to generate thin films or volumetric data storage media. Because the optical clarity of the thin films or cuvettes used for these applications is critical for performance, the polymerization of protein/polymer suspensions must be uniform and avoid gravity-induced gradients and sedimentation. A microgravity environment was explored to create more homogenous polymer suspensions for the optical memory application. Three Space Shuttle flights were carried out to examine if microgravity polymerization might enhance the homogeneity of the protein-based photoactive memory medium used in our branched-photocycle optical three-dimensional memory (Birge et al. 1999). The motivation behind these studies followed from the observation that polymerization of the BR and polyacrylamide matrix in gravity resulted in a density gradient such that the amount of protein increased slightly at the bottom of the cuvette due to gravitational forces.

The apparatuses that were used in these microgravity studies are shown in Figure 15.7. The first two flights were carried out using the "old design," which was improved for the final flight (Space Transportation System [STS]-69) as shown in the "new design." Both apparatuses were automated and were contained during the flights in the Commercial Generic Bioprocessing Apparatus. The key observation of these studies was that microgravity does indeed improve homogeneity of the protein/polymer cuvettes to the point that whatever inhomogeneities were remaining were not causing any diminution in optical storage density.

15.4 Future Directions in LEO Commercialization

In the sections above, we have described the advantages of LBL deposition in microgravity and provided examples of a number of technologies that could benefit from an improved manufacturing paradigm supported by LEO resources. Over the past two decades, advances in tissue engineering, regenerative medicine, and advanced materials have demonstrated the full potential of learning from basic scientific analyses in LEO and translating those outcomes to viable products that provide economic value and benefits to life on Earth. We are still in the early stages of demonstrating the full potential of microgravity on the LBL manufacturing process; however, potential enhancements hold promise in improving the homogeneity, orientation, and stability of multilayered thin films for numerous applications. In addition to our review of several potential LBL-mediated biomedical applications, we have highlighted several applications of BR that require oriented multilayers of the protein on a scaffold, including the design of photovoltaic cells, water-splitting devices, chemical sensors, and, specifically, LambdaVision's artificial retina technology. BR has been at the forefront of biotechnology development dating back to its discovery in the 1970s. Due to the inherent stability and unique photochemistry of the protein, BR continues to represent a valuable and versatile biomaterial that can pave the way for future technology development, infrastructure establishment, and product discovery.

LambdaVision's current needs of pilot-scale production are well suited for the near-term goals of National Aeronautics and Space Administration (NASA's) commercial LEO development strategy, in which we will continue to leverage ISS capabilities and spaceflight opportunities to explore commercial development in LEO. While technology-related manufacturing applications in microgravity are under consideration, capabilities to support medical device and implantable manufacturing applications have yet to be adequately addressed. Specifically, an infrastructure for GMP production in LEO must be established to generate safe and effective biomedical products and to fully meet FDA standards. Similar to physical property alternations in microgravity that facilitate improved technology products, changes in both physical and biological processes in microgravity have the potential to reshape the current paradigm of production for patient therapies involving artificial retinas, LBL-mediated drug products, and other materials used in a clinical setting.

Our artificial retina technology offers a unique test case for developing a product in microgravity. By targeting an orphan disease first, we will have the opportunity to meet a reasonably low demand for pilot-scale GMP production before establishing steady-state production for larger markets and alternative technology sectors. We will continue to leverage ISS capabilities and spaceflight opportunities to explore commercial development in LEO, and as we look ahead to developing the infrastructure for GMP in LEO, our manufacturing will be congruous with NASA's transition to using commercial allocations attached to the ISS or to using free flyers from commercial partners. On a larger scale, LambdaVision's commercial efforts in microgravity will demonstrate that it is feasible to manufacture a drug or therapy using a LEO platform, and this achievement will continue to inspire new research and commercial product development that can foster a thriving LEO economy.

References

Ahari, H., Bedard, R.L., Bowes, C.L. et al. (1997). Effect of microgravity on the crystallization of a self-assembling layered material. *Nature* 388: 857–860.

Andres, C.M. and Kotov, N.A. (2010). Inkjet deposition of layer-by-layer assembled films. *Journal of the American Chemical Society* 132: 14496–14502.

Becker, A.L., Johnston, A.P., and Caruso, F. (2010). Layer-by-layer-assembled capsules and films for therapeutic delivery. *Small* 6: 1836–1852.

Berson, E.L. (1996). Retinitis pigmentosa: unfolding its mystery. *Proceedings of the National Academy of Sciences of the United States of America* 93: 4526–4528.

Birge, R. R. and Rangarajan, R. (2015). Protein-based photovoltaics and methods of use. US Patent 9,023,989, filed 19 February 2008 and issued 5 May 2015.

Birge, R.R., Gillespie, N.B., Izaguirre, E.W. et al. (1999). Biomolecular electronics: protein-based associative processors and volumetric memories. *Journal of Physical Chemistry B* 103: 10746–10766.

Bogomolni, R.A., Baker, R.A., Lozier, R.H. et al. (1976). Light-driven proton translocations in *Halobacterium halobium*. *Biochimica et Biophysica Acta (BBA) - Bioenergetics* 440: 68–88.

Borges, J. and Mano, J.F. (2014). Molecular interactions driving the layer-by-layer assembly of multilayers. *Chemical Reviews* 114: 8883–8942.

Bunkin, F.V., Vsevolodov, N.N., Druzhko, A.B. et al. (1981). Diffraction efficiency of bacteriorhodopsin and its analogs. *Soviet Technical Physical Letters* 7: 630–631.

Chen, Z. and Birge, R.R. (1993). Protein based artificial retinas. *Trends in Biotechnology.* 11: 292–300.

Dag, O., Ahari, H., Coombs, N. et al. (1997). Does microgravity influence self-assembly? *Advanced Materials* 9: 1133–1149.

Decher, G. and Hong, J.D. (1991). Buildup of ultrathin multilayer films by a self-assembly process: 2. Consecutive adsorption of anionic and cationic bipolar amphiphiles and polyelectrolytes on charged surfaces. *Berichte der Bunsengesellschaft für physikalische Chemie* 95: 1430–1434.

Delucas, L.J., Smith, C.D., Smith, H.W. et al. (1989). Protein crystal growth in microgravity. *Science* 246: 651–654.

Flaxman, S.R., Bourne, R.R.A., Resnikoff, S. et al. (2017). Global causes of blindness and distance vision impairment 1990–2020: a systematic review and meta-analysis. *Lancet Global Health* 5: e1221–e1234.

Friedman, D.S., O'colmain, B.J., Munoz, B. et al. (2004). Prevalence of age-related macular degneration in the united states. *Archives of Ophthalmology* 122: 564–572.

Gentile, P., Carmagnola, I., Nardo, T. et al. (2015). Layer-by-layer assembly for biomedical applications in the last decade. *Nanotechnology* 26: 21.

Hammond, P.T. (2012). Building biomedical materials layer-by-layer. *Materials Today* 15: 196–206.

Hampp, N. (2000). Bacteriorhodopsin as a photochromic retinal protein for optical memories. *Chemical Reviews* 100: 1755–1776.

Harbaugh, J. (2020). On-orbit servicing, assembly, and manufacturing (OSAM-2). *National Aeronautics and Space Administration* https://www.nasa.gov/mission_pages/tdm/archinaut/index.html (accessed 13 February 2020).

He, J.-A., Samuelson, L., Li, L. et al. (1998). Oriented bacteriorhodopsin/polycation multilayers by electrostatic layer-by-layer assembly. *Langmuir* 14: 1674–1679.

Hong, F. (1994). Retinal proteins in photovoltaic devices. In: *Molecular and Biomolecular Electronics* (ed. R.R. Birge), 527–560. Washington, DC: American Chemical Society.

Humayun, M.S., Prince, M., De Juan, E. et al. (1999). Morphometric analysis of the extramacular retina from postmortem eyes with retinitis pigmentosa. *Investigative Ophthalmology & Visual Science* 40: 143–148.

Jackson, M.B. and Sturtevant, J.M. (1978). Phase transitions of the purple membranes of *Halobacterium halobium*. *Biochemistry* 17: 911–915.

Johnson, K.E., Gakhar, S., Risbud, S.H. et al. (2018). Development and characterization of titanium dioxide gel with encapsulated bacteriorhodopsin for hydrogen production. *Langmuir* 34: 7488–7496.

Kwon, C.H., Ko, Y., Shin, D. et al. (2018). High-power hybrid biofuel cells using layer-by-layer assembled glucose oxidase-coated metallic cotton fibers. *Nature Communications* 9: 4479.

Lanyi, J.K. and Luecke, H. (2001). Bacteriorhodopsin. *Current Opinion in Structural Biology* 11: 415–419.

Lee, S.-S., Hong, J.-D., Kim, C.H. et al. (2001). Layer-by-layer depositied multilayer assemblies of ionene-type polyelectrolytes based on the spin-coating method. *Macromolecules* 34: 5358–5360.

Lilley, S., Letissier, P., and Robbins, J. (2004). The discovery and characterization of a proton-gated sodium current in rat retinal ganglion cells. *Journal of Neuroscience* 24: 1013–1022.

Marc, R.E., Jones, B.W., Watt, C.B. et al. (2003). Neural remodeling in retinal degeneration. *Progress in Retinal and Eye Research* 22: 607–655.

Martin, J.R., Howard, M.T., Wang, S. et al. (2021). Oxidation-responsive, tunable growth factor delivery from polyelectrolyte-coated implants. *Advanced Healthcare Materials* 10: e2001941.

Miyasaka, T. and Koyama, K. (1992). Rectified photocurrents from purple membrane Langmuir-Blodgett films at the electrode-electrolyte interface. *Thin Solid Films* 210–211: 146–149.

Oesterhelt, D. and Stoeckenius, W. (1971). Rhodopsin-like protein from the purple membrane of *Halobacterium halobium*. *Nature New Biology* 233: 149–152.

Oesterhelt, D., Brauchle, C., and Hampp, N. (1991). Bacteriorhodopsin: a biological material for information processing. *Quarterly Reviews of Biophysics* 24: 425–278.

Overbey, E.G., da Silveira, W.A., Stanboly, S. et al. (2019). Spaceflight influences gene expression, photoreceptor integrity, and oxidative stress-related damage in the murine retina. *Scientific Reports* 9: 13304.

Pomerantseva, E. and Gogotsi, Y. (2017). Two-dimensional heterostructures for energy storage. *Nature Energy* 2: 17089.

Ranaghan, M.J., Shima, S., Ramos, L. et al. (2010). Photochemical and thermal stability of green and blue proteorhodopsins: implications for protein-based bioelectronic devices. *Journal of Physical Chemistry B* 114: 14064–14070.

Richardson, J.J., Björnmalm, M., and Caruso, F. (2015). Multilayer assembly. Technology-driven layer-by-layer assembly of nanofilms. *Science* 348: aaa2491.

Richardson, J.J., Cui, J., Bjornmalm, M. et al. (2016). Innovation in layer-by-layer assembly. *Chemical Reviews* 116: 14828–14867.

Schaaf, P., Voegel, J.C., Jierry, L. et al. (2012). Spray-assisted polyelectrolyte multilayer buildup: from step-by-step to single-step polyelectrolyte film constructions. *Advanced Materials* 24: 1001–1016.

Scheffler, K., Bilz, N.C., Brueckner, M. et al. (2020). Enhanced uptake and endosomal release of LbL microcarriers functionalized with reversible fusion proteins. *ACS Applied Bio Materials* 3: 1553–1567.

Shah, N.J., Macdonald, M.L., Beben, Y.M. et al. (2011). Tunable dual growth factor delivery from polyelectrolyte multilayer films. *Biomaterials* 32: 6183–6193.

Stuart, J.A., Tallent, J.R., Tan, E.H.L. et al. (1996). Protein-based volumetric memories. *Proceedings of Nonvolatile Meory Technology Conference* 6: 45–51.

Tamaki, Y. (2007). Novel drug delivery system for age-related macular degeneration using nanotechnology. *Nanobiotechnology* 3: 107–115.

Tang, T., Weng, T., Jia, H. et al. (2019). Harnessing the layer-by-layer assembly technique to design biomaterials vaccines for immune modulation in translational applications. *Biomaterials Science* 7: 715–732.

Turvey, M.E., Uppu, D., Mohamed Sharif, A.R. et al. (2019). Microneedle-based intradermal delivery of stabilized dengue virus. *Bioengineering & Translational Medicine* 4: e10127.

Váró, G. and Keszthelyi, L. (1983). Photoelectric signals from dried oriented purple membranes of *Halobacterium halobium*. *Biophysical Journal* 43: 47–51.

Wagner, G. and Linhardt, R. (1991). Liquid-liquid diffusion profiles in microgravity experiments, and crystal growth of the membrane protein bacteriorhodopsin. *Journal of Crystal Growth* 110: 114–118.

Wagner, N.L., Greco, J.A., Ranaghan, M.J. et al. (2013). Directed evolution of bacteriorhodopsin for applications in bioelectronics. *Journal of the Royal Society Interface* 10: 20130197.

Wang, Y., Liu, Y., Cheng, Y. et al. (2011). Coupling electrodeposition with layer-by-layer assembly to address proteins within microfluidic channels. *Advanced Materials* 23: 5817–5821.

Weleber, R.G. (2005). Inherited and orphan retinal diseases: phenotypes, genotypes, and probable treatment groups. *Retina* 25: S4–S7.

Wilson, W.C. and Boland, T. (2003). Cell and organ printing 1: protein and cell printers. *Anatomical Record* 272A: 491–496.

Winter, J.O., Stuart, C.F., and Rizzo, J.F. III, (2007). Retinal prostheses: current challenges and future outlook. *Journal of Biomaterials Science, Polymer Edition* 18: 1031–1055.

Wong, W.L., Su, X., Li, X. et al. (2014). Global prevalence of age-related macular degeneration and disease burden projection for 2020 and 2040: a systematic review and meta-analysis. *Lancet Global Health* 2: e106–e116.

Wu, L., Lu, Z., and Ye, J. (2019). Enzyme-free glucose sensor based on layer-by-layer electrodeposition of multilayer films of multi-walled carbon nanotubes and cu-based metal framework modified glassy carbon electrode. *Biosensors and Bioelectronics* 135: 45–49.

16

3D Bioprinting Aboard the International Space Station Using the Techshot BioFabrication Facility

Carlos C. Chang and Eugene D. Boland

Techshot, Inc., 7200 US-150, Greenville, IN 47124, USA

16.1 Introduction

The BioFabrication Facility (BFF) is the first 3D bioprinter installed and operated on the International Space Station (ISS) that is capable of precision printing and culturing. The BFF is part of a larger plan to manufacture complex tissue-like structures in microgravity, with long-term goals of fabricating organs in space. Microgravity provides a unique third solution to printing complex organ structures, as minimal gravity removes the need for scaffolding structures to support complex tissue shapes. The BFF is a platform for researchers to print organ-like tissues and begin proving viability for human organ fabrication in space. BFF is used in conjunction with existing Techshot bioreactor cassettes, which hold the printed tissues for several weeks after initial printing, allowing the tissues to cohesively form on a cellular level. During this incubation period, the cassettes are housed inside the Techshot Advanced Space Experiment Processor (ADSEP). In time, the BFF payload could become a part of a larger system capable of manufacturing whole, fully functioning human organs from existing patient cells to reduce the burden of organ shortages for transplant.

16.2 Engineering a 3D Bioprinter for Use in Microgravity

3D bioprinting in microgravity presents a number of unique challenges compared to typical terrestrial operations that require specific engineering modifications of existing terrestrial technologies. The Techshot BFF incorporates a quad-head 3Dn BioAssembly Tool developed by nScrypt, Inc. (Orlando, FL). To reduce acceleration loads on the construct during printing, a gantry platform (3-axis motion applied to head rather than printing platform) with precision linear motors has also been developed. The linear drives of the BFF have a resolution of 100 nm, repeatability of 500 nm, and accuracy of 1 μm to assure printing detail at a physiologically significant size. This motion control system features high-velocity drives with a large print area. The *Z*-axis has 50 mm travel and velocities of greater than 400 mm/s from a

In-Space Manufacturing and Resources: Earth and Planetary Exploration Applications, First Edition. Edited by Volker Hessel, Jana Stoudemire, Hideaki Miyamoto, and Ian D. Fisk.

single motor. The Y-axis has dual motors and is capable of greater than 750 mm/s while supporting the full gantry with 140 mm of travel. The X-axis is also capable of greater than 750 mm/s and has a varying range of motions depending on the number of print heads used. With a single head, it possesses a maximum travel of 180 mm but with all four heads, the range is limited to a maximum of 105 mm (single head – 180 × 140 × 50 mm; quad head – 105 × 140 × 50 mm). As the printheads are arranged along the X-axis, range is determined by the region where all pumps can overlap the same point allowing for full dispensing capability. To limit imparted vibration and motion artifacts, the drives are limited to $0.5 \times G$ acceleration although each is capable of higher values. Filtered and humidified air is circulated within the print volume to maintain a consistent environment for hydrogel printing.

Dispensing is controlled by 4 nScrypt Model 100, Generation two SmartPumps. These digitally controlled microdispensers operate through positive pressure and patented valving to enable near-perfect starts and stops without over extrusion (drips) or under extrusion (voids). The pumps can handle materials from 1 centipoise (cP) to over 1 million cP with inks that contain cells, flake-loaded pastes, or simple hydrogels. Volumes as low as 100 pl can be dispensed and line widths as small as 20 μm can be printed through a variety of nTip ceramic cones or blunt needles. Material is fed into the SmartPumps through a Luer fitting in the valve body connected to a gas-driven dispensing syringe. For typical operations, a 10 ml volume is chosen and held within a cooled bracket on the face of the pump. The print chamber is illustrated in Figure 16.1 (original BFF version without thermal control of the SmartPumps).

The analogy of a pancake is helpful to explain our approach to 3D bioprinting. The pancake batter is much like the bioink. All the necessary components are present, but no one wants a pile of batter on their plate. Batter must be cooked to reach its optimal properties like a printed construct must be conditioned to reach biological significance. This is where BFF starts to differentiate itself from other bioprinters whether in space or in terrestrial applications. To call our conditioning chamber a bioreactor may be an understatement as many people equate bioreactors with large, stirred vessels for amplifying cell cultures. Techshot's system provides queues much like stem and progenitor cells receive during development of tissues *in vivo* in the womb. These signals are thermal, chemical, mechanical, and electrical. The proper balance, timing, and intensity of these signals will coax the printed cells, proteins, growth factors, and other nutrients to fuse into a cohesive solid. This solid could be as simple as a tissue-analog monoculture or as complex as a neo-organ.

16.3 Bioink Development

Extensive planning goes into each BFF experiment design and execution. The BFF system itself is outfitted with four independently controlled dispensing systems (referred to as SmartPumps). Planning begins by determining the number of SmartPumps that will be used and how many hydrogels (referred to as Bioinks) will be utilized to fabricate the structures of interest.

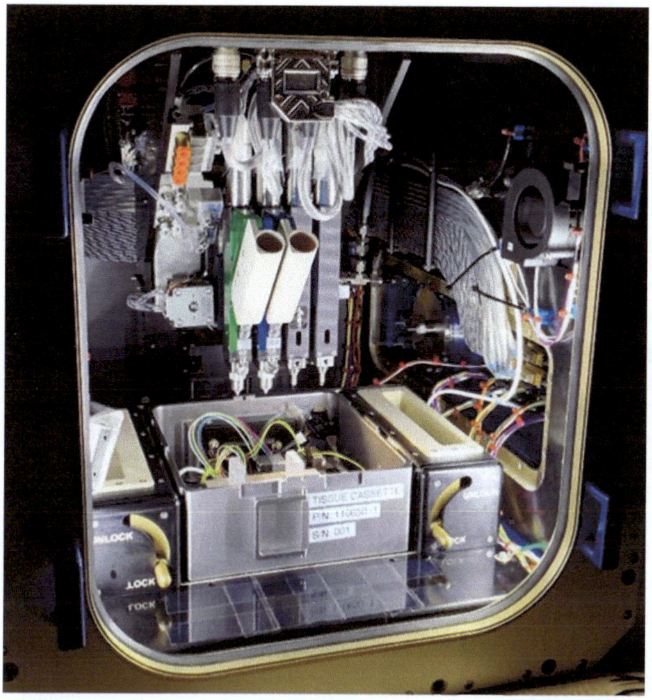

Figure 16.1 Interior of the Techshot BioFabrication Facility print volume. Two syringe holders are mounted on the face of the leftmost SmartPumps and a bioreactor cassette is installed to receive the printed construct. Source: Techshot.

Bioinks tend to be cell specific, containing both viable cells and cell-specific extracellular matrix (ECM) components to support cell development. For our initial experiments, we were interested in fabricating cardiac tissue–like structures utilizing four bioinks. The first three bioinks contained cells and ECM specific to cardiac, nerve, and vascular components. The fourth bioink was an acellular dissolvable hydrogel used to assist construct shape.

The cell-specific bioinks were each made with decellularized tissue (Figure 16.2). Tissues were aseptically minced, decellularized, and then treated with peracetic acid for viral inactivation. Tissue samples were then frozen, lyophilized, and cryomilled to a fine powder. The powder was further digested via enzymes, neutralized, and finally reconstituted into a hydrogel at concentrations between 3 and 6 mg ECM per ml.

For bioprinting activities, the hydrogel materials must meet a few requirements. First and foremost, the hydrogels must support cellular growth and development. Ground experiments were conducted to verify the hydrogels supported cell growth. Each cell type was cultured within its cell-specific hydrogel. Cells were first cultured in ideal conditions; hydrogels were suspended with the hydrogels, polymerized, and cultured within well plates in standard tissue culture incubators. Following successful culture, cells were then incorporated into bioinks and loaded into the bioprinter

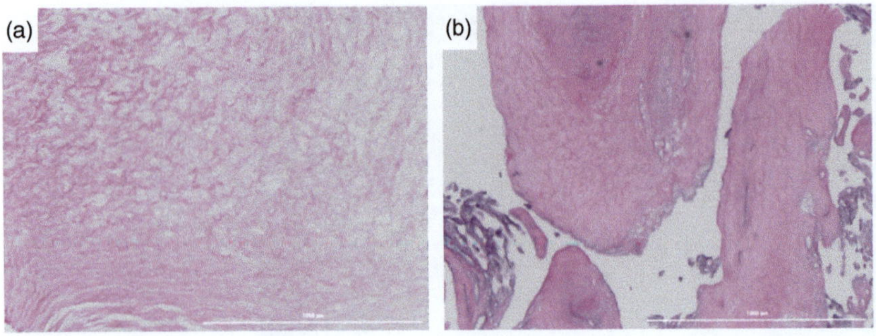

Figure 16.2 Images of decellularized human tissue sections. Hematoxylin and Eosin (H&E)-stained human umbilical cord (a) and placenta (b) tissues were decellularized and treated to remove any remaining viral or bacterial material within the tissue. The pink color illustrates the maintenance of protein structure within the processed tissue. The lack of dark-stained nuclei within the samples suggests efficient removal of cellular nucleic material. Scale bar in both images is 1000 μm. Source: Techshot.

for dispensing into simple patterns. Those patterns were again polymerized and cultured within standard tissue culture incubators.

Just prior to the printed culture evaluations, hydrogel polymerization time and viscosity were evaluated. Hydrogels were mixed and their polymerization rates were evaluated at different temperatures. In particular, we investigated polymerization rates at 4 and 20 °C. Looking forward to on-station activities, these were two temperatures of focus. The acellular ink component was planned to be stored at 4 °C. Once mixed with the cells, there would be a slow increase of the ink temperature to 20 °C – the operational temperature of BFF. In preliminary studies (Figure 16.3) it was important to determine the amount of time that the mixed hydrogels would be stable, maintaining a low viscosity compatible with dispensing at 20 °C.

16.4 Print Testing in Zero-G

To reduce cost and validate concepts during the development of a future ISS payload, alternate methods of obtaining short durations of microgravity are available to a researcher. One such available method, which was chosen by the Techshot-led space bioprinter team, was a parabolic flight provided by the commercial company Zero-G. During the parabolic flight, Zero-G is capable of producing a microgravity environment for 20–30 seconds as the specialized aircraft reaches the apex of its parabolic ascent. This maneuver is repeated up to 24 times which allowed sufficient print time and repetition to be a platform suitable to conduct the experimental printing of human tissue in a microgravity environment.

To complete this, Techshot partnered with nScrypt, an Orlando, FL-based Digital Manufacturing company, to modify a commercial-off-the-shelf (COTS) gel printer into a configuration amenable to printing a cell suspension material in a microgravity environment. In addition to the engineering effort to meet the physical constraints and requirements of a microgravity flight, printing processes and controls

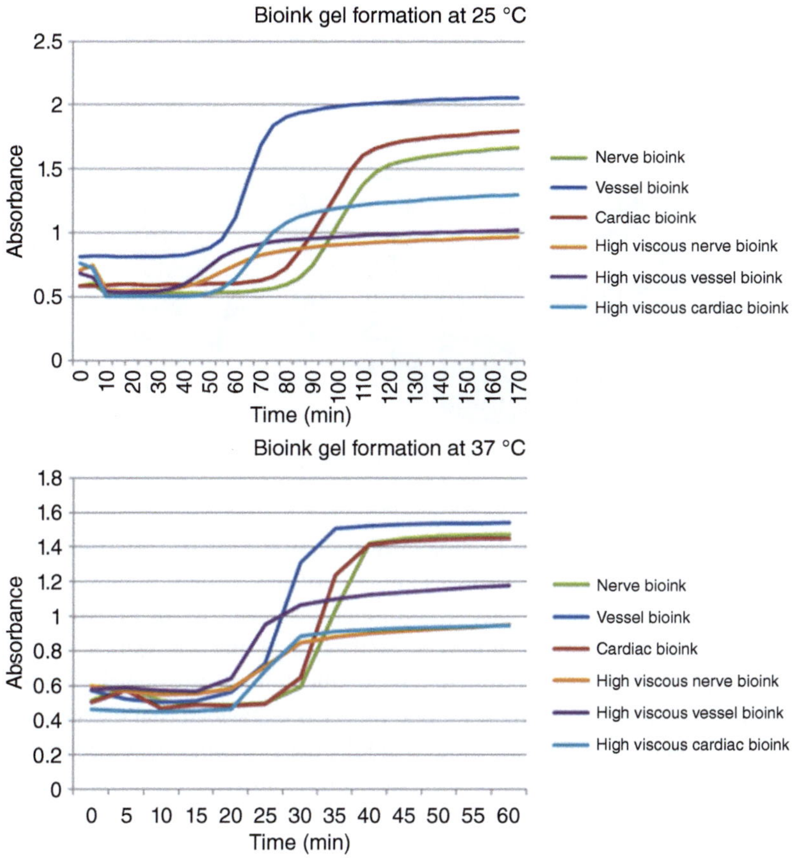

Figure 16.3 Hydrogel tests – viscosity versus time and temp for six variations of bioinks without exogenous cross-linkers. Source: Techshot.

unique to the space environment were developed. Upon completion of this effort, the first biological printer designed to print in a microgravity environment had been developed and readied for flight.

In June of 2016 Techshot flew the first bioprinter on a Zero-G parabolic flight from Orlando, FL, and printed the first neonatal heart ventricle using adult human stem cells (Figure 16.4).

During this flight various geometries to replicate vascular structures were printed with gels containing suspensions of human cells. The gel viscosities used to fabricate these vascular structures could only be printed in a microgravity environment as the mass and intricacy of the structure would collapse when subjected to a 1 g environment. Successful printing of these structures proved the viability and capability of the technology for longer-term microgravity research available on the ISS.

Ground-breaking biological goals accomplished through this research include optimization of gel viscosity for printing structures in microgravity, selection of cell size and type for tissue viability and printing of a nonviable neonatal-sized

Figure 16.4 Techshot Zero-G Flight of the BioFabrication Facility (BFF). Source: Techshot https://youtu.be/tAzcB_3dVzE.

heart (ventricles only). In addition to biological accomplishments, this research enabled the development of print processes and algorithms unique to fabricating gel-based structures in a microgravity environment, as well as demonstrating surface-tension-based layer fusion not seen with typical ink viscosities when printing on the ground. Furthermore, the mechanical hardware developed to conduct printing on the parabolic flight formed the basis of payload that eventually flew to the ISS. Techshot was now positioned to take the next iterative step which is further refinement to become a payload facility for conducting research on the ISS. Additional process automation, cell-ink mixing, syringe installation, and print assembly cleaning methods were needed to transfer a technical operation on the ground and parabolic flight to the time/training constraints of ISS operations.

Mostly due to the nature of the cyclic printing on the parabolic flight, a great deal of personnel interaction was required for each print. Looking forward to on-station activities, it was important to minimize the time and technical requirements to operate BFF. Modifications were made to syringe mixing of bioink components, syringe installation into BFF, and print assembly cleaning.

In terrestrial development activities, ink components were measured with digital pipettes in a biosafety cabinet just prior to printing activities. On the ISS, these typical "simple" operations are not logistically possible. Instead, we worked to prepare pre-loaded syringes of cells and ECM that could be connected. Safety considerations also required the incorporation of leak-proof connectors. Two revisions of connectors were incorporated into BFF.

The first set of connectors were stainless steel locking sleeve connectors with internal valves. During ground tests, the connectors were utilized in multiple successful test prints. In preliminary tests on the ISS, a few issues were presented. The flow path through the connectors proved to provide too small an internal opening, and the tortuous path around spring-loaded valves led to higher resistance and shear on the dispensed bioinks. The combination of these two issues caused clogging within BFF SmartPumps. In response to these issues, a second set of

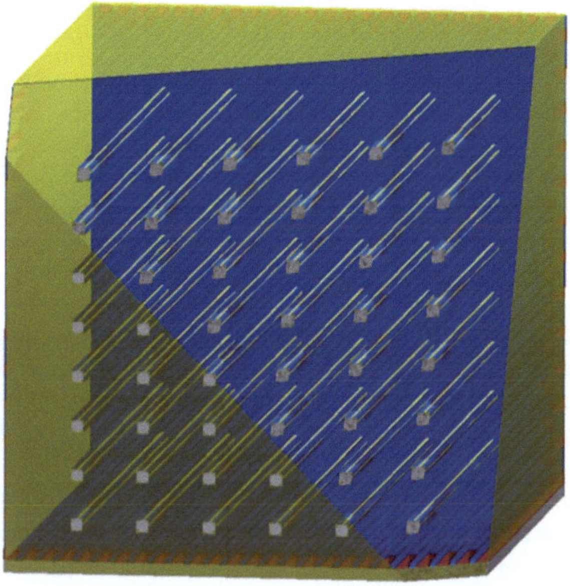

Figure 16.5 Cut-away view of tissue construct. Rods through the construct are sacrificial ink aligned to needle array within bioreactor chamber. Rods are wrapped in vascular cells/inks and nerve cells/inks. The bulk of the construct is cardiomyocyte ink/cells. Source: Techshot.

connectors were introduced. The second connectors were a simpler design. The internal fluid path diameter was much larger permitting lower resistance. In addition, the safety valves were internal and external silicone needleless septa, providing a fluid path of much lower resistance. There were far fewer clogging issues after changing to connector system 2.

Installation of the syringes into the BFF system was also streamlined. The first locking sleeve connectors did not present any issues during ground testing. However, on the ISS, the sleeves would often not engage, delaying prints and preventing proper BFF operation. The change to the second connectors introduced quarter-turn Luer connectors that were much more reliable.

Once installed, the operation of the printer could be conducted in microgravity. Our ISS printing activities were divided into two projects. The first project was the fabrication of a vascularized tissue-like construct. The second was the fabrication of a 3D meniscus-like structure. Both structures utilized all four SmartPumps and relied on materials with multiple viscosities.

The vascularized tissue construct, depicted in Figure 16.5, was designed as a rectangular prism containing an array of internal parallel channels. Each channel would be surrounded by vascular cells with cardiac and nerve cells distributed through the rest of the volume.

Thinking about how materials would interact in the absence of gravity, in particular low-viscosity hydrogels, surface interactions would be the dominant driving force of pattern structure. This thinking led to dedicating one bioink as a higher-viscosity, degradable hydrogel. This "structural" hydrogel would be used to provide a 3D framework to shape the lower-viscosity cell-suspended hydrogels. Once the cell-containing hydrogels polymerized, the structural hydrogel would be rinsed away.

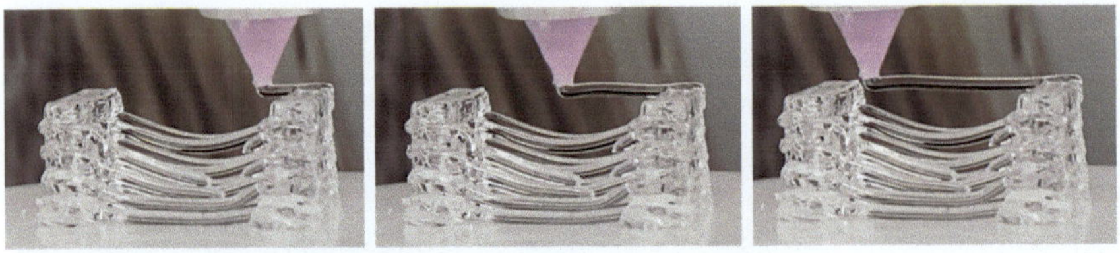

Figure 16.6 A collection of images illustrating the optimized dispensing of a free-standing channel of our high-viscosity support hydrogel. This example shows a channel of approximately 300 µm spanning a distance of approximately 30 mm. Source: Techshot.

In ground studies, we optimized the viscosity of the structural hydrogel to permit dispensing of an outer box to contain the low-viscosity cellular bioinks. Print parameters were also optimized to permit the printing of freestanding channels within the structure (Figure 16.6). The internal channels served the two important purposes in the print. First, the additional internal surfaces provided guides to help the shape and direct flow of the low-viscosity inks once they left the print tip. Secondly, once the cell containing hydrogels polymerized, these channels were made to dissolve, leaving behind an array of hollow channels to perfuse the surrounding developing construct (Figure 16.7).

Dispensing of the low-viscosity hydrogels was also scripted with a focus on surface interactions. To ensure materials left the print tip and remained within the pattern, the initiation of each low-viscosity material started in a corner of the layer. In that way, material would immediately leave the print tip and contact at least two surfaces. These surfaces provided a large interface to help hold the hydrogel in place as additional material was added to the structure.

Although a great deal of focus was placed on developing hydrogels and optimizing printing, another key aspect of bioprinting that also needed to be addressed was cleaning the print tips after a session. On the ground, SmartPumps are typically disassembled after each print run, components are soaked in a variety of solvents and sterilization fluids, permitted to dry, and then aseptically reassembled prior to the next run. On station, typical cleaning fluids such as 70% ethanol and acetone present

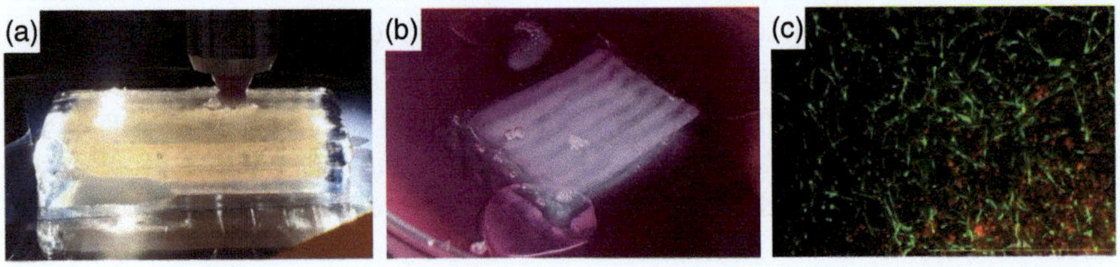

Figure 16.7 (a) Image of ground printing of test vascular construct. As the low-viscosity cell-containing hydrogels are printed, the high-viscosity (clear) structural material maintains its shape permitting the hydrogels to polymerize. (b) image of polymerized vascular construct after transfer to an incubator. The internal channels can be seen within the construct as the structural material is rinsed away. (c) A representative image of vascular cells growing within the printed construct after three days of culture. Viable cells (green) are visibly extending out into the matrix, beginning the remodeling process. The scale bar is 300 µm. Source: Techshot.

too much of a safety risk to be used. In addition, disassembly and reassembly of each SmartPump presented too much of a logistical challenge to be feasibly incorporated into mission operations. New cleaning methods needed to be developed.

Ground evaluations were conducted to evaluate alternate cleaning solutions. Efforts focused on two different solvents, both approved for use onboard the ISS. First, investigations evaluated the use of benzalkonium chloride (BZK), a disinfectant with a history of use in past missions. BZK was found to be an effective solvent for cleaning the smart pump components and the flow path. However, during the first test prints in microgravity, it was found that over time, due to cohesion and surface wetting residual fluid remained in the flow path between prints. There was concern that this remaining fluid could negatively impact cell viability in subsequent prints. In response, cleaning fluids were changed to a 4% hydrogen peroxide solution. The new cleaning solution adequately cleared residual bioinks and disinfected the fluid path. As an added benefit, any residual fluid would break down into water and oxygen, posing a trivial threat to cell viability in subsequent print runs.

An additional concern with SmartPump cleaning was accommodating the waste cleaning fluid. Unlike in the lab, small disposable containers cannot be easily placed into and removed from the BFF print chamber for waste collection. Likewise, the low-viscosity cleaning solutions present a few challenges. If permitted to simply eject from the print tip, the fluid could wet the print tips, forming large masses of fluid. Depending on printer movement, these fluid volumes could then dislodge and float within the print chamber – potentially becoming a safety and/or contamination concern. To avoid this scenario, each print cassette was designed with a dedicated waste port. This port contained a needless septum that created a seal around the print tip minimizing fluid leaks during dispensing. Once the tip was disengaged from the port, the septum then sealed preventing fluid from leaving the waste port.

16.5 Launching to ISS

In preparation for rocket launch to the ISS, a great deal of energy went into selecting, evaluating, and testing component packaging as described below. Coordinating with both NASA Safety and Operations teams, experiment components were organized to optimize efficient packing and sample grouping for on-station printing activities.

16.6 Containers and Containment

A major overarching NASA safety requirement is the testing of all sample packaging to ensure no leaks occur. All samples are inspected prior to leaving our facility. They are inspected again at turnover to our NASA colleagues. Once delivered, all samples are inspected on the ISS prior to being stowed in their respective temperature-controlled areas.

For bioprinting activities, samples were contained within a number of different containers including multiple sizes and styles of syringes, gas-permeable bags, and sterilization pouches. All syringes were pressure tested to ensure they could withstand the expected forces exerted during launch. Similarly, gas-permeable bags (used to store medium) were both pressure tested and evaluated to withstand puncture. To minimize contamination, all utilized materials were purchased presterilized (preferred) or sterilized prior to use. When possible, materials were autoclaved. However, most utilized polymers could not withstand 121 °C without adverse material warping, melting, or cracking. These materials were typically gas sterilized via ethylene oxide exposure. We also took the extra step to gas sterilize all utilized sterilization pouches prior to packing any components for use.

16.7 Bioink Packaging

On the ground, combining bioinks is a fairly standard procedure. Hydrogels are prepared ahead of time and are typically maintained in a refrigerator or within a biosafety cabinet, stored in an insulated container loaded with ice. Hydrogel components are maintained at 4 °C until needed. Meanwhile, the cells of interest are retrieved from liquid nitrogen storage and then thawed for use. Once ice crystals are no longer visible, the cells are added to the base hydrogel. Once mixed, the cell-loaded hydrogel is transferred to a syringe and then installed in the bioprinter for use.

Onboard the ISS, this "simple" process is far more complicated. First, storage facilities are spread around the multi-module facility. For cooled samples, separate Minus Eighty Degree Laboratory Freezer (MELFI) and General Laboratory Active Cryogenic ISS Experiment Refrigerator (GLACIER) systems are utilized to maintain specimens between −160 and +2 °C. Typical operational temperatures are −95, −35, and +2 °C. Each is maintained by separate units that may be located in different modules within the ISS. Ambient temperature (20 °C) storage areas are distributed throughout the ISS modules. From a logistics standpoint, one crew member is assigned to each experiment each day. Experiments often run in parallel, so their time is meticulously scheduled with minute increment resolution for their entire mission. It is extremely important to minimize the time required to set-up and execute each experiment.

With time management in mind, it was imperative to group as many like-temperature items together as possible. Minimizing the number of "kits" to be retrieved during experiment preparation. As part of the scheduled experiment prep-time, crew members float around station to collect all the items required for the experiment. Once gathered and placed in the work area, the experiment can begin.

For printing activities, all the cells (stored at −95 °C) were packaged together for each run (three frozen cell syringes per bag). Likewise, syringes of cold hydrogels were packaged together for storage at +2 °C (four hydrogel syringes per bag). Cells and hydrogels were all double bagged for shipment, providing two levels of containment in the event of a leak. With space and storage at a premium, we were permitted to send two extra sets of both cell kits and hydrogel kits. All of the other room-temperature components were packaged together and maintained in standard ambient storage.

Due to limited space in the "powered" containers that transport temperature-controlled samples to the ISS, only the bare minimum of materials is permitted. For instance, a connector was required to attach each cell-syringe to the hydrogel-syringe, permitting mixing of the two components prior to use. To minimize contamination, the connectors were individually packaged within sterile pouches and grouped together in sets of three (corresponding to the three cell-containing bioinks of each print). The connectors and packaging would have increased the size of the overall packaging of the bioinks. As a result, this connector was included with the room-temperature components.

An additional interesting consideration of packaging is related to orientation during launch. Both SpaceX and Northrup Grumman rockets were loaded in a horizontal orientation during our initial flights, and then moved to a vertical orientation for launch. Depending on packaging, the potential for fluid separation, or mechanical components, packed items may need to be loaded onto the rocket considering the launch vector. In our systems, the print cartridges were packed so the print surface would be perpendicular to the launch vector. With the initiation of Commercial Crew flights carrying astronauts, SpaceX rockets are now loaded vertically, reducing constraints related to horizontal loading pre-flight.

16.8 In-Flight

In-flight crew activities and how materials behave in microgravity provide a substantial impact on how experiments are organized. For example, many experiments that require the use of cell culturing start with a frozen cell vial that is thawed. On the ground, cells are transferred from liquid nitrogen storage and immediately thawed in a water bath. The cell vial may or may not be manually swirled for a few minutes until ice crystals are no longer visible. Cells are then transferred to a conical tube and gently reconstituted in medium. After a gentle mixing (often by a few conical tube inversions) the cells suspension is centrifuged, the extra medium is removed, and then cells are collected and added to their destined container or culture vessel for use.

On the ISS, this process is far more complicated. Starting with frozen cells, there are safety requirements that prevent direct handling of extremely cold samples (typically below −35 °C). For warming samples, there are no water baths on station. Samples are typically thawed by setting them out in an ambient space. For frozen cell samples in 3 ml of freeze medium, this can take approximately 45 minutes. Cell and medium transfers all must be conducted within liquid-tight containment fields or boxes. The available centrifuge on station may be accessible, but the operation time is fairly long and also requires repackaging of samples for transport to and from the instrument.

16.9 Bioprinting Preparation

Past experience with planning experiments on the ISS informed how we approached sample preparation and loading of the BFF for on-orbit operations. Operations began

Figure 16.8 Image of Astronaut Koch working with BFF during on-station operations. Source: NASA.

the day before printing. Crewmembers placed a custom heating block in a 37 °C incubator. This heating block was fabricated to accommodate three frozen cell syringes. It was determined that the use of this heating block would thaw the cell samples in under seven minutes (far better than the 45 minutes of ambient warming).

The morning of printing, crew members first expanded the outer containment bag fitted to the front of BFF. Once locked in place and ready to use, the crew member retrieved the frozen cell kit, cooled hydrogel syringe kit, ambient stored printing cassette, experiment kit bags, cleaning wipes, and the pre-warmed heat block from the. All items were placed inside the outer bag and sealed (Figure 16.8).

The initial steps involved placing the frozen cells into the heat block. Cells thawed in the block as cleaning procedures were conducted – all exterior and surfaces of BFF were wiped down. In addition, the printing cassette was opened and locked into place. At this time, cells were checked to evaluate their thawing status. Once thawed, cell syringes were connected to the cold hydrogel ink syringes and slowly mixed. Care was taken to mix slowly to minimize shearing cells and also to minimize the formation of bubbles. Once mixed, the cell-loaded bioink syringes were loaded into BFF and quick visual and tug checks verified BFF was set-up and ready to print.

16.10 Bioprinting in Microgravity on the ISS

During 3D bioprinting experiments on the ISS the entire bioink-dispensing process was controlled and monitored from our ground facility. BFF was designed with multiple cameras that permitted observation of print tip location. Each print session began with facility health and safety tests, camera checks, and operational tests of each pen tip. *XYZ* coordinates and dispensing were checked to ensure no anomalies had been introduced in the previous cleaning activities. Once positions and movement had been verified, print files were uploaded to the printer.

The printer was designed to run once loaded in an automated fashion without the need for crewmember interaction. Once the outer door was closed, the facility was under our remote control. One of the most challenging aspects of microgravity printing was the need for perfection. On the ground, if a small error occurs in the beginning of a print – such as an errant drop of ink falls from a print tip, a bubble in a support ink causes a pattern abnormality, maybe the ink is slow to dispense causing a pattern gap – the print can be stopped, the stage can be wiped, and the print can be restarted. On station, with the limited crew time, crew safety concerns, and limited supplies that can be launched to station, errors, and print restarts were not possible.

To maximize opportunities for success under these constrained parameters, the entire printed construct was broken up into scripts of approximately seven-layer thicknesses. These built-in pauses permitted us to monitor how the materials were printing, facilitating (if needed) minor modifications of print pressures, speeds, and position. In the rare occasion that some defect was noticed during the print, it could be addressed before continuing with additional layers. Altogether, our cardiac construct printed in microgravity on the ISS was approximately 100 layers, fabricating a volume of approximately 31 × 23 × 27 mm. Once printed, approximately one hour of time was allocated to permit polymerization of the print (Figure 16.9).

Computer commanding scripts were designed and tested on the ground prior to flight using a system that was nearly identical to the flight model. All printing components were identical between the two systems. The ground system commanding routines also incorporated data downlink time consideration to mimic slight video delays from the ISS to mimic remote printing. From a preparation standpoint, during ground testing, procedures were defined and the printing activities had been developed, tested, and were ready for use. However, every print session was unique in one way that could not be easily predicted – loss of signal (LoS) timepoints during our assigned experiment time. The ISS communicates with ground assets through a network including ground stations and geostationary satellites known as the Tracking and Data Relay Satellite System (TDRSS). Due to line of sight constraints during orbit of the ISS there are gaps in coverage that can occur.

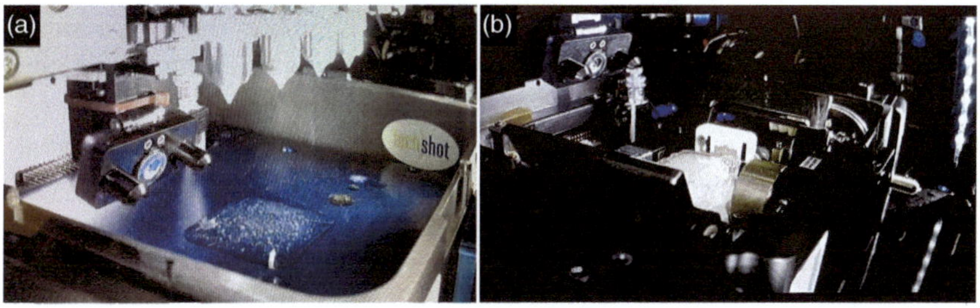

Figure 16.9 Example images of 3D bioprinted constructs on ISS utilizing the Techshot BioFabrication Facility (BFF). (a) SpX-18 printing of low-viscosity material. The printed structure exhibited a thickness of a few millimeters in the *z*-direction. (b) SpX-19 printing of both low- and higher-viscosity bioinks. The higher-viscosity inks were used to create a border, permitting the fabrication of thicker (~20 mm) structures in the *z*-direction. Source: Techshot.

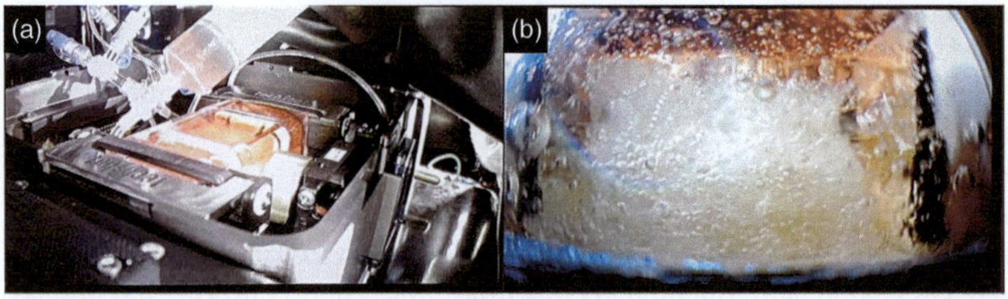

Figure 16.10 Images of (a) Astronaut Koch (not shown) filling the bioreactor with media following bioprinting. (b) side image of sealed bioprinted construct following media addition. Source: NASA.

Lasting anywhere from a few minutes to potentially an hour, LoS presented a major challenge to printing activities. The BFF system was designed to run prints autonomously. Once the print was loaded and started to run, BFF could complete the print without any need for user intervention. For safety purposes, we tried to minimize the time the printer would be active without our ability to monitor it remotely. This often led to creative organization of printer-setup and subsequent print loading and printing.

16.11 Post-Print Activities and Conditioning on ISS

A unique aspect of our BFF system was the direct printing of our pattern into a bioreactor. Once bioprinting activities had completed and the print had time to polymerize, we relied on crew members for a number of crucial steps to complete the print activities. The bioreactor was designed for simple operation and access to all the port. First the lid was placed over the print then sealed. Next, media was slowly added to the bioreactor chamber and a few media lines were primed to remove air from the system (Figure 16.10). Due to the fragility of the newly printed construct, a crew member slowly added media to the sealed bioreactor volume. As it filled, some bubbles formed. We incorporate both a manual bubble retrieval tool and bubble traps within the media circulation system to remove bubbles from the system. Lastly, the bioprinting cassettes were removed from BFF and carefully loaded into our powered incubator system to support construct conditioning.

During the incubation time for construct conditioning, media kits were prepared to replace spent media. Operations were coordinated to exchange media every 4 or 5 days during the 30-day mission utilizing crew member support to access the print cartridge and replace the media.

16.12 Return of 3D Bioprinted Tissues from the ISS

Live return capability of tissue printed on orbit is provided via a custom rack which contains three cassettes and fits within the Microgravity Experiment Research

Locker/Incubator (MERLIN). MERLIN controls temperature and provides power to maintain feed/circulation pumps, physical conditioning (stretch), and electrical conditioning (pacing). After recovery at a NASA space center (typically the Kennedy Space Center for current SpaceX returns), the cassettes can be returned to investigators at the recovery center or transported to either the Techshot facility or the principal investigator (PI's) facility within a thermally controlled potable incubator/refrigerator while maintaining internal power similar to the return from ISS in MERLIN. Because of the time constraints for culturing and conditioning, we typically work backward from return to schedule on-orbit operations. Additionally, chemical fixation and/or on orbit thermal preservation are also available if live return is not desired. Fixation or preservation on orbit may be preferred for capturing mid-stage growth conditions or for prevention of any gravitational readaptation when the capsule returns.

16.13 Future Considerations

Based on initial successful BFF demonstrations performed on the ISS we are continuing to develop optimize procedures and designs that facilitate robust and repeatable procedures for 3D bioprinting of tissues constructs and potentially future organs on the ISS. We envision our initial success with 3D bioprinting in microgravity will be further enhanced by capabilities such as edge computing on additional crewed platforms under development by commercial space companies that address remote communication delays, newer facilities requiring less maintenance and therefore more availability of crew time, as well as flexible form factors for design that address current design constraints on ISS become available. While there is still much to be learned, the opportunity to provide 3D bioprinting in microgravity as a new capability presents exciting possibilities for the terrestrial 3D bioprinting industry, patients in need of scarce tissues and organs for transplantation, and technologies required to support NASA's future exploration goals.

Part VII

Space Construction

17

Beyond the ISS: The World's First Commercial Space Station

Christian Maender, Jason Aspiotis, Rachel Clemens, Anjali Gupta, Divya Panchanathan, David Zuniga, and Jana Stoudemire

Axiom Space, 1290 Hercules Avenue, Houston, TX 77058, USA

17.1 Historical Context

Space stations have been in orbit for over 50 years, beginning with the Salyut 1 launched by the Soviet Union in 1971 [1]. National Aeronautics and Space Administration (NASA)-built Skylab achieved spaceflight in 1973 [2] and while it orbited Earth until 1979, it was inhabited by humans for less than a year. In this brief period, numerous experiments were conducted to understand the effects of microgravity on human physiology and materials. Mir, operated by the Soviet Union (and eventually Russia), followed in 1986, remained in orbit for 15 years until 2001 [3], and played a key role in informing the assembly of the International Space Station (ISS). The ISS, an international collaboration between five space agencies representing 15 countries, has been a home for humanity in space for over 20 continuous years [4]. Throughout the 50 years of humans living in low-Earth orbit (LEO), astronauts have studied and learned about living and working in the unforgiving environment of space, and how to leverage microgravity to benefit research and manufacturing efforts. Each successive space station has demonstrated enhancements and improvements in structural design, technology, communications, research capabilities, and critical life-support systems.

By the end of this decade the ISS will have spent nearly 30 years in space providing a long-term test bed for microgravity research and space exploration technology development. Despite outliving its expected lifespan, the ISS is eventually slated to be decommissioned. To ensure there is no gap in human access to the LEO environment, commercial space is stepping in to enable a new era of human spaceflight, to build space stations, and to develop emerging markets in space.

17.2 The World's First Commercial Space Station

With the aerospace industry experiencing massive growth (projected in 2020 to reach US$1.4 trillion within a decade [5, 6]) and organizations all over the world carving

In-Space Manufacturing and Resources: Earth and Planetary Exploration Applications, First Edition. Edited by Volker Hessel, Jana Stoudemire, Hideaki Miyamoto, and Ian D. Fisk.

out their place in LEO, Axiom has established their mission to build the world's first commercial space station to serve as the successor to the ISS.

Axiom Space Inc., a privately held venture-backed American enterprise, was founded in 2016 by Michael T. Suffredini and Kam Ghaffarian. Prior to starting Axiom, Suffredini was NASA's International Space Station program manager from 2005 to 2015 [7], and Ghaffarian, both an engineer and entrepreneur, sold his company, Stinger Ghaffarian Technologies, Inc., a large NASA contractor, to KBR – (formerly Kellogg Brown & Root) in 2018 [8]. Together they have decades of expertise in designing, building, and operating a space station, crewed and uncrewed missions, and spaceflight logistics.

But why construct a commercial space station? The first reason is to maintain US leadership in space. With the recent launch of the Tiangong station by China [9], and the lessons learned from not having a transition plan after the end of the Space Shuttle Program, having an ISS replacement is a priority for NASA and the nation. The seamless transition of activities from the ISS to an improved and upgraded replacement platform like the Axiom Station maintains continuous U.S. presence in LEO, especially for continued research and emerging in-space manufacturing opportunities. It also optimizes costs, permitting NASA and international space agencies to concentrate their efforts and focus agency budgets on deep space exploration, including a sustainable presence on the Moon and sending the first humans to Mars. Current cost projections to build the world's first commercial space station are significantly lower than the costs needed to build, operate, and maintain the ISS. It is estimated that Axiom's cost to build four modules for the Axiom Station is US$3 billion [10] versus US$150 billion to build and US$4 billion to operate the ISS [11].

Furthermore, other contributing commercial companies in the aerospace industry such as SpaceX, Northrop Grumman, and Boeing are at the forefront of providing new commercial launch capabilities yielding decreased transportation costs to LEO for cargo and crew. For comparison, NASA's cost to send one of its astronauts to the ISS today is approximately US$55 million [12] per seat using SpaceX commercial services versus US$80 million [13] per seat on the Russian Soyuz spacecraft saving the U.S. government US$20 billion to US$30 billion in 2020 [11, 12].

Another important reason for building a commercial space station in LEO is to support, foster, and grow human presence in space via the emerging economic activities enabled by that increased human presence. There is a latent demand for more human spaceflight from both countries that participate in the ISS partnership but more importantly from countries who do not have access to the ISS but wish to join the league of spacefaring nations fielding spaceflight and microgravity research programs. This demand is augmented by strong international interest from a growing population of private citizens with the means and the intent to pursue a life-changing experience in microgravity. Thanks to the ISS, the continuous human presence in LEO has yielded a rich set of research results demonstrating that microgravity can have interesting effects on all types of physical and biological processes. In turn, these discoveries unlock the potential for a diversified set of applications in LEO leveraging the microgravity environment for industrialized research and in-space manufacturing opportunities. Commercial platforms in LEO are well positioned to foster these emerging economic opportunities providing accommodations in orbit that can scale as markets mature.

On a mission to improve life on Earth and foster possibilities beyond it, Axiom Station will be a thriving home for living and working in space that will orbit approximately 250 miles above Earth, traveling at 17,500 mi / hr. Not only offering residential facilities for next-generation space explorers, it will offer (i) a training ground for professional astronauts, (ii) a laboratory for research and discovery, (iii) a production facility for developing products that benefit from the extreme environment of space and microgravity, (iv) an opportunity for innovation and invention for entrepreneurial first movers, (v) a near-Earth platform for space-environment materials testing and validating technologies for deep-space missions, (vi) a place for global collaboration, and (vii) a beacon of inspiration and home for the next generation of private and professional space explorers, and much more. This diversified strategy will, in turn, ensure availability of LEO access for national interests and maintain a continuous and important US presence in LEO during and after the ISS lifetime.

17.3 Construction and Assembly of the Axiom Station

Construction of a space station in LEO is not a trivial undertaking. Like any building project in an austere or inhospitable environment, it has significant project challenges due to the site location and the complex supply chain to get components and materials to the site. Analogous to construction of an offshore petroleum well, an operations platform must be assembled in pieces in port and then transported to the production site for assembly and activation of regular operations. In the case of Axiom Station, each component of the station (modules) must be individually assembled and tested on the ground and then launched into orbit to be connected into a complete station. Each component must be built to both survive the launch environment and tolerate the inhospitable environment of LEO for the entirety of its useful life. Each component must be carefully designed to interface successfully with various other modules such that pressure, power, thermal control, and communications can all be maintained.

Axiom Station will be constructed by launching a series of modules from the ground into space and then assembling those modules together as each one arrives in orbit. Each module is capable of fully independent spacecraft operations which ensures that the modules can navigate and fly to the assembly location to be robotically grappled and installed onto the growing space station. The initial Axiom Station configuration will consist of four large habitable modules along with several other components (Figure 17.1) described in detail below.

17.3.1 First Module (AxH1)

Axiom's first module, the Axiom Habitat-1 (AxH1), will attach to the ISS in 2024 as the first piece of a fully commercial space station. Axiom, partnered with NASA,

Figure 17.1 Exterior view of the final assembly of the Axiom Station before departure from the International Space Station. Source: Axiom Space.

will initially build Axiom Station attached to the ISS Node 2 ensuring continuity of access to existing ISS customers. This first module includes crew quarters for up to four crew and research and manufacturing capabilities that allow for iterative experimentation. The module will provide approximately 75 m^3 in living space and volume dedicated to research and manufacturing payloads and experimentation. Other features include an external platform to expose experiments to the extreme conditions of space and several Earth-facing (nadir) windows. This additional volume for both research and habitation provides opportunities not only to advance the pace of commercial research, but for advancing NASA's human exploration research in preparation for lunar missions.

17.3.2 Second Module (AxH2)

A second module, Axiom Habitat-2 (AxH2), is scheduled to attach to AxH1 between 2025 and 2026 and will offer additional crew capacity, expand the capabilities for research, and will be equipped with advanced infrastructure to enable pilot-scale manufacturing for Axiom customers and partners. Additionally, this module will be home to Axiom Station's Earth Observatory, a large glass-paneled module featuring the largest windows in space, offering up to six crew an unparalleled view to observe and study Earth. The addition of AxH2 enables more opportunities for human exploration research (e.g. long-duration studies on astronaut health in isolation) and closed ecosystem studies (e.g. *in situ* resource utilization and recycling systems) to prepare for exploration class missions back to the Moon and to Mars (Figure 17.2).

17.3.3 Third Module (AxRMF)

The Axiom Research and Manufacturing Facility (AxRMF) will provide state-of-the-art manufacturing capabilities for materials, technology, and biomedical products such as cleanroom and current Good Manufacturing Practice (cGMP) facilities within the approximately 133 m^3 dedicated volume. This module will offer

Figure 17.2 Viewing windows from the Earth Observatory (top) and crew quarters on a module of the Axiom Station. Source: Axiom Space.

an expansion of research capabilities, but the focus for this facility is to enable quality-controlled, human-enabled but semi-autonomous manufacturing, using in-line sensors and remote control capabilities from the ground. While continuous crew support will be available for research and pilot-scale manufacturing on the AxH1 and AxH2 modules, the semi-automated scaled manufacturing systems of the AxRMF should only require minimal crew interaction for startup, shutdown, and product recovery procedures. The launch of the AxRMF will also include a new external Axiom Robotic arm capable of supporting manipulation and operation of research and on-orbit servicing, assembly, and manufacturing (OSAM) payloads on the outside of the Axiom Station.

17.3.4 The Power and Thermal Tower

The final component of a fully independent Axiom Station is the Power Thermal Facility (AxPTF) which will provide oversized power and heat rejection capabilities, along with a crew airlock and extravehicular activity (EVA) suits to support spacewalks for up to three crew at a time. The power generation system will include upgraded solar panels and modern batteries capable of generating power for a four-module station in excess of what is currently available on the ISS and provide a surplus for module expansion in the future.

17.3.5 Common Systems

Axiom Station modules will all contain modern spacecraft systems that support interoperability and flexibility of operations. For example, each module will be equipped with an independent, regenerative propulsion system and its own guidance, navigation, and controls enabling the module to attach and detach from the main assembly as a free-flying vehicle that can transition away from the habitation modules for more complex operations, or eventually to serve as a transport system for high-Earth orbit (HEO) activities, such as docking with NASA's Gateway. The

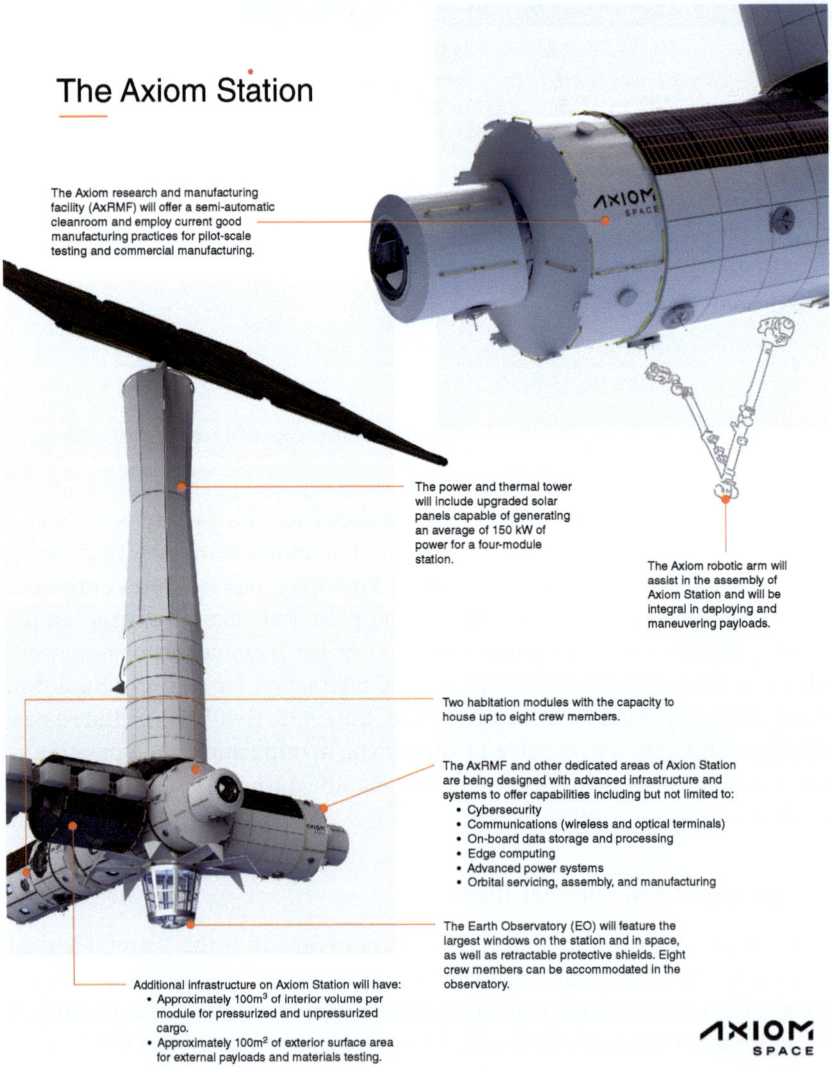

Figure 17.3 Features and capabilities for in-space research and manufacturing on Axiom Station once completed. Source: Axiom Space.

regenerative propulsion system will use the Sabatier process which converts exhaled carbon dioxide captured from each module into methane used as propellant for the propulsion system. Environmental control and life-support systems (ECLSS) will be available in all modules providing redundancy in safe human operations and enable large crews to operate long term in LEO. Modules will also be equipped with significant compute power enabling customers to virtualize processing operations and data analysis. In 2028, when the build is complete, Axiom Station will separate from the ISS to become the world's first commercial space station serving as the hub for a future "city in space" (Figure 17.3).

17.4 Preparing for the Axiom Station: A New Era of Human Spaceflight

The 2020s marked a historic return to spaceflight for American astronauts launching to the ISS from American soil on American rockets. NASA's Commercial Crew Program [14] in partnership with American aerospace manufacturers such as SpaceX and Boeing are beginning regular crew rotations supporting missions aboard the ISS.

NASA's SpaceX Demo-2 [15] flight test on May 30, 2020, was the first mission to transport crew to the ISS using a spacecraft designed, built, and operated by a private company. Since this historic launch, the agency's Commercial Crew Program has conducted subsequent missions, including NASA's SpaceX Crew-1 through Crew-4 [16] putting commercial crew spaceflights into full operation. Axiom is leveraging these new capabilities in commercial human spaceflight to prepare for the docking and operations of future Axiom modules to the ISS by providing full-service spaceflight missions for private astronauts to the ISS, and later to the Axiom Station. The first private astronaut mission conducted by Axiom Space, Ax-1, was successfully completed on 25 April 2022, opening the door to private citizens living and working in 'future cities in space' [17].

Axiom's early private astronaut missions include 10- and 30-day missions to the ISS. Each mission includes a crew of four astronauts consisting of a professional Axiom astronaut and three private astronauts. All astronauts undergo NASA-level training which is scaled to the mission's objectives and tailored to the spaceflight expertise of the private astronauts. Each crew receives an appropriate amount of training to certify their safe and independent operations on the ISS. Currently, Axiom is the only private company equipped to provide astronaut training using the same facilities and personnel used by NASA. As the Axiom Station comes online, Axiom will fly longer and more complex missions to the Axiom segment attached to the ISS, culminating in regular 180-day missions performed on the Axiom Station providing a permanent crewed presence.

Considering the myriad of functions that the Axiom Station will support, Axiom plans to support an equal variety of personnel as needed to fulfill the roles created by those functions. For example, a commercial company may sponsor a mission specialist on a 180-day rotation to oversee its manufacturing process aboard the station. Or a dedicated career astronaut in the role of an instructor may reside on Axiom Station to provide focused training for students or astronauts in training directly from the orbiting spacecraft. A research scientist from a pharmaceutical or technology company might periodically visit the station to conduct critical research to collect proprietary and sensitive data for the development of a novel therapeutic or an advanced material that benefits humankind here on Earth. A university or non-profit foundation may consider sponsoring a member of its faculty, graduate or research programs to serve as a mission specialist to facilitate research, teach from space, or serve as a requirement for an unprecedented curriculum. Rather than the traditional model of career astronauts, sponsored by national space agencies, funded by their respective nations or unions, the path to visiting or residing in space on Axiom Station will be more diverse than the historical past.

17.4.1 Building a Robust and Sustainable Commercial Space Economy in Low-Earth Orbit

The current space economy is already an important part of our everyday world. Products that require space such as cell phones, Global Positioning System (GPS), and weather forecasting tools are used every day. Space touches our daily lives in ways so common it can be overlooked. In 2020, the global space economy revenues were US$371 billion, with US$271 billion of the total generated from the satellite, telecommunications, and remote-sensing industries (satellite services US$117.8 billion and ground equipment US$135.3 billion), and non-satellite government space budgets with commercial human spaceflight generating US$100.7 billion [18]. In the same year, there were 114 total orbital launches of which 41 were commercial launches [19]. The number of launches and launch providers continues to grow at a rapid pace providing more routine access to LEO at reduced costs. Bank of America estimates that the global space industry could generate revenue of more than US$1.4 trillion or more in 2030 [5, 6].

While the current space economy takes advantage of LEO for established markets such as data, communications, and Earth observation, commercial access to the microgravity environment is driving the creation of two new market sectors: one for the discovery and manufacturing of advanced materials, and another for the discovery and manufacturing of biomedical products. Microgravity is the most promising environment for innovation and problem solving since the internet and the unique conditions that exist in LEO provide opportunities to accomplish a multitude of transformations in space that cannot be achieved here on Earth.

17.5 Emerging Market Sectors: In-Space Manufacturing of Advanced Materials

Most of the knowledge in the traditional materials industries like metals, alloys, ceramics, and plastics was created by studying the formation of these materials in bulk using methods like forming, heat treatment, and quenching, among others.

These traditional manufacturing methods have been optimized to achieve superior properties by changing parameters like temperature, pressure, and gas phases. The advent of additive manufacturing has disrupted the advanced materials sector and created new methods of material formation by layering small quantities of various raw materials which can lead to fundamentally different structural properties when compared to the bulk processing of the same material. Manufacturing in microgravity will create yet another disruption in the field of material science due to the changes in the way atoms, molecules, and larger particles arrange themselves without the limitations gravity induces on the material formation properties.

For example, both buoyancy, a major cause of convection (the bulk movement of molecules within a fluid), and hydrostatic pressure (force exerted by a fluid at equilibrium at a given point within the fluid) are dependent on gravity. These forces are functionally absent in the microgravity environment as observed on the ISS. These effects on fluid movement also influence how substances interact at the boundary layer (or interface) between a liquid and a solid or a liquid and a gas. Many aspects of the interaction at these boundaries are easier to study in microgravity. Potential applications include studies focused on fluid-gas boundaries in foams and gels used in consumer care products, drug formulation and delivery, and advanced manufacturing processes.

Containerless processing of materials becomes possible in microgravity and can help avoid defects introduced by the contact of the material with the container or crucible. Amorphous alloys (non-crystalline structures) are an example where it is necessary to remove nucleation sites for crystals in the melt during the cooling process, and this could be possible through containerless processing. The extreme environment of space itself can become an advantage to manufacturing where extreme temperatures and near-vacuum conditions can be leveraged for materials processing applications.

A commercial space station provides ample space to exploit this environment and paves the way for material science innovation that might change the way we think about improving existing materials and creating novel materials by leveraging this unique environment. Although there is still much research to be done to capitalize on the properties of microgravity to the maximum extent for advanced materials manufacturing, construction of commercial platforms is an inflection point for using space systemically to change current paradigms for advanced materials manufacturing.

The scope of materials research on the ISS has thus far focused on individual projects to develop next-generation production methods, improve understanding of mechanisms involved in material transformations, advance fundamental materials discovery, or test processes or manufacturing methods of novel design and synthesis pathways. There is increasing emphasis on LEO-based applied research and development seeking to demonstrate space-based materials manufacturing and production activities that enable new business growth and capital investment, represent scalable and sustainable market opportunities, and produce recurring economic value while enabling new avenues to in-space construction using materials developed in orbit. To date in-space production application projects enabled by the ISS National

Lab include 57 payloads, several patents, and $13.3 million in committed sponsored program funding [20].

Some examples of in-space manufacturing applications that are being explored include the following.

17.5.1 Optical Fibers and Exotic Glasses

Gravity's impact on material production affects many properties associated with defects in the current terrestrial processes. For example, during solidification, materials may form voids (gas-filled pockets, cracks, and phase distortions [glasses, amorphous, or misaligned crystals]) that affect material durability, strength, and performance. Lack of sedimentation and buoyancy in microgravity allows glass fibers, doped with heavy metals to enhance performance, to be manufactured free from gravity-induced defects.

17.5.2 Industrial Crystals

Crystals in a gravity environment tend to be smaller and less uniform because crystallization is an exothermic reaction (reactions that release energy to its surrounding usually in the form of heat). This heat, along with concentration gradients, leads to currents that move molecules away from the growing crystal. Reinforcing effects resulting from sedimentation and convection lead to crystals growing down terrestrially. Microgravity frees crystallization processes from these Earth-bound constraints. Additionally, in microgravity the same molecule can potentially layer upon itself in a regular pattern (crystallize) in different ways to create different macromolecular structures called forms. Microgravity enables manufacturers to manipulate crystal growing conditions that may yield novel forms with new potential applications across a wide variety of industries including semiconductor and electronics. Several applications are the subject of continued studies on the ISS. Given the immense potential in this area, several commercial space companies have designed and built industrial commercialization facilities [21, 22].

17.5.3 Metal Alloys

The weightless environment enables the formation of an alloy's structure to be driven by diffusion, as described by Brownian motion, rather than gravity-driven sedimentation. A diffusion-driven process encourages molecules to be distributed evenly in the material and results in a more uniform microstructure. When an alloy, for example, solidifies on Earth, gravity-driven phenomena such as convection, sedimentation, and hydrostatic pressure can significantly impact the material's internal microstructure. Non-uniform densities in materials can cause the alloying elements in an alloy to separate during phase transitions. Microgravity offers opportunities to promote uniformity within a material's microstructure because of its independence from convection, sedimentation, and hydrostatic pressure.

17.5.4 Thin-Layer Deposition

Thin-layer deposition is the process of creating and depositing thin-film coatings on a substrate material. Coating can be made from many different materials including metals, oxides, and compounds. Different coatings have different characteristics; some are durable and scratch-resistant or transparent, and some can decrease the conductivity of electricity or transmission of signals. Thin-layer coatings are used in a variety of optical, electrical, and health-related industries. In microgravity the absence of sedimentation and buoyancy allows dominant surface tension effects to result in more uniform and atomically and molecularly precise layering for products like semiconductors and sensors [23]. Given the significant potential number of applications, this area is being explored further.

17.6 Emerging Market Sectors: In-Space Production Biomedical Products

The last 20 years of continuous human presence in space have provided opportunities to examine the effects of spaceflight on the human biology. It is widely known that the prolonged exposure to microgravity and the harsh environment of space (radiation, temperature, vacuum, etc.) results in health effects that resemble aging such as muscle deterioration, bone loss, reduced cardiopulmonary function, and immune dysregulation among other changes [24]. While these changes normalize within a short time from when astronauts return to Earth, the ability to leverage a unique environment that produces these changes for study at the cellular level provides an unprecedented opportunity for accelerating understanding of disease pathways and potential new therapeutic targets.

State-of-the-art capabilities for iterative experiments with real-time on-orbit analysis provided by the Axiom platform, along with increasing access to LEO through the growing number of launches and launch providers, stand to change the paradigm for understanding disease, developing novel treatments, and addressing significant bottlenecks that exist in the manufacturing of the growing number of new therapies that have been developed in the last decade.

The pharmaceutical industry has transitioned from pills, to biologics, to cell and gene therapies. In each of these areas, microgravity stands to provide unique opportunities for discovery and potential manufacturing advantages that currently limit the full commercial potential of these therapies. A key component to accessing this emerging market is establishing cGMP [25] used in the production of products intended for human use. All pharmaceutical, biologics, and medical devices are regulated by the U.S. Food and Drug Administration (FDA) to ensure quality of the products produced. The cGMP requirements were established to be flexible to allow companies to use modern technologies and innovative approaches to achieve higher quality through continuous improvement. Accordingly, the “C” in cGMP stands for “current,” requiring companies to use technological systems that are up to date to comply with regulations. Importantly, these are minimum requirements, and many manufacturers of products intended for human use implement comprehensive, modern quality systems, controls, and risk management approaches that exceed these minimum standards. Axiom intends to establish these capabilities and dialogue with the FDA to facilitate development of the market sector for the space economy.

Areas where characteristics of the microgravity environment may provide benefit to discovery and in-space manufacturing include the following.

17.6.1 Flow Chemistry (Active Pharmaceutical Ingredient Production)

As the demands of chemically complex and targeted drugs coming to market increase, more pharmaceutical companies consider adopting a continuous-flow process (versus traditional batch processing) for finished-dose tablets, as well as active pharmaceutical ingredients [26]. While the pharmaceutical industry adoption of these methods has slowed compared to other industries, the benefits microgravity provides for improving photochemistry and cryogenic/exothermic reactions at a macromolecular level, that are already being studied for the synthesis of organic compounds in microgravity [27], could help accelerate this process and generate an increased number of potential novel classes of therapeutics. Flow chemistry is uniquely suited for the synthesis of organic chemicals in spaceflight because chemicals and solvents remain completely enclosed in tubing throughout the investigation.

17.6.2 Biologics

Monoclonal antibody (mAbs) Keytruda® (pembrolizumab) studied on the ISS, where sedimentation and convection currents are minimized, demonstrated crystalline suspensions produced in microgravity had lower viscosity and were more uniform than ground controls [28]. Identification of conditions for producing crystalline suspensions of homogenous particle size and distribution could enable drug delivery via subcutaneous injection versus intravenous dosing, and have important implications for drug purification and storage of therapeutic mAbs like pembrolizumab. There are past examples of new crystal forms being generated on the ISS that were used for drug targeting and discovery [29, 30], but the application of uniform crystals as a therapeutic was not a focus, so it was not explored and is currently being evaluated for applications in both biomedical and technology products.

17.6.3 Layer-by-Layer Deposition

Improved uniformity of a layer-by-layer deposition process used to create artificial retina that uses a light-activated protein from bacteria (bacteriorhodopsin) to restore vision is currently being evaluated in studies on the ISS [31]. Microgravity provides the opportunity for improvements in the layering process of the protein and polymer that could lead to reduction in materials needed for production, lowering manufacturing costs, and improving the quality of the implant. The layer-by-layer process being developed by LambdaVision has wide applications to other biological and technology-related products.

17.6.4 Drug Delivery

Current limitation in the cell and gene therapy fields, and more recently mRNA (messenger RNA) vaccines, includes challenges related to delivery of genes [32] or mRNA [33] therapies into the cell. Creation of uniform nanoparticle delivery systems in microgravity may provide benefits that could help to advance the availability of these therapies to patients in need.

17.6.5 Regenerative Medicine

In recent years, the possibility of using the unique microgravity environment of the ISS has helped accelerate progress in the fields of cell biology, physiology, and tissue engineering. Completed, ongoing, and future studies in microgravity related to regenerative medicine include improved understanding of multiple challenges in the field such as: (i) stem cell proliferation and differentiation, (ii) biomarker and growth factor regulation, (iii) vascularization during thick tissue formation, (iv) cell–matrix interactions and the mechanical loading environment, (v) bioreactor engineering and design, (vi) bioink and scaffold materials, and (vii) tissue preservation and transport to and from the ISS.

In space, several changes take place in cells, including changes in cell aggregation (the physical contact between cells leading to their 3D structural organization into tissues) and matrix deposition [34, 35]. Microgravity also profoundly changes the physics of fluid movement and the interface of nanoparticles and cells. These cellular and physical changes provide opportunities for discoveries not possible on Earth, which may advance the regenerative medicine fields of stem cell biology and organ bioengineering.

The first long-duration stem cell culture of cardiac myocytes (the cells within the heart tissue) [36], evaluation of mesenchymal stem cells for clinical applications [37], tissue chip studies on multiple tissue types in collaboration with National Institutes of Health (NIH), National Center for Advancing Translational Science (NCATS) [38], and tissue engineering projects focused on mechanobiology in collaboration with the National Science Foundation (NSF) [39], and 3D bioprinting demonstrations [40, 41], have been conducted on the ISS in support of advancing the field of regenerative medicine. Additionally, two teams from the Wake Forest Institute of Regenerative Medicine (WFIRM) were awarded first and second places in the NASA Centennial Vascular Tissue Challenge (VTC) competition for successfully generating vascularized liver tissue maintaining functionality for 30 days [42]. The first place team is currently working with the ISS U.S. National Laboratory to conduct a future mission on the ISS.

Specific areas of promise for regenerative medicine that are currently continuing to be developed include the following.

17.6.6 Disease modeling

Using induced pluripotent stem cells (iPSCs) in microphysiological systems (MPS or “tissue chips”) and organoid models provides the opportunity to study human disease models for preclinical safety testing, development, and testing of therapeutics, and personalized medicine applications. Studying these human analog models in microgravity provides an opportunity to understand disease mechanisms at a cellular level that could provide breakthrough insights into new therapeutic options. Additionally, these models are of high interest to NASA for the potential development of countermeasures for astronauts. In addition to the direct human correlation of these systems to human health, they offer the opportunity to be constructed from iPSCs of astronauts to potentially allow development of personalized countermeasures prior to spaceflight.

17.6.7 Stem Cells

Given early directional work from the Space Shuttle Program, and continued confirmation from long-duration stem cell cultures on the ISS that stem cells proliferate at an increased rate while maintaining a state of pluripotency, microgravity may provide an opportunity to address the terrestrial manufacturing challenge of producing these therapies at sufficient scale.

17.6.8 3D Bioprinting

Microgravity may be an ideal environment for 3D bioprinting of complex scaffold geometries for regenerative medicine, medical devices, and ultimately tissue structures. Given the absence of gravity, placement of bioinks in a true three-dimensional configuration may allow the creation of structures that provide new opportunities for materials used in surgical medical devices, medical implants, antibiotic or cell-growth promoting coatings on non-porous surfaces of artificial joints, and scaffold materials that facilitate integration of cells and vascularization for complex tissues and organs.

17.7 Conclusion

Axiom envisions their first commercial space station as the cornerstone of a future "city in space" built on future economic activity in LEO centered around human-enabled applications. The Axiom Station will support diverse global market segments such as national security, LEO services, biomedical and advanced materials manufacturing, and exploration systems testing providing a variety of services addressing current and future customers from around the world. This foundational in-space infrastructure will not only sustain U.S. leadership in human spaceflight and space exploration, but also advance international science objectives, foster technology and research capabilities, and development of future in-space manufacturing applications in LEO. Ultimately it is the growth of new, stable, and sustainable economic markets in LEO, catalyzed by the construction of Axiom Station, that will drive the need for significant additional in-space construction of infrastructure required to support all the new economic activity and facilitate value creation in an expanded global economy that includes space.

References

1 National Geographic Resource Library. (2020). The History of Space Exploration (24 January), https://www.nationalgeographic.org/article/history-space-exploration/#:~:text=We%20human%20beings%20have%20been,known%20as%20the%20Cold%20War (accessed 2 November 2021).
2 NASA (2014). Mai, T. (Page Ed.). May 1973- Skylab Launched (3 October), https://www.nasa.gov/directorates/heo/scan/images/history/May1973.html (accessed 2 November 2021).
3 NASA (2001). Mir Space Station. https://history.nasa.gov/SP-4225/mir/mir.htm (accessed 2 November 2021).
4 ISS National Laboratory (2021). History and Timeline of the ISS. https://www.issnationallab.org/about/iss-timeline/ (accessed 2 November 2021).
5 Sheetz, M. (2020). Bank of America expects the space industry to triple to $1.4 trillion market within a decade. *CNBC* (4 October). https://www.cnbc.com/2020/10/02/why-the-space-industry-may-triple-to-1point4-trillion-by-2030.html (accessed 6 November 2021).
6 AXIOS (2021) Space industry continues its growth in 2021 (26 October). https://www.axios.com/space-industry-growth-2021-cb0ffa9c-9621-4552-97e9-4cfbd0b1dc86.html (accessed 6 November 2021).
7 Reichhardt, T. (2018). This Group of NASA Veterans Wants to Build Their Own Space Station. *Smithsonian Institution. airspacemag.com* (September). https://www.airspacemag.com/space/13_sep2018-michael-lopez-alegria-180969922/ (accessed 6 November 2021).
8 Wilkers, R. (2018). In SGT deal, KBR further transforms its government services business. *Washington Technology* (26 April). https://washingtontechnology.com/articles/2018/04/26/kbr-sgt-deal-closes.aspx (accessed 6 November 2021).
9 Jones, A. (2021). China's Tiangong Space Station. Space.com (24 August). https://www.space.com/tiangong-space-station (accessed 6 November 2021).
10 Economist (2021). Placing Perches in the Sky: Private space stations will soon be in orbit. And with them industry (27 October). https://www.economist.com/science-and-technology/private-space-stations-will-soon-be-in-orbit/21805944 (accessed 6 November 2021).
11 Petrova, M. (2019). The International Space Station will soon have a replacement but it won't come from NASA. *CNBC* (26 August). https://www.cnbc.com/2021/08/26/heres-why-the-next-space-station-will-be-commercial.html (accessed 4 November 2021).
12 McCarthy, N. (2020). Why SpaceX is A game Changer for NASA. *Statista* (8 June). https://www.statista.com/chart/21904/estimated-cost-per-seat-on-selected-spacecraft/ (accessed 29 August 2021).
13 Wikipedia (2021). Soyuz-2 (30 October). https://en.wikipedia.org/wiki/Soyuz-2 (accessed 29 August 2021).
14 NASA (2021). Heiney, A. (Page Ed.). Commercial Crew Up to the Minute (26 October). https://www.nasa.gov/exploration/commercial/crew/index.html (accessed 6 November 2021).

15 NASA (2020). Potter, S. (Page Ed.). NASA Astronauts Launch from Americ in Historic Test Flight of SpaceX Crew Dragon. Press release (30 May). https://www.nasa.gov/press-release/nasa-astronauts-launch-from-america-in-historic-test-flight-of-spacex-crew-dragon (accessed 6 November 2021).

16 NASA (2022). Commercial Crew Program https://www.nasa.gov/exploration/commercial/crew/index.html (accessed 26 April 2022).

17 Gohd, C. (2022). Axiom Space, SpaceX change the landscape of private spaceflight with Ax-1 mission. Space.com (25 April). https://www.space.com/axiom-space-spacex-ax1-splashdown-private-spaceflight (accessed 26 April 2022).

18 Bryce Technologies (2021). SIA State of the Satellite Industry Report (June) (accessed 22 August 2021).

19 Bryce Technologies (2021). 2020 Orbital Launches Year in Review (June) (accessed 22 August 2021).

20 Savin, K. (2021). In-Space Production Application with a focus on Crystal Production, presented as part of NASA Commercial Space Lecture Series (28 April). https://www.nasa.gov/sites/default/files/atoms/files/space_portal_ken_savin.pdf (accessed 29 August 2021).

21 ISS360: The ISS National Lab Blog (2021) Redwire to Launch a New Industrial Crystal Growth Platform on Northrop Grumman CRS-15 (18 February). https://www.issnationallab.org/iss360/redwire-industrial-crystal-growth-platform-ng-15/ (accessed 29 August 2021).

22 Techshot (2021). Macromolecular Crystal Growth Facilities. https://techshot.com/aerospace/technology/macromolecular-crystal-growth/ (accessed 29 August 2021).

23 ISS360: The ISS National Lab Blog (2021). In-Space Production Applications: Thin-layer Deposition. https://www.issnationallab.org/ispa-thin-layer-deposition/#:~:text=Thin%2Dlayer%20deposition%20in%20microgravity%20provides%20opportunities%20to%20produce%20enhanced,portable%2C%20and%20easier%20to%20scale (accessed 29 August 2021).

24 Afshinnekoo, E. et al. (2020). Fundamental biological features of spaceflight: advancing the field to enable deep space exploration. *Cell* 183 (5): 1162–1184.

25 U.S. Food and Drug Administration Facts About Current Good Manufacturing Practices (CGMP). https://www.fda.gov/drugs/pharmaceutical-quality-resources/facts-about-current-good-manufacturing-practices-cgmps (accessed 29 August 2021).

26 American Chemical Society (2019). Drug Companies Warming up to Continuous Manufacturing. *Chemical and Engineering News (C&EN)* (1 May). https://phys.org/news/2019-05-drug-companies.html (accessed 26 August 2021).

27 NASA Mission Page: Flow Chemistry Platform for Synthetic Reactions on ISS. https://www.nasa.gov/mission_pages/station/research/experiments/explorer/Investigation.html?#id=8172 (accessed 26 August 2021).

28 Reichert, P., Prosise, W., Fischmann, T.O. et al. (2019). Pembrolizumab microgravity crystallization experimentation. *NPJ Microgravity* 5: 28. https://doi.org/10.1038/s41526-019-0090-3.

29 Space, B.D. (2021). Space BD is appointed by JAXA as a private partner for High-Quality Protein Crystal Growth Experiment Services on the ISS Kibo Contributing to life science R&D areas such as drug discovery. Press release (10 May). https://space-bd.com/en/news/20210510.php (accessed 29 August 2021).

30 Ng, J.D., Baird, J.K., Coates, L. et al. (2015). Large-volume protein crystal growth for neutron macromolecular crystallography. *Acta Crystallographica Section F: Structural Biology and Crystallization Communications* 71 (Pt 4): 358–370. https://doi.org/10.1107/S2053230X15005348. Epub 2015 Mar 30. PMID: 25849493; PMCID: PMC4388167.

31 Smith, A.W. (2020). LambdaVision Receives $5 Million Award from NASA for Additional ISS Research. *ISS360: The ISS National Lab Blog* (10 April). https://www.issnationallab.org/iss360/lambdavision-space-tango-5-million-award-from-nasa-additional-retinal-research/ (accessed 26 Aug 2021).

32 Strachan, J. (2021). Advanced Therapies: New Year New Challenges. *The Medicine Maker* (12 March). https://themedicinemaker.com/manufacture/what-is-the-greatest-challenge-facing-cell-and-gene-therapy-in-2021 (accessed 26 Aug 2021).

33 Cao, Y. and Gao, F.G. (2021). mRNA vaccines: a matter of delivery. *Lancet* 32: 100746 (1 February), https://www.thelancet.com/journals/eclinm/article/PIIS2589-5370(21)00026-2/fulltext. (accessed 26 August 2021).

34 Blaber, E.A. et al. (2015). Microgravity reduces the differentiation and regenerative potential of embryonic stem cells. *Stem Cells and Development* 24: 2605–2621.

35 Baio, J. et al. (2018). Spaceflight activates protein kinase C alpha signaling and modifies the developmental stage of human neonatal cardiovascular progenitor cells. *Stem Cells and Development* 27: 805–818.

36 Wnorowski, A. et al. (2019). Effects of spaceflight on human induced pluripotent stem cell-derived cardiomyocyte structure and function. *Stem Cell Reports* 13: 960–969.

37 Huang, P. et al. (2020). Feasibility, potency, and safety of growing human mesenchymal stem cells in space for clinical application. *NPJ Microgravity* 6: 16.

38 Low, L.A. and Giulianotti, M.A. (2020). Tissue chips in space: modeling human diseases in microgravity. *Pharmaceutical Research* 37: 1–6.

39 ISS360: The ISS National Lab Blog (2020). Three Tissue Engineering Projects Awarded From Joint National Science Foundation and CASIS Solicitation to Leverage the Space Station, Press release (19 October) (accessed 25 June 2021).

40 ISS360: The ISS National Lab Blog (2019). 3D Printer for Human Tissue Now Available for Research Onboard the ISS National Laboratory (13 August). https://www.issnationallab.org/iss360/3d-printer-for-human-tissue-now-available-for-research-onboard-the-iss-national-laboratory/ (accessed 25 June 2021).

41 Parfenov, V.A. et al. (2020). Magnetic levitational bioassembly of 3D tissue construct in space. *Science Advances* 6: 1–13.

42 NASA (2021). NASA Highlights Winners of Challenge to Engineer Human Tissue. Press release (7 June).

18

Leveraging Open Innovation to Incentivize Advances in Additive Construction in Space and on Earth

Monsi Roman[1], *Christopher Frangione*[2], *and Ademir Vrolijk*[3]

[1] *National Aeronautics and Space Administration, Marshall Space Flight Center, Huntsville, AL, USA*
[2] *Frangione & Associates Innovation Advisors, 28 First Street, Rumson, NJ, USA*
[3] *Engineering Management and Systems Engineering, The George Washington University, Washington, DC., USA*

18.1 Background and Overview

Long-duration human presence on other planets is quickly moving from science fiction to programmatic goal. The National Aeronautics and Space Administration (NASA) has embarked on a new era of lunar exploration: Artemis, named after Apollo's twin sister [1]. The goal is to explore more of the lunar surface than ever before, and establish a sustainable, long-term presence by the end of the decade. This will prepare us to take the next giant leap—sending astronauts to Mars and beyond. Teams across the agency are currently working on problems that the Artemis Generation astronauts will face as they start this journey. But one of them stands out when planning for long-duration stays: where will our astronauts live and work?

Despite the exciting developments by the launch industry, the cost of landing one kilogram on a planetary surface is still restrictively high. As such, bringing all materials to build a habitat with us is not feasible: it is costly and takes up volume that could be used for more essential purposes. Instead, NASA has been studying ways to sustainably and responsibly use resources available on other planets. One of these avenues is to innovate on tools that can use resources found *in situ* to construct the needed infrastructure; in particular, leveraging the strides made by additive manufacturing and applying these to planetary additive construction.[1]

NASA subject matter experts are currently working on this problem [3], but they are not the only ones that can make strides toward this goal. The study of innovation has long known that cooperation across boundaries, across organizations, or across knowledge areas, can drive innovation. Across multiple industries, breakthrough ideas have come from unexpected corners [4], and the addition of their "distant expertise" can have a big impact on the focal organization and its industry [5, 6].

1 Planetary additive construction can be thought of as 3D printing roads, berms, shelters, and other infrastructure on a planetary surface [2].

In-Space Manufacturing and Resources: Earth and Planetary Exploration Applications, First Edition.
Edited by Volker Hessel, Jana Stoudemire, Hideaki Miyamoto, and Ian D. Fisk.

Academics, businesses, and individuals – some outside of the space industry – are currently facing problems that are related to additive construction on other planetary bodies. Leveraging these overlaps might lead to similar breakthroughs. But finding what those overlaps are and who is best suited to lend their expertise can be a daunting task.

That's where prize challenges come in. These bridge the innovation gap between NASA and the nation: catalyzing sources of innovation inside and outside of the traditional aerospace industry. In this chapter, we describe how NASA's 3D Printed Habitat (3DPH) Challenge raised awareness of planetary additive construction and attracted technology solutions from outside of the space industry. Participants' backgrounds included architecture, construction, material science, mechanical engineering, and 3D printing, to name a few. These new players invested at least 1.5 times the NASA prize purse to explore habitat designs, feedstock materials, and printing processes and systems that could set the stage for further technology developments.

The rest of this chapter is organized as follows: first, we introduce prize challenges and describe their use across NASA, followed by a description of what the 3DPH Challenge set out to do. Next, we describe how the challenge reached out to new communities, attracted their participation, and gathered new and useful solutions as well as millions of dollars of spending on planetary additive construction. We conclude with a summary and a look ahead on how challenges can shape other areas of space construction.

18.2 Historical Context: What are prize challenges?

The use of challenges to accelerate innovation and research and development (R&D) goes back over 300 years to when the British Parliament launched the Longitude Prize in 1714 [7, 8]. Facing heavy losses of crews and cargo, they set a cash prize to be paid to anyone who could demonstrate a method to calculate the position of their ship at sea within a certain accuracy. The day's leading scientists, including Sir Isaac Newton himself, believed that the solution to this problem was in the field of astronomy. Instead, a "country bumpkin" clockmaker [7, p. 72] applied his expertise and came up with the solution – the first marine chronometer.

Here, as with challenges generally, the prize challenge (i) highlighted to the world how important the problem was, and drew both effort and (new) attention to it; and (ii) enticed new solvers with relevant expertise to create a solution to the problem [9, 10, 11]. We elaborate on these two features below.

First, challenges engage and excite a broad audience. The public profile of a prize helps shine a spotlight on a problem that needs to be solved. Bringing awareness to the problem by making it public through the prize serves multiple purposes including (i) helping recruit teams that might be able to solve the problem from all walks of life and backgrounds and (ii) engaging the general public to understand the problem and to have them take action – whether that be participating in the prize or taking other actions in their lives that can help solve the problem. To solve the grandest challenges, we need to engage the public so that they take action or demand others take action.

Second, challenges democratize innovation. Prizes inspire teams from around the world to compete to achieve a specific goal. These teams come from all walks of life and often they are not the current "experts" or industry incumbents. Some solvers are from tangential fields and have solutions that could be modified to solve the problem at hand, while others possess little to no experience at all.

To solve our grandest challenges, we need to pull from a variety of solvers because we need all the ideas we can get. Non-industry individuals may have some of the most innovative ideas to solve the problem, and the democratized nature of a challenge will allow those ideas to be considered. Using a traditional grant or contract, an organization would be very unlikely to find such innovators. The focus in a grant or contract most often falls on the known players or experts with the "best" resumes. But a prize does not care if someone has 10 days of experience or 10 years of experience: if they can solve the challenge presented to them, then they win. This allows prizes to attract diverse individuals to participate (and a prize administrator can market to those demographics they want to participate in the prize). Most likely, an organization would not risk giving a grant or contract to an innovator with little or no experience in a field. To solve our grandest challenges, we need to democratize innovation – encouraging anyone from anywhere with any background to help us solve them.

Organizations can use challenges to spur all types of innovation. Challenges can range from ideas challenges (e.g. ideas for innovative ways to solve a particular problem or for new products or services) to data science challenges (e.g. identifying Alzheimer's bio markers) to creativity challenges (e.g. the best ways to communicate scientific information to a particular audience) to technology challenges (e.g. creating a way to create valuable products out of carbon dioxide) to many more.

Because of their benefits, in the years since the Longitude at Sea prize, the use of challenges to solve problems (both local and global) and expedite R&D is experiencing a renaissance, with many organizations and governments around the world launching challenges to accelerate innovation. Many well-known innovations have resulted from challenges, including canning as a form of food preservation (a competition launched by Napoleon in 1795); modern plastics (the Billiard Ball Prize launched in 1863); transatlantic flight (a competition won by Charles Lindbergh in 1927); and educating kids in rural areas via software on a tablet (the Global Learning XPRIZE launched in 2014). From a US government perspective, NASA has been both one of the pioneers and one of the biggest users of prizes and challenges. Section 18.3 summarizes prizes and challenges at NASA and is followed by a section describing the 3DPH Challenge.

18.3 Prize Challenges at NASA

NASA recognizes the power of prize challenges and has a portfolio of programs that leverage this innovation tool [12, 13, 14]. To this end, NASA formed an Early-Stage Innovations and Partnerships (ESIP) group responsible for creating policy and objectives as a basis for the planning and formulation of national innovation initiatives in space technology and direct the planning and implementation of programs and activities under their authority. Part of the ESIP portfolio includes the Prizes, Challenges, and Crowdsourcing program: it supports many of NASA's public-facing prizes and competitions that incentivizes the generation of ideas and/or products for the agency. It elicits these from diverse, multi-disciplinary, and nontraditional sources, including individual citizens as well as those in academia, industry, and other government agencies.

Problems on the scale and complexity of the Longitude at Sea prize – which are of interest to NASA and the nation – are hosted by NASA's Centennial Challenge Program [13]. In 2005, NASA received legislative authority[2] to use appropriated funds to conduct public prize competitions, and established the Centennial Challenges Program with prizes for specific achievements in alignment with the Vision for Space Exploration. The first competition was opened in 2005, and since then, the program has conducted 22 challenges in technology development areas including propulsion, robotics, communication and navigation, human health, destination systems, science instrumentation, nanotech, materials and structures, and aerodynamics. Seventeen of the twenty-two challenges have produced winners and awarded more than $10 million. The length of the competitions has varied from one year for less complex challenges to five years for competitions requiring complex hardware demonstrations. The prize purse offered for challenges under the Centennial Challenge Program has historically ranged from $50,000 to $5,000,000, but the NASA Prize Authority allows the use of bigger prize purses if funding is available. This prize purse could also be provided and/or augmented by funding provided by entities outside of NASA, if the right prize authority is used.

18.4 NASA's Additive Construction Challenge: 3D-Printed Habitat

The 3DPH Challenge focused the public's attention on an important NASA problem: designing and constructing housing solutions that used additive construction technology for extended stays on other planets [15, 16, 17]. Across its three phases, 3DPH asked teams to (i) design a Martian habitat that leveraged additive construction systems, (ii) develop and test a printing material that could apply to the Martian surface, and (iii) develop and test additive construction systems that could autonomously print at large-scale structures. Collectively, these formed an ambitious goal at the time: the technology to autonomously 3D-print large vertical structures – what NASA needed to shelter astronauts living and working on Mars – had not been developed.

Planning for the challenge started in 2014, and its final stage ran in 2019. Its partners included America Makes in Phase 1 and Bradley University in Phases 2 and 3. In turn, Bradley University successfully secured the sponsorship of Caterpillar, Bechtel, Brick and Mortar Ventures, the American Concrete Institute, and the United

2 51 U.S.C. §20144: Prize authority

Figure 18.1 Overview, timeline, and winners of 3DPH Challenge. Source: Roman, NASA.

States Army Corps of Engineers (USACE) Engineer Research and Development Center (ERDC). A summary of the timeline, goals and winners of Phases 1, 2, and 3 can be found in Figure 18.1. A short summary of each phase follows below.

Phase 1: The first phase of the 3DPH Challenge was an architectural design competition. It challenged teams to develop state-of-the-art architectural concepts that took advantage of the capabilities offered by additive construction. This phase was completed in 2015 and offered $50,000 in prizes.

Phase 2: The second phase focused on development of materials and printing systems for planetary construction. It asked teams to develop a feedstock consisting of materials found on the Martian surface and trash recyclables which would otherwise be nuisance materials on space missions. The rules required the former to be at least 70% of the feedstock by weight. Printed samples of these feedstocks were tested for their material strength (ultimate load); printers were tested for their ability to print small-scale structures with high overhang. Caterpillar's Edwards Demonstration & Learning Center in Peoria, Illinois, hosted the final head-to-head portion of this phase. Phase 2 was completed in 2017 and offered $1,100,000 in prizes.

Phase 3: The final phase of 3DPH emphasized the scale-up and autonomous operation of the printing systems. It covered these areas across two sub-challenges: a virtual challenge and a construction challenge. In the former, teams created a high-fidelity building information model (BIM) of their habitat, taking some Martian environmental conditions into account. The virtual challenge also asked teams to simulate their construction sequencing and flow in 4D BIM models.

In the latter, teams printed several large-scale structures, which were subjected to a range of tests. These included materials samples subjected to strength tests; a structural foundation subjected to freeze/thaw and impact testing; a habitat element subjected to a hydrostatic test; and a 1 : 3 scale habitat subjected to impact, pressure, and crush testing at the Edwards Center. The construction challenge also asked teams to autonomously place penetrations and interfacing elements in their printed structures. Both the virtual and construction challenges of Phase 3 were completed in 2019, and offered participants $200,000 and $1,800,000 in prizes, respectively.

18.5 How 3DPH Raised Awareness of Planetary Additive Construction

The 3DPH Challenge team made sure that the message of the challenge was widely broadcast in hopes of reaching individuals and organizations that might provide solutions to the problem. Instead of a narrow focus on space-players, the team strategically pursued collaboration with non-profit partners with connections to participants with civil engineering, architecture, construction, and 3D printing knowledge and that could solicit sponsors if needed to execute the competition [11]. The NASA team designed a competition that clearly connected NASA technology needs with Earth benefits like environmentally sustainable and affordable construction of beautiful and safe structures. The result of this connection and the active participation of the NASA team in out-of-discipline (non-aerospace) conferences/activities attracted the attention of new groups.

The NASA team had a communication strategist as part of the group during the design and execution of the competition. The communication expert helped engage the media and the public in new ways by creating a visually attractive campaign that included very short videos for social media campaigns designed to attract a larger community of participants. In addition, the NASA team designed a competition that required participants to provide Mars habitats architectural concepts that resulted in a series of innovative, realistic-looking designs that successfully inspired the public to visualize what a colony of sustainable habitats on Mars might look like. Many of those designs are commonly used now by the media.

A strong media presence can lead to invaluable outcomes, including inspiring future participants, discovering new technology solutions or applications, and fostering public support for future space endeavors.

The result of this strategy was engagement with the challenge at levels comparable to NASA's public interest of the 2019 total solar eclipse. Phase 3's final head-to-head level live broadcast attracted local, national, and international attention. Locally, the challenge was covered by Peoria's local NBC affiliate, Peoria Journal Star, and WMBD (a local radio station). Approximately 2000 people visited the Edwards

Center to learn about the value of NASA challenges and viewed various science-, technology-, engineering-, and mathematics-themed exhibits. All middle- and high-school students from the area were invited to the testing facility during the competition and were presented with ways in which science, technology, engineering, and mathematics could be used to revolutionize the way we can build affordable houses in the future. Nationally, results of the competition, but especially Phase 3, were featured in more than 382 media publications, including Business Insider, Fox News, Mashable, Popular Mechanics, Popular Science, Diversity in Action, CNN Cable News Network, Popular Science, and USA Today. Other notable guests included writers from Popular Science and IEEE Spectrum. A film crew for YouTube Premium's Age of A.I. also attended to gather material for their eight-part documentary that explores the latest in technology. Finally, Phase 3 also received international media exposure in countries including Switzerland, Peru, Mexico, Singapore, Vietnam, and Germany. The estimated reach from traditional media coverage was between 113.5 and 170 million people.

18.6 How 3DPH Attracted Nontraditional Participants

The 3DPH Challenge attracted more participating teams than any other Centennial Challenge, nearly all outside of the space industry. A total of 200 teams registered for the 3DPH: Phase 1 saw 167 teams (though only the 30 top scoring teams were considered for awards); Phase 2 had 8 teams; and Phase 3 saw 18 and 7 in its virtual and construction sub-challenges, respectively. These numbers included a noticeable amount of international teams despite not being eligible for a prize award.[3] The challenge accepted registrations from Chile, Hong Kong, Germany, the United Kingdom, Russia, Poland, Canada, South Korea, Singapore, Saudi Arabia, and Australia. Notably, teams from South Korea and Singapore won top awards in Phase 2.

Furthermore, there was a large diversity in the backgrounds of the teams. First, teams spanned the range of industry (33%), academic (22%), international (23%) and other independent competitors, like hobbyists (22%). Second, while the focus of the challenge was printing on Mars, only two teams traced directly to the space industry. Several other teams were space-adjacent: team members had existing experience in the (aero)space industry but were not actively working in it at the time of their participation. Instead of the space industry experience, a survey of participants showed 3D printing, architecture and design, civil engineering, construction management, electrical engineering, material science, mechanical engineering, and venture capital experience. Notably, more than half of the respondents who were new to additive construction said they were still in the industry after well past the end of the challenge.

These data show that the outreach that the challenge conducted to different communities was successful in attracting participants from those backgrounds. The 3DPH became a platform for NASA to engage with different kinds of individuals and organizations outside the space industry [11].

3 Centennial Challenges' legal authority prevents prize awards to international teams.

18.7 How 3DPH Advanced Planetary Additive Construction

The influx of participants brought with it a variety of solutions and funding that NASA would not have seen through any other innovation mechanism [11]. Challenges allow for a portfolio of solutions because a collection of teams will compete for the prize – each with their own unique solution to address the problem. Not all solutions will be equally good. However, having a large solution set allows for many opportunities to address the problem. Across all prize awards across all phases of the 3DPH there were at least two teams that presented compliant solutions; in some cases, that number was as high as 100. In contrast, had 3DPH been a traditional grant or contract, NASA would likely only hire one or two companies. This would result in one or two solutions – not the hundreds incentivized by the challenge, with their related approaches and processes.

Participants in 3DPH invested considerable amounts of funding on the planetary additive construction problem, most from outside the space industry. Recognizing that challenges can leverage investment by the participating teams, we surveyed teams about the costs incurred during their participation. However, collecting cost data is very difficult. Often, this information is confidential. Costing can also be calculated very differently: some include the value of staff time while others do not. In an effort to get anecdotes of the spend, respondents were asked an open-ended question of how much they spent to participate. How they estimated this amount was left up to them to encourage their response. Table 18.1 below summarizes the responses of 25 teams: in total, these respondents *alone* invested upwards of $4,500,000, at least 1.5 times the prize purse offered by NASA.

These participants' investments were put to good use – their solutions made significant advancements related to planetary additive construction [11, 15]. These included diversity and innovation in viable designs for planetary habitats; demonstrations of new material compositions for additive construction in both planetary and Earth contexts; and demonstrations of processes and equipment for autonomous, large-scale, vertical construction, which included control algorithms for depositing material in three dimensions instead of two.

First, participants presented a range of habitat designs and construction methods [18]. The designs varied on their construction materials, printers, and their processes, and the shapes and layout of the habitats. For example, one team chose

Table 18.1 Summary of self-reported investments in solutions across 3DPH for 25 respondents.

	Phase 1*	Phase 2*	Phase 3*
Average investment ($k)	7.7	84.8	237.4
Investment range ($k)	0.3–15	6–200	1.5–800
Total ($k)	15.3	509	4035.5

Note: *Phase 1: 2 respondents, Phase 2: 6 respondents, and Phase 3: 17 respondents.

water-ice as their main construction material for the habitat. Their decision was driven by water's properties to shield against galactic cosmic rays and solar particle radiation, as well as evidence of the potential hazards of perchlorates in the Martian soil. Some teams distributed their materials gathering and construction tasks across an array of robots; some concentrated these functions in one multipurpose robot. One team with a single robot design included a lander structure that encloses their printer: providing a pressurized, thermally controlled print environment for processing of the extracted materials into feedstock and fabrication of their habitat module(s). Habitat architectures ranged from vertical towers with multiple stories, to various independent spaces connected through corridors, to a large, comprehensive single-story dwelling. A taxonomy of these design features has been created for future reference [19].

Second, participants in 3DPH also made advancements in the materials used to print these habitats, in both polymer concrete as well as cementitious concrete families [18, 20]. A notable example of the former was a thermoplastic and basalt fiber blend, and associated printing process, where the material solidifies in free space as it is printed. This demonstrated the ability to freeform print a cellular matrix in midair – without formwork – and produce building components that can take on a variety of shapes. Another team developed a biopolymer and basalt fiber blend that could be produced on Mars through a series of chemical reactions. This blend also has a low coefficient of thermal expansion, attains its full strength instantaneously, and is recyclable. With respect to the latter, one team developed a dry geopolymer binder composed of minerals indigenous to both Mars and Earth. The developed mix required only a small amount of water to be added when printing, which could be obtained from ice found on Mars. On Earth, this geopolymer binder could dramatically decrease the emission of greenhouse gases if used instead of Portland cement. Finally, one team demonstrated the production of Portland cement from Mars indigenous materials. Their insistence on using cementitious concrete while remaining competitive under the indigenous material rules drove them to produce this proof of concept. The team produced Mars Portland cement that combined basalt and limestone, both indigenous to Mars, with gypsum. Their work showed that Portland cement can, conceptually, be produced on Mars.

Third, participants also advanced the printing processes and equipment needed to construct these habitat designs [16]. Phase 3 in particular saw several demonstrations of both gantry and robotic arm printers. Guided by their choice in printer, teams tackled large-scale printing in different ways: some used separate hydraulic systems to print vertically (e.g. using a forklift) or horizontally (e.g. using a retractable boom), while others relied on the fixed components of their printer to complete the task. Many teams in this phase successfully used two robots working in tandem to complete the required structures: one to print the structure's layers and one to correctly place building components. Finally, participants made significant strides in addressing the lack of software and control algorithms for depositing material in a non-two-dimensional layer. Teams created new tool path algorithms that allowed them to optimize material usage: tailoring the material deposition to where it is most needed (i.e., in the direct load path).

3DPH participants continued their contributions to planetary additive construction even after their prize awards [11]. After Phase 1, one team collaborated with personnel at Langley Research Center on two feasibility studies of a deployable ice habitat. After Phase 2, feedstocks created for the challenge were tested in NASA additive construction printers at Kennedy Space Center, as well as on the International Space Station.

The majority of follow-on activity happened after Phase 3. Similar to Phase 2, two teams had their feedstocks – a polymer concrete and a cementitious concrete – tested on-orbit. Additionally, several teams were awarded follow-on contracts/partnerships with NASA to continue the work that they had started under the challenge. These included a partnership with Marshall Space Flight Center to further develop a large-scale 3D printer for the Moon or Mars, as well as high-fidelity lunar habitation concepts; a partnership with Kennedy Space Center to develop a lunar feedstock, construct a printer suited to that feedstock, and print test samples in vacuum conditions; a partnership with Johnson Space Center to print a full-scale Mars habitat analog for habitation experiments. These partnerships totaled more than $40 million of funding; it is unlikely that these large expenditures would have been offered to teams without the work carried out within the 3DPH [11].

18.8 Conclusions and Future Outlook

Additive construction truly is a disruptive technology. As construction becomes increasingly automated and less labor-intensive, additive construction will be at the forefront in facilitating changes in which structures, both on Earth and on other planetary surfaces, are traditionally designed and fabricated. This technology can sustainably produce a set of reasonably priced habitats in weeks, rather than months, even in remote areas, capable of sustaining earthquakes and hurricanes. This can be a game-changer.

This chapter has summarized how the 3DPH contributed to additive construction through two aspects of challenges: (i) the challenge raised awareness of a specific problem/niche sector/nascent industry – participation in the challenge pushed the acceptance of additive construction as a viable extraterrestrial and terrestrial construction technique and (ii) elicited new participants to (bring their expertise to bear and) solve a complex problem. The solvers' nontraditional expertise was brought to bear, and they showed hardware development and operation in the additive construction context; a feat that is not trivial. It has also enabled significant inroads into understanding the critical parameters associated with development of a robust additive construction system that with appropriate modifications, can be used on the Moon and Mars as well.

NASA hopes to successfully apply the lessons learned from the 3DPH challenge in the design of new competitions and continue to harness the power of crowdsourcing. To that end, the Centennial Challenge Program is working to engage the public in the development of additional technologies needed to support the infrastructure needed for a sustainable human presence on the moon and beyond.

Its new competitions will incentivize technology development of reliable power grids, equipment for the excavation of needed resources, and the development of nutritious food, among others.

The Centennial Challenges Program is dedicated to encouraging innovation and imagination through its organic approach to utilizing the great talents this nation and the world have to offer, while also capturing the public imagination, engaging communities not historically involved in the development of aerospace technologies, and attracting greater public attention to these endeavors. This program exemplifies the values that have formed the bedrock of the culture at NASA since the beginning: innovation, imagination, and a passion for exploration.

References

1 NASA. (2020). The Artemis Plan. NASA's Lunar Exploration Program Overview. (September), https://www.nasa.gov/sites/default/files/atoms/files/artemis_plan-20200921.pdf (accessed 6 November 2021).

2 Labonnote, N., Rønnquist, A., Manum, B., and Rüther, P. (2016). Additive construction: state-of-the-art, challenges and opportunities. *Automation in Construction* 72: 347–366. https://doi.org/10.1016/j.autcon.2016.08.026.

3 Fiske, M., Edmunson, J.E., Weite, E. et al. (2018). The disruptive technology that is additive construction: system development lessons learned for terrestrial and planetary applications. In: *2018 AIAA SPACE and Astronautics Forum and Exposition*. Orlando, FL: American Institute of Aeronautics and Astronautics https://doi.org/10.2514/6.2018-5127.

4 Jeppesen, L.B. and Lakhani, K.R. (2010). Marginality and problem-solving effectiveness in broadcast search. *Organization Science* 21 (5): 1016–1033. https://doi.org/10.1287/orsc.1090.0491.

5 Lifshitz-Assaf, H. (2018). Dismantling knowledge boundaries at NASA: the critical role of professional identity in open innovation. *Administrative Science Quarterly* 63 (4): 746–782. https://doi.org/10.1177/0001839217747876.

6 Szajnfarber, Z. and Vrolijk, A. (2018). A facilitated expert-based approach to architecting "openable" complex systems. *Systems Engineering* 21 (1): 47–58. https://doi.org/10.1002/sys.21419.

7 Sobel, D. (1995). *Longitude: The True Story of A Lone Genius Who Solved the Greatest Scientific Problem of His Time*. Walker Publishing.

8 Spencer, R.W. (2012). Open innovation in the eighteenth century. *Research-Technology Management* 55 (4): 39–43. https://doi.org/10.5437/08956308X5504074.

9 Gustetic, J.L., Crusan, J., Rader, S., and Ortega, S. (2015). Outcome-driven open innovation at NASA. *Space Policy* 34: 11–17. https://doi.org/10.1016/j.spacepol.2015.06.002.

10 Szajnfarber, Z., Zhang, L., Mukherjee, S. et al. (2020). Who is in the crowd? Characterizing the capabilities of prize competition competitors. *IEEE*

Transactions on Engineering Management 1–15. https://doi.org/10.1109/TEM.2020.2991370.

11 Vrolijk, A., Roman, M.C., and Szajnfarber, Z. (2022). *Who Provides What and When: Mapping the Benefits Stemming from Innovation Contests.* Research-Technology Management 65 (1): 29–38. https://doi.org/10.1080/08956308.2022.1993683.

12 Kaminski, A.P., Buquo, L., Hemmings, S.N., and Roman, M.C. (2019). How public-private collaboration through challenges and prize competitions is enabling NASA to shape the future of space exploration and innovation. In: *Proceedings from 70th International Astronautical Congress (IAC).* Washington, DC, United States.

13 NASA (2019). NASA's centennial challenges overview. *STMD: Centennial Challenges* https://www.nasa.gov/directorates/spacetech/centennial_challenges/overview.html.

14 Kaminski, A.P. (2012). Can the demos make a difference? Prospects for participatory democracy in shaping the future course of US space exploration. *Space Policy* 28 (4): 225–233. https://doi.org/10.1016/j.spacepol.2012.09.004.

15 Vrolijk, A. and Szajnfarber, Z. (2021). Exploring how prize challenges complement an organization's innovation efforts. In: *Proceedings from CESUN 2020 Engineering Systems Symposium*, CESUN Conference, West Lafayette, IN. https://doi.org/10.5703/1288284317288.

16 Mueller, R.P., Prater, T.J., Roman, M.C. et al. (2019). NASA centennial challenge: three dimensional (3D) printed habitat, phase 3. In: *Proceedings from 70th International Astronautical Congress (IAC).* Washington DC, USA.

17 Mueller, R.P., Howe, S., Kochmann, D. et al. (2016). Automated additive construction (AAC) for Earth and space using *in situ* resources. In: *Proceedings of the Fifteenth Biennial ASCE Aerospace Division International Conference on Engineering, Science, Construction, and Operations in Challenging Environments (Earth & Space 2016).* Reston, Virginia, USA: American Society of Civil Engineers. Available at: http://oro.open.ac.uk/45865/ (accessed 9 November 2021).

18 Roman, M., Yashar, M., Fiske, M. et al. (2020). 3D-printing Lunar and Martian habitats and the potential applications for additive construction. In: *Presented at the 2020 International Conference on Environmental Systems*, Lisbon, Portugal. Available at: http://ttu-ir.tdl.org/handle/2346/86486. (accessed 9 November 2021).

19 Roman, M.C., Eberly, E.A., Mueller, R.P., and Deutsch, S. (2016). NASA centennial challenge: three dimensional (3D) printed habitat. In: *Proceedings of Earth and Space 2016: Engineering for Extreme Environments*, 333–342. Reston, VA: American Society of Civil Engineers.

20 Prater, T., Kim, T., Roman, M.C., and Mueller, R.P. (2018). NASA's centennial challenge for 3D-printed habitat: phase II outcomes and phase III competition overview. In: *Proceedings of the 2018 AIAA SPACE and Astronautics Forum and Exposition.* American Institute of Aeronautics and Astronautics https://doi.org/10.2514/6.2018-5405.

Part VIII

Space Policy, Law, and Economics

19

The Impact of the Artemis Accords on Resource Extraction

Melissa de Zwart

Flinders University, Jeff Bleich Centre, Adelaide, 5001, Australia

19.1 Introduction

Recent years have witnessed a renewed interest in crewed missions to the Moon and beyond. China and Russia have announced that they will establish a base on the Moon (Jones 2021) and have formally invited both states and international organizations to be involved in their International Lunar Research Station (ILRS), which will focus on projects including *in situ* resource utilization, water and mineral extraction, manufacturing, and effects of low gravity on biology. This project is generally similar in scope to the Artemis project lead by NASA announced in 2020, a bold vision to return humans to the Moon by 2024. The Artemis project also encompasses a Lunar Gateway, the return of the first woman and the first person of color to the Moon, and the use of the Moon as a base in anticipation of a mission to Mars.

This commitment by multiple nations to extend human space activities to the Moon and beyond reflects a major shift away from the focus of the past few decades. After the conclusion of the Apollo program the US Space Shuttle program was devoted to projects in Low-Earth Orbit. The most successful space collaboration of this period has been the International Space Station (ISS), which orbits the Earth at a distance of only 250 mi. The ISS is operated pursuant to the International Space Station Intergovernmental Agreement (IGA 1988; IGA 1998; Yakovenko 1999) supported by four Memoranda of Understanding between NASA, the European Space Agency (ESA), the Canadian Space Agency (CSA), Roscosmos, and the Japanese Space Agency (JAXA), and is an exemplar of international co-operation in outer space.

More recent projects however highlight the growth in importance of commercial actors in space. The successful crewed mission to the ISS, launched on 30 May 2020 which delivered astronauts to the ISS in a SpaceX Dragon capsule, was the first launch of American astronauts from US soil since the demise of the Space Shuttle program in 2011. Since this historic flight, SpaceX has conducted 7 human spaceflight operations including 4 Commercial Crew missions for NASA and the first private astronaut mission conducted by Axiom Space on April 8, 2022. It is

In-Space Manufacturing and Resources: Earth and Planetary Exploration Applications, First Edition.
Edited by Volker Hessel, Jana Stoudemire, Hideaki Miyamoto, and Ian D. Fisk.

unsurprising therefore that NASA has announced that the Artemis project will involve commercial partners.

The Artemis Accords provide a new model for participation in space activities. Premised on the involvement of national space agencies and commercial participants the Artemis Accords articulate a set of 10 principles which provide the context for further development of relevant projects. One of those principles directly espouses the importance of resource extraction and utilization. Others relate to the general principles which will support and underpin those activities. It is contemplated that these principles will provide the general parameters for the conclusion of more specific project agreements. In this way, the Accords themselves serve as a framework within which there may be a spectrum of participation pathways. As will be discussed further below, Australia's participation in the Artemis Accords as well as its membership of the Moon Agreement raises specific questions regarding how these two regimes are reconcilable.

Questions arise regarding the compatibility of these principles with international space law, as well as how they may be implemented and enforced in a far more complex space environment from that of the 1960s when much of the relevant space law was created. Do the Artemis Accords provide an accurate articulation of space law or provide strong guidance regarding how useful principles of space law may be further developed?

This chapter will therefore identify the relevant principles of international space law and the apply those laws to the proposals of the Artemis Accords, with particular attention to their capacity to support or impede the development of *in situ* resource extraction and use.

19.2 International Space Law

International Space law comprises five major United Nations (UN) Space Treaties, although not all five have attracted the same number of signatories. The primary or foundational treaty is the Treaty on Principles Governing the Activities of States in the Exploration and Use of Outer Space, including the Moon and Other Celestial Bodies (OST) (Outer Space Treaty 1967). That treaty is supplemented and expanded by the Rescue and Return Agreement (1968); the Liability Convention (1972); the Registration Convention (1975) and the Moon Agreement (1979) (collectively the "UN Space Treaties"). As at June 2021 the Outer Space Treaty has 111 ratifications (UNOOSA 2021). It is considered to embody a number of fundamental principles which reflect customary international law, such as the principle of non-appropriation of outer space, but it does not address in any detail many of the questions confronting users of space today. Nor does it fully contemplate the major shift to commercial operators in the modern space environment. It should be noted that the context surrounding the conclusion of the OST was the potential threat of use and placement of nuclear weapons in outer space. Understanding the Cold War origins of that treaty explains the focus on ensuring access to outer space, the emphasis on scientific activities for the benefit of all of humanity, the prohibition of placement of weapons of mass destruction in space, and the perception that the only actors in space would be nation states (Article IV).

This section will address the key principles of international space law relevant to resource extraction and use.

> **Article I** of the OST requires that:
> The exploration and use of outer space, including the Moon and other celestial bodies, shall be carried out for the benefit and in the interests of all countries, irrespective of their degree of economic or scientific development, and shall be the province of all mankind.
> Outer space, including the Moon and other celestial bodies, shall be free for exploration and use by all States without discrimination of any kind, on a basis of equality and in accordance with international law, and there shall be free access to all areas of celestial bodies.
> There shall be freedom of scientific investigation in outer space, including the Moon and other celestial bodies, and States shall facilitate and encourage international co-operation in such investigation.
>
> **Article II** provides:
> Outer space, including the Moon and other celestial bodies, is not subject to national appropriation by claim of sovereignty, by means of use or occupation, or by any other means.
> **Article III** requires:
> States to undertake activities in the exploration and use of space in accordance with international law, including the UN Charter, "in the interest of maintaining international peace and security and promoting international co-operation and understanding."
> Important for understanding the unique nature of responsibility of states for activities in outer space is **Article VI** which provides:
> States Parties to the Treaty shall bear international responsibility for national activities in outer space, including the Moon and other celestial bodies, whether such activities are carried on by governmental agencies or by non-governmental entities, and for assuring that national activities are carried out in conformity with the provisions set forth in the present Treaty. The activities of non-governmental entities in outer space, including the Moon and other celestial bodies, shall require authorization and continuing supervision by the appropriate State Party to the Treaty. When activities are carried on in outer space, including the Moon and other celestial bodies, by an international organization, responsibility for compliance with this Treaty shall be borne both by the international organization and by the States Parties to the Treaty participating in such organization.

This Article essentially imposes responsibility upon States for the space activities of private operators and obliges States to ensure compliance of those operators with the obligations of the OST. This Article derives from a compromise between the views of the USSR that no private activities should be permitted in space and the US view that private enterprise should be allowed to flourish in space as elsewhere, intending to impose an obligation upon a State to be responsible for activities of any non-State entities (von der Dunk 2011, pp. 1–2). However, there is significant uncertainty regarding the scope of what is encompassed within the concept of "national activities" and hence the scope of the obligation under this Article. Various interpretations may be applied which look to concepts of nationality, attribution with respect to activities undertaken, and concepts of jurisdiction under international law. Further, this interpretation is influenced by the approach to the requirements of **Article VII** with respect to international liability for damage. This uncertainty has resulted in States adopting various interpretations in their domestic laws with respect to their obligations under **Article VI** (von der Dunk 2011, pp. 4–12).

Later treaties address specific issues in greater detail, but reflect a dwindling level of support and engagement, evidenced by a reduced number of ratifications. The least successful of these treaties is the Moon Agreement. Australia is one of only 18 signatory states.

The Moon Agreement is of particular relevance to *in situ* resource extraction and utilization activities.

The Moon Agreement explicitly contemplates the advent of space mining and provides in **Article 11** a number of core principles which affect the development of such activities. Firstly, it states that "the Moon and its natural resources are the common heritage of mankind (sic)" and the Moon "is not subject to national appropriation by any claim of sovereignty, by means of use of occupation, or by any other means." It further elaborates that "neither the surface not the subsurface of the Moon, nor any part thereof or natural resources in place, shall become the property of any State" nor any other institution or person. It affirms that "States Parties have the right to exploration and use of the Moon without discrimination of any kind, on the basis of equality and in accordance with international law and the terms of this Agreement." Most importantly, this section obliges signatories to establish an international regime to "govern the exploitation of the natural resources of the Moon as such exploitation is about to become feasible." This obliges parties to inform the UN, as well as the public and international scientific community "to the greatest extent feasible and practicable, of any natural resources they may discover on the Moon." Finally, the main purposes of the contemplated international regime are expressed to include:

(a) The orderly and safe development of the natural resources of the Moon;
(b) The rational management of those resources;
(c) The expansion of opportunities in the use of those resources;
(d) An equitable sharing by all States Parties in the benefits derived from those resources, whereby the interests and needs of the developing countries, as well as the efforts of those countries which have contributed either directly or indirectly to the exploration of the Moon, shall be given special consideration.

Other provisions deal with related sharing of information and samples, as well as inspections and visits to the installations of another party.

As noted above, the Moon Agreement has not been enthusiastically received by the international community. As von der Dunk (2020, pp. 55–56) and Tronchetti (2009, 99–110) both explain, the elaboration of the principle of "the common heritage of mankind" in the 1982 UN Convention on the Law of the Sea, generated concerns regarding the sharing of resources and resulted in both regimes being rejected by many states. Until recently this need for such a regime has of course been highly theoretical. However, recent developments have brought a new urgency to the question of resource extraction and use. Several countries have now enacted domestic laws permitting and regulating space mining activities, including the US Commercial Space Launch Competitiveness Act, the Luxembourg Law on the exploration and utilization of space resources, and the UAE Federal Law on the Regulation of the Space Sector (von der Dunk 2020, p. 57).

In an Executive Order in April 2020, the Trump Administration went so far as to explicitly repudiate the Moon Agreement as not reflecting customary international law, stating that the "US does not consider the Moon Agreement to be an effective or necessary instrument to guide nation States regarding the promotion of commercial participation in the long-term exploration, scientific discovery, and use of the Moon, Mars, or other celestial bodies." (Executive Order on Encouraging International Support for the Recovery and Use of Space Resources, 6 April 2020).

19.3 The Artemis Accords

The Artemis Accords were released by NASA as a set of high-level principles on 15 May 2020. The release followed growing speculation in early 2020 that the Trump Administration was developing a "legal blueprint for mining on the Moon" (Roulette 2020). This speculation had been fueled by the Executive Order noted above, which stated:

> Americans should have the right to engage in commercial exploration, recovery, and use of resources in outer space, consistent with applicable law. Outer space is a legally and physically unique domain of human activity, and the United States does not view it as a global commons. Accordingly, it shall be the policy of the United States to encourage international support for the public and private recovery and use of resources in outer space, consistent with applicable law.
>
> (Executive Order 2020)

That Executive Order affirmed the principles which had been set out in the US Commercial Space Launch Competitiveness Act 2015. **Title IV** of that Act, entitled "Space Resource Exploration and Utilization" provided three key provisions: (i) The recognition of the rights of a US citizen "engaged in commercial recovery of an asteroid resource or a space resource" to the resource so obtained; (ii) a mandate for a US regime which required authorization and supervision of such activities in conformity with US obligations under international law; and (iii) A requirement for the President to promote and facilitate through appropriate Federal agencies, commercial space resources exploration and extraction and to promote the rights of US citizens to engage in such activities in accordance with international obligations (von der Dunk 2017).

The Executive Order reflected that: "Successful long-term exploration and scientific discovery of the Moon, Mars, and other celestial bodies will require partnership with commercial entities to recover and use resources, including water and certain minerals, in outer space." Importantly therefore (and acting within the ambit of **Title IV**) the Executive Order foreshadowed the need for some alternative road rules for the commercial use of space, setting the stage for the release of the Artemis Accords: "With numerous countries and private sector players conducting missions and operations in cislunar space, it's critical to establish a common set of principles to govern the civil exploration and use of outer space."

The Artemis Accords are articulated as a series of principles that must be agreed to by any State that wants to participate in the Artemis program. In other words, other national space agencies that join NASA in the Artemis program will do so by executing:

> bilateral Artemis Accords agreements, which will describe a shared vision for principles, grounded in the Outer Space Treaty of 1967, to create a safe and transparent environment which facilitates exploration, science, and commercial activities for all of humanity to enjoy.
>
> (NASA 2020a)

The approach of the Artemis Accords, using bilateral agreements to determine the rules of the joint venture, clearly builds upon the success of the ISS IGA. The original signatories to the Artemis Accords were Australia, Canada, Italy, Japan, Luxembourg, United Arab Emirates, United Kingdom, and the United States of America, with New Zealand, Ukraine, South Korea, and Brazil joining in the following months. While initially indicating that it was open to participation in the Artemis Accords (Sheldon 2020) Russia has now declared that the Artemis Gateway project is too "US-centric" despite the fact that it builds upon the IGA (Foust 2020b) and, as noted above, is pursuing the development of the ILRS Moon base with China. NASA is legally prohibited from collaboration with any Chinese entity using government funding in the absence of direct Congressional approval under the Wolf Amendment.

The Artemis Accords consist of ten principles, although it is envisaged that these principles will be developed in further detail in the terms of any pursuant bilateral agreement. Each of these principles will be addressed in turn below. Commentators (Foust 2020a, Johnson 2020) have observed that the principles fall roughly into two extremes: existing norms that are broadly repetitive of the provisions of the Outer Space Treaty and those that extend well beyond its provisions and attempt to establish new norms of behavior.

The provisions which may be described as broad reaffirmations of fundamental space law principles are:

Section 3: Peaceful Purposes: This section requires signatories to affirm that the activities under the Accords shall be "exclusively for peaceful purposes and in accordance with relevant international law." The concept of peaceful purposes appears four times in the OST, most notably in **Article IV**. While the exact scope of the concept remains open to conjecture, it is accepted as a fundamental tenet of international space law and its presence in the Artemis Accords is uncontroversial. Similarly, the requirement to conduct activities in accordance with international law is a fundamental provision of the OST (**Article III**).

Section 4: Transparency: The section requires States to provide "transparency in the broad dissemination of information regarding their national space policies and space exploration plans" and the sharing of scientific information resulting from activities conducted pursuant to the Accords with the public and international scientific community on a good-faith basis and consistent with **Article XI** of the OST. It has been recognized since the dawn of the space age that clear communication about space activities is a necessary pre-condition of the safe use of space for peaceful purposes.

Section 6: Emergency Assistance: **Article V** OST requires parties to render assistance to astronauts and the Rescue and Return Agreement prescribes obligations regarding rendering assistance to personnel of a spacecraft who are in distress. This section of the Accords reinforces these obligations by requiring signatories to "commit to taking all reasonable efforts to render necessary assistance to personnel in outer space who are in distress, and acknowledge their obligations under the Rescue and Return Agreement." While seeming a simple restatement of the current state of international law with respect to astronauts, in practice the actions required by Artemis may in fact go well beyond the Space Treaties.

It is acknowledged that with increased activities in outer space, it is no longer clear that all personnel operating in the space environment will be astronauts. Private space operators, including SpaceX, Blue Origin, Virgin Galactic, and Axiom Space, all use the term "astronaut"; however, they also make use of terms such as commercial crew and passengers. By referring to all "personnel in outer space who are in distress" this seems to go well beyond any technical definition of astronauts or personnel of a spacecraft, as would be logical if those personnel are located on the Moon or celestial body.

Section 7: Registration of Space Objects: The Registration Convention requires the launching state of a space object to register that object and, where there are two or more launching states, to determine which of them shall register that object. This section explicitly requires parties to the Accords to determine the appropriate means and state of registration for any project conducted pursuant to the Accords.

Section 8: Release of Scientific Data: A fundamental concept as outlined in **Article I** of the OST is the encouragement of international cooperation. The UN Space Treaties also contain numerous mentions of visits to facilities, observation, consultations, and provision of certain information to the UN and other States; however, fundamentally concepts of openness and transparency have been anathema to the space race. Technological superiority has been the key advantage in both civilian and military contexts (de Zwart 2019, p. 341), while the sharing of scientific data, such as the US Air Force Space Surveillance Network which tracks about 26 000 space objects, has been an important asset in space security (Garber and Herron 2020). Unsurprisingly, therefore, while this section asserts that the signatories "are committed to the open sharing of scientific data" and "plan to make the scientific results obtained from the cooperative activities under these Accords available to the public and the international scientific community" in a timely manner, the Accords will require signatories to "coordinate with each other in advance regarding the public release of information that relates to the other Signatories" activities under these Accords in order to provide appropriate protection for any proprietary and/or export-controlled information." Notably the commitment to share scientific data openly is not intended to apply to private sector operations unless being conducted on behalf of an Agency.

Significantly new principles include:

Section 5: Interoperability: Acknowledging the specific needs of collaborative activities, as well as the growing use of commercial providers in the space domain, this section requires signatories to commit to use current interoperability standards for space-based infrastructure, as well as to establish standards where they do not currently exist or are inadequate. This principle is fundamental to developing a major space operation between multiple parties. Russia has even indicated that, despite the fact that it will not be participating in the Lunar Gateway project, it wants to ensure that standard interfaces are used so Russian spacecraft may dock with the Gateway (Foust 2020b).

Section 9: Preserving Outer Space Heritage: The preservation of the Apollo Moon Landing site has attracted significant interest in recent years (NASA 2011). This section actively commits signatories to "preserve outer space heritage" including "historically significant human or robotic landing sites, artifacts, spacecraft, and other evidence of activity on celestial bodies in accordance with mutually developed standards and practices." Further, signatories agree to contribute to multilateral efforts to preserve space heritage. The key contention here is not the question of historic value of these sites and artifacts but rather the quarantining of certain areas of the Moon – placing them off limits for other users. It also raises questions regarding whether protection of such sites may then amount to a *de facto* appropriation of the relevant area.

Section 10: Space Resources: This provision makes the most significant contribution to the debates regarding the legitimacy of space mining under the UN Space Treaties and international law. The section explicitly states that the "utilization of space resources can benefit humankind by providing critical support for safe and sustainable operations." Notably, the section affirms that "the extraction and utilization of space resources, including any recovery from the surface or subsurface of the Moon, Mars, comets, or asteroids, should be executed in such a manner that complies with the Outer Space Treaty and in support of safe and sustainable space activities." Essentially this leaves it up to project participants to determine how to comply with the OST.

Reflecting the current state of international debate on the legality of resource extraction, the section continues: "the extraction of space resources does not inherently constitute national appropriation under **Article II** of the Outer Space Treaty, and that contracts and other legal instruments relating to space resources should be consistent with that Treaty." Thus the Accords skirt the question of whether resource extraction is legal, but rather maintain that it does not amount to national appropriation, which is explicitly prohibited by **Article II** OST (Nelson 2020). Finally, the signatories agree "to use their experience under the Accords to contribute to multilateral efforts to further develop international practices and rules applicable to the extraction and utilization of space resources, including through ongoing efforts at the COPUOS (Committee on the Peaceful Uses of Outer Space)."

One of the major criticisms of the Artemis project is the fact it did not emerge from the COPUOS process (Tronchetti and Liu 2021 p. 246; Nelson 2020). Therefore the encouragement of participants to use their Artemis experience to contribute to ongoing debates and work of COPUOS is an explicit acknowledgement of the importance of that process. It should also be noted that the Hague International Space Resources Governance Working Group has developed and adopted a set of guiding principles in the Building Blocks for the Development on an International Framework on Space Resource Activities (Hague International Space Resources Governance Working Group 2019). Thus there are already other projects which endeavor to address the gap in law making in this area.

Section 11: Deconfliction of Space Activities: The key provision of the OST dealing with the management of competing uses of outer space is **Article IX**, which articulates concepts of "due regard" and "harmful interference." This section requires signatories to acknowledge and reaffirm their commitment to these provisions, in particular, of the OST and to act in accordance with the requirements with respect to these concepts under **Article IX** OST.

Signatories also affirm their commitment to the United Nations Guidelines for the Long-term Sustainability of Outer Space Activities (UNCOPUOS 2019).

Critically, this section states that signatories will "contribute to multilateral efforts to further develop international practices, criteria, and rules applicable to the definition and determination of safety zones and harmful interference." The concept of "safety zones," while being one well recognized in aviation law, is not currently recognized in space law, despite there being a 200 km "keep out" zone in place around the ISS. However, this section reflects a principle which may assist in ensuring that competing uses of a celestial body, such as the surface of the Moon for resource extraction activities, may be conducted with some degree of mutual co-operation and understanding.

Section 11 states:

> In order to implement their obligations under the Outer Space Treaty, the Signatories intend to provide notification of their activities and commit to co- ordinating with any relevant actor to avoid harmful interference. The area wherein this notification and coordination will be implemented to avoid harmful interference is referred to as a "safety zone." A safety zone should be the area in which nominal operations of a relevant activity or an anomalous event could reasonably cause harmful interference."

Signatories are required to apply principles of reasonableness of size, scope, and duration, considering the nature of operation, environmental conditions, requirements of notice in respect of safety zones. Further, signatories are required to coordinate with one another regarding notification of operations being undertaken within a safety zone, subject to protections for proprietary and security (export-controlled) information.

Notably, safety zones are explicitly identified as applicable in the context of resource extraction activities. Mining activities may create a significant hazard for personal working on or under the Moon's surface. The Hague Space Resources Working Group, for example, provides that the international framework "should permit States and international organisations responsible for space resources activities to establish a safety zone, or other area-based safety measure, around an area identified for a space resource activity as necessary to assure safety and to avoid any harmful interference" with that activity (Hague Space Resources Working Group 2019).

Commentators have considered how zones could be established and exercised in compliance with the principle of non-appropriation under **Article II** OST. Johnson, for example, argues that the concept of keep out zones is justified under the terms of the OST as their practical implementation provides some real context for the concept of due regard under **Article IX** (Johnson 2020). Zones as articulated by the Artemis Accords could clearly and practically establish how various uses of the Moon could co-exist where there is adequate co-operation in the creation and recognition of various usage zones. The extent, size, and duration of the zones would also be important to determining their character (Schingler 2020). Further questions arise as to whether zones must be limited to physical or safety matters or could be extended to economic or exclusive "keep out" zones. Johnson notes that "it would be contrary to the Outer Space Treaty to create blanket "keep out" zones. Rather, these Artemis safety zones will have to be articulated carefully to partners, and to the international community, in a way that stresses that Artemis activities are not asserting property rights over the boundaries of such zones. Instead, they should be explained as the physical manifestation of due regard amongst Artemis partners" (Johnson 2020).

Section 12: Orbital Debris: Recognizing that the problem of space debris is of increasing concern for the continued access to and use of outer space, this section of the Accords requires signatories to plan for the mitigation of orbital debris "including the safe, timely, and efficient passivation and disposal of spacecraft at the end of their missions." Further, signatories commit to limit the creation of harmful debris "through normal operation, break-up in operational or post- mission phases, and accidents and conjunctions." The problem of orbital congestion has been recognized as a major issue for space sustainability. The increase in launch of smaller satellites has generated exponentially more objects in space, posing greater risks of collision and hazards for larger space objects, such as the International Space Station. In 2020, the ISS was maneuvered three times to avoid space debris, prompting then NASA Administrator Jim Bridenstine to Tweet:

> The @Space_Station has maneuvered 3 times in 2020 to avoid debris. In the last 2 weeks, there have been 3 high concern potential conjunctions. Debris is getting worse! Time for Congress to provide @CommerceGov with the $15 mil requested by @POTUS [President of the United States] for the Office of Space Commerce.
>
> (Bridenstine 2020)

The growing recognition of the increasing threat posed by the proliferation of space debris to continued access to space, especially to prized orbits and management of the radiofrequency spectrum prompted UNCOPUOS to establish a Working Group on the Long Term Sustainability of Outer Space Activities, which produced a set of voluntary, non-binding guidelines encouraging the adoption of national regulatory frameworks, the supervisions of national space activities, promotion and sharing of space debris monitoring information, and the investigation of measures to manage space debris (Martinez 2018, pp. 15–16). Those guidelines were adopted by UNCOPUOS in 2019 (UNCOPUOS 2019).

19.4 Issues for Australia as a Signatory to the Moon Agreement

As noted above, Australia is unusual as not only is it one of the few signatories of the Moon Agreement but it is also a partner in the Artemis Accords, through the Australian Space Agency.

Australia is supporting NASA's plans to return to the Moon and on to Mars through its Moon to Mars initiative. Key aspects of that initiative include the Supply Chain Program, the Demonstrator Program and the Trailblazer Program "Australia's flagship mission supporting NASA's Moon to Mars activities that demonstrates and develops Australia's strengths, ingenuity and capabilities, while inspiring the nation." (Australian Space Agency 2021, p. 3). One of the strengths identified as aligned with the key priority areas identified in Australia's Civil Space Strategy (Commonwealth of Australia 2019) is *In situ* Resource Utilisation (ISRU). With express reference to the opportunity provided by Artemis to leverage key Australian expertise in remote asset management and field robotics, the proposed Exploration Mission includes activities of regolith collection and sampling.

Some have claimed that "Australia's simultaneous membership of both instruments is problematic both from a diplomatic and a legal standpoint" (Tronchetti and Liu 2021, p. 243). Tronchetti and Liu assert that the Artemis Accords and the Moon Agreement articulate and reflect a very different and indeed, inconsistent, approach to regulation of activities on the Moon and beyond. Unlike the Moon Agreement, which is a multilateral treaty, the Artemis Accords reflects a much narrower set of interests, developed in accordance with the US stance both on the Moon Agreement and the extraction and use of resources in space. It should however be observed that the high-level principles outlined in the Accords are intended to "establish a common vision via a practical set of principles, guidelines, and best practices to enhance the governance of the civil exploration and use of outer space" (Artemis Accords, p. 2). The details of common activities and programs are to be developed and agreed in separate agreements and instruments.

Tronchetti and Liu further note that the Artemis Accords are premised on the concept that extraction and use of space resources is permitted by international space law, whereas the Moon Agreement anticipates that "States Parties to this Agreement hereby undertake to establish an international regime… to govern the exploitation of the natural resources of the Moon as such exploitation is about to become feasible" (Moon Agreement, **Article 11.5**). However, again, this overlooks the provisions of **Article 4.2** which states that states will be guided by the "principle of cooperation and mutual assistance in all their activities concerning the exploration and use of the Moon." It goes on to state that "International cooperation in pursuance of this Agreement should be as wide as possible and may take place on a multilateral basis, on a bilateral basis or through international intergovernmental organizations." This appears to well and truly leave open the door for initiatives such as the Artemis Accords, which provide an umbrella set of principles, pursuant to which it is anticipated that further agreements between two or more States or agencies will be concluded. Such instruments and consequent activities are expressed to be intended to be "carried out in accordance with the legal obligations applicable to each Signatory" with a consequent obligation for each Signatory to ensure that entities acting on its behalf comply with the principles of the Accords (NASA 2020a, p. 2). Thus, Australia is able to participate in Artemis in accordance with its obligations under international law, including the Moon Agreement. More concerning is the assertion by Tronchetti and Liu that **Article 11** obliges Australia to participate in the establishment of an international regime as the exploitation of the natural resources of the Moon is about to become feasible. They state that not only has Australia *de facto* opted out of such a process but failed to consult with other members of the Moon Agreement regarding their intention to do so, questioning even if the decision to join the Artemis Accords places Australia in breach of its international obligations (Tronchetti and Liu 2021 p. 248).

In fact, Australia has, through the Australian Space Agency, been clear that it sees no disparity or inconsistency in its membership of both the Moon Agreement and participation in the Artemis Accords. Australia is committed to meeting its international obligations under the Outer Space Treaties and broader principles of international law.

19.5 Recent Developments: Purchase of Lunar Regolith

Reflecting the continuing uncertainty regarding the legal framework regarding ISRU, in September 2020 NASA moved ahead with issuing a Request for Quotations for the Purchase of Lunar Regolith and/or Rock Materials from Contractor. The purpose of that project is to test commercial technical capability and the appropriate legal mechanisms for the extraction and transfer of materials on the Moon. Four companies have been selected under that request for tender worth US$25,001 to establish a precedent for ownership and use of space resources: Lunar Outpost ($1 following arrival of a lander to the lunar South Pole in 2023); ispace Japan (US$ 5,000 following arrival in 2022 of a lander to Lacus Somniorum on the Moon's northeastern near side); ispace Europe (US$ 5,000 following arrival in 2023 of a lander to the lunar South Pole); and Masten Space Systems (US$ 15,000 following arrival in 2023 of a lander to the lunar South Pole). Payment is made by instalments of 10% of their total proposed price upon award, 10% upon launch, and the remaining

80% upon successful completion. The payment is expressed to be exclusively for the lunar regolith and not with respect to associated services. Each company will collect a small amount of lunar regolith from the specified location and is required to provide imagery to NASA evidencing the collection of the material, including data identifying the collection location. Upon receipt of that evidence an "in-place" transfer of ownership of the lunar regolith to NASA is deemed to have occurred. At present there are no processes for retrieval of the materials and the regolith will remain in place for future use as part of the Artemis program (NASA 2020b). The procurement arrangements require the successful companies to undertake performance of the contracts in "full compliance with the Registration Convention, Article II, and other provisions of the Outer Space Treaty, as well as in accordance with NASA's other relevant international obligations" (NASA 2020b).

19.6 Conclusions: Where to from Here for International Space Law?

What do the Artemis Accords mean for the future development of international law affecting space mining activities? Does the fact that they have been negotiated outside of the UNCOPUOS process diminish their likelihood of success?

Rather than adopting a negative approach that Artemis has been undertaken externally to UNCOPUOS, it may be more productive to view the Accords as an effort to influence and advance the lawmaking process through the influence of State practice. At the 60th Session of the COPUOS Legal Subcommittee, in May 2020, Agenda Item 14: Potential Legal Models for Activities in Exploration, Exploitation and Utilization of Space Resources the US Head of Delegation, Gabriel Swiney, confirmed the US view that utilization of space-based resources (including commercial use) is consistent with the "four main United Nations Space treaties" and presented the following views:

> [T]he Outer Space Treaty does not provide a comprehensive international regime for space resource utilization activities. At this stage, the United States sees neither a need nor a practical basis to create such a regime. We do, however, see a need to ensure that all nations engaged in space resource activities share a common set of fundamental beliefs: in the rule of law, in transparency, and in peaceful purposes. NASA's Artemis Accords underscore these critical principles, and form a starting point for the US Government's future work on space resources, which we hope to share with other likeminded allies and partners.
>
> (UNCOPUOS 2021)

The Accords provide a framework for the exploration of the next frontier of space. This step will inevitably require the development of a legal regime around the extraction and use of resources. This means that meaningful and frank discussions need to be had now to ensure that there is a common understanding regarding the laws that will guide those activities. It is simply not sensible to wait for another large multilateral treaty nor to continue to bemoan the impossibility of ever reaching one again.

While not all nations may agree with the fundamental tenets of the Artemis Accords, the Accords at least represent an articulation of rules that are up for debate, there is an attempt at a common understanding of the guiding principles and there is an invitation to be part of the future agreement of the rules guiding humanity's next steps to Mars. It is clear that commercial operators will not delay space activities, awaiting clarification of the legal rules of the road. It is to be hoped that Artemis will provoke a genuine engagement with the next set of legal and ethical principles to guide humanity into space.

References

Australian Space Agency (2021). *Moon to Mars initiative – Trailblazer Program Consultation paper,* 8 April 2021, https://consult.industry.gov.au/space/moon-to-mars-trailblazer-program/user_uploads/trailblazer-consultation-paper-3.pdf (accessed 19 June 2021).

Bridenstine, J (2020). https://twitter.com/JimBridenstine/status/1308540671725318144 (accessed 18 June 2021)

Commonwealth of Australia (2019). *Australian Civil Space Strategy 2019 – 2028*, April 2019, https://publications.industry.gov.au/publications/advancing-space-australian-civil-space-strategy-2019-2028.pdf (accessed 19 June 2021).

von der Dunk, F.G. (2011). The origins of authorisation: Article VI of the outer space treaty and International Space Law. In: *National Space Legislation in Europe: Issues of Authorisation of Private Space Activities in the Light of Developments in European Space Cooperation, Studies in Space Law*, vol. 6, 3–28. Leiden: Martinus Nijhoff Publishers.

von der Dunk, Frans (2017). Asteroid Mining: International and National Legal Aspects. *Michigan State International Law Review*, 26:1 83–101 at 94–95.

von der Dunk, F.G. (2020). *Advanced Introduction to Space Law*. Cheltenham, UK: Edward Elgar.

Executive Order on Encouraging International Support for the Recovery and Use of Space Resources, 6 April 2020.

Foust, J. (2020a). What's in a name when it comes to an "accord"? (13 July 2020) *The Space Review*, https://www.thespacereview.com/article/3987/1 (accessed 13 June 2021).

Foust, J. (2020b). Russia skeptical about participating in lunar Gateway, (12 October 2020) *Space News*, https://spacenews.com/russia-skeptical-about-participating-in-lunar-gateway/ (accessed 13 June 2021).

Garber, S. and Herron, M. (2020). How has traffic been managed in the sky, on waterways, and on the road? Comparisons for space situational awareness (part 1) (8 June 2020) *The Space Review* https://www.thespacereview.com/article/3961/1 (accessed 13 June 2021).

IGA (1988). *Agreement among the Government of the United States of America, Governments of Member States of the European Space Agency, the Government of Japan, and the Government of Canada on Cooperation in the Detailed Design, Development, Operation, and Utilization of the permanently Manned Civil Space Station*, 29 September 1988 (entered into force on 30 January 1992).

IGA (1998). *Agreement among the Government of Canada, Governments of Member States of the European Space Agency, the Government of Japan, the Government of the Russian Federation, and the Government of the United States of America concerning Cooperation on the Civil International Space Station*. 29 January 1998 (entered into force 28 March 2001) (IGA).

Johnson, C. (2020). The Space Law Context of the Artemis Accords (Part 2) (28 May 2020) *Space Watch Global*, https://spacewatch.global/2020/05/spacewatchgl-feature-the-space-law-context-of-the-artemis-accords-part-2/ (accessed 13 June 2021).

Jones, A. (2021). China, Russia open moon base project to international partners, early details emerge. *Space News* (26 April 2021) https://spacenews.com/china-russia-open-moon-base-project-to-international-partners-early-details-emerge/ (accessed. 13 June 2021).

Liability Convention (1972). *Convention on International Liability for Damage Caused by Space Objects*, opened for signature 29 March 1972, 961 UNTS 187 (entered into force 1 September 1972).

Martinez, P. (2018). Development of an international compendium of guidelines for the long-term sustainability of outer space activities. *Space Policy* 43: 13–17. https://doi.org/10.1016/j.spacepol.2018.01.002.

Moon Agreement (1979). *Agreement Governing the Activities of States on the Moon and Other Celestial Bodies*, opened for signature 18 December 1979, 1363 UNTS 3 (entered into force 11 July 1984).

NASA (2011). NASA's recommendation to space-faring entities: how to protect and preserve the historic and scientific value of US Government Lunar Artifacts (11 July 2011).

NASA (2020a). The Artemis Accords: Principles for cooperation in the civil exploration and use of the Moon, Mars, Comets and Asteroids for Peaceful Purposes, signed 13 October 2020.

NASA (2020b). NASA Selects Companies to Collect Lunar Resources for Artemis Demonstrations, (4 December 2020) https://www.nasa.gov/press-release/nasa-selects-companies-to-collect-lunar-resources-for-artemis-demonstrations (accessed 18 June 2021).

Nelson, J.W. (2020). The Artemis Accords and the future of International Space Law. *Insight* 24 (31): 1–7.

Outer Space Treaty (1967). Treaty on Principles Governing the Activities of States in the Exploration and Use of Outer Space, including the Moon and Other Celestial Bodies (1967, January 27). 18 U.S.T. 2410, 610 U.N.T.S. 205 (entered into force 10 October 1967).

Registration Convention (1975). *Convention on Registration of Objects Launched into Outer Space*, opened for signature 14 January 1975, 1023 UNTS 15 (entered into force 15 September 1976).

Rescue and Return Agreement (1968). *Agreement on the Rescue of Astronauts, the Return of Astronauts and the Return of Objects Launched into Outer Space*, opened for signature 22 April 1986, 672 UNTS 119 (entered into force 3 December 1968).

Roulette, J. (2020). Trump Administration drafting 'Artemis Accords' pact for space mining. *Science News* https://www.reuters.com/article/us-space-exploration-moon-mining-exclusi-idUSKBN22H2SB (accessed 19 June 2021).

Schingler, J.K. (2020). Imagining safety zones: implications and open questions. *The Space Review* https://www.thespacereview.com/article/3962/1 (accessed 19 June 2021).

Sheldon, J. (2020). Lunapolitics: Russia's Roscosmos Signals Openness to US-proposed Artemis Accords, SpaceWatch.global. https://spacewatch.global/2020/05/lunapolitics-russias-roscosmos-signals-openness-to-us-proposed-artemis-accords/ (accessed 13 June 2021).

The Hague Space Resources Working Group (2019). *Building Blocks for the Development of an International Legal Framework on Space Resources Activities*. https://www.universiteitleiden.nl/binaries/content/assets/rechtsgeleerdheid/instituut-voor-publiekrecht/lucht--en-ruimterecht/space-resources/bb-thissrwg--cover.pdf (accessed 18 June 2021).

Tronchetti, F. (2009). *The Exploitation of Natural Resources of the Moon and Other Celestial Bodies: A Proposal for a Legal Regime*. Leiden, The Netherlands: Brill.

Tronchetti, F. and Liu, H. (2021). Australia's signing of the Artemis Accords: a positive development or a controversial choice? *Australian Journal of International Affairs* 75 (3): 243–251. https://doi.org/10.1080/10357718.2021.1893268.

UNCOPUOS (2019). *Guidelines for the Long-term Sustainability of Outer Space Activities*, June 2019, A/74/20, para 163 and Annex II). https://www.unoosa.org/oosa/en/oosadoc/data/documents/2019/a/a7420_0.html (accessed 13 June 2021).

UNCOPUOS (2021). 60th Session of the COPUOS Legal Subcommittee – Agenda Item 14: Potential Legal Models for Activities in Exploration, Exploitation and Utilization of Space Resources, as delivered by U.S. Head of Delegation, Gabriel Swiney, June 1, 2021 https://vienna.usmission.gov/2021-copuos-lsc-u-s-on-the-exploration-exploitation-and-utilization-of-space-resources/ (accessed 22 June 2021).

UNOOSA (2021). Status of International Agreements relating to Activities in Outer Space, https://www.unoosa.org/oosa/en/ourwork/spacelaw/treaties/status/index.html (accessed 13 June 2021).

Yakovenko, A. (1999). The intergovernmental agreement on the International Space Station. *Space Policy* 15 (2): 79–86. https://doi.org/10.1016/S0265-9646(99)OOD11-9.

de Zwart, M. (2019). Outer space. In: *New Technologies and the Law in War and Peace* (ed. W. Boothby), 337–358. UK: Cambridge University Press.

20

Space Resources: Physical Constraints, Policy Choices, and Ethical Considerations

Martin Elvis[1], *Alanna Krolikowski*[2], *and Tony Milligan*[3]

[1] *Center for Astrophysics, Harvard and Smithsonian, 60 Garden Street, Cambridge MA 02138, USA*
[2] *Missouri University of Science and Technology (Missouri S&T), Department of History and Political Science and Center for Science, Technology, and Society, 500 W 14th Street, Rm 122, Rolla MO 65409, USA*
[3] *King's College London, Cosmological Visionaries Project, Virginia Wolf Building, 22 Kingsway, London WC2B 6NR, UK*

20.1 Introduction and Outline

For over 50 years, technology has dominated the discussion of human expansion into space. This is for the very good reason that without proven technology there can be no space exploration. The question of "can it be done?" is now all but answered in the affirmative. New questions of "should it be done?" or at least "how should it be done?" are coming to the fore. In this chapter, we discuss such questions in the light of an important assumption about current actions and their later impact. For this, we use the familiar idea of a founder effect: "even a few score, initial colonizers can mean much more for the cultural geography of a place than the contributions of tens of thousands of new immigrants a few generations later" (Zelinksky 1972, pp. 13–14). Adapted to the context of humanity's expansion into space, the formulation becomes:

> **Founder effect**: while we cannot hope to dictate terms to future generations, decisions implemented at the beginning of humanity's exploitation of space resources can have disproportionate and lasting impact.

This effect is likely to operate as a significant societal constraint upon future patterns of economic and social development. The upside of this effect is a special kind of influence upon future events, allowing us to shape our projects in ways that might encourage them to be continued by others. For example, responsible practices adopted early can create enduring precedents that shape future conduct for many generations. The downside of the founder effect is that beginnings cannot be made again. They are irreversible or difficult to reorient in ways that later

In-Space Manufacturing and Resources: Earth and Planetary Exploration Applications, First Edition. Edited by Volker Hessel, Jana Stoudemire, Hideaki Miyamoto, and Ian D. Fisk.

actions are not. Mistakes made now have a disproportionate likelihood of rippling indefinitely and unfavorably into unintended future outcomes. This downside reflects physical constraints as well as social ones. The future expansion of a space economy will face constraints of both sorts. Here, we consider only a few that we regard as important and foreseeable. The substantial constraints of our human physicality (which is difficult to accommodate in space) will be set aside in order to focus upon the limiting factor of finite resources. Section 20.2 will consider such physical constraints upon the emergence of competitive markets, but also as a factor pushing human space expansion toward a systematic presence at multiple interconnected places, rather than merely at the sites with the material resources of more direct interest. Given the potential of these resources, Section 20.3 addresses the policy questions that now confront governments, highlighting the ways in which the decisions made in the near term will shape whether, when, and how markets for space resources emerge over the medium to long terms. Section 20.4 will then tentatively frame these possible policy responses within a larger ethical dilemma concerning our obligations to current and future generations: the dilemma of having to continually trade off *near-term justice* against *the interests of future generations*. Near-term justice may be enabled by economic expansion in space, but at a cost to those future generations who live closer to the edge of resource depletion and, ultimately, exhaustion.

20.2 Economic Prospects and Strategic Resources

Over the past decade the idea that there could be a large and dynamic space economy involving humans working, and possibly living, in space has revived after long pause and several false starts (Sonter 1997; Lewis 1996). Such an economy was widely anticipated during the Apollo era and formed part of the economic rationale for the US Congress to fund NASA's Space Shuttle program, as a way to make spaceflight routine and thereby help bring the cost of space flight down to the point where commercial activity could flourish (NASA 1976). The anticipated reduction in costs, and the economic flourishing it promised, failed to materialize. There are, however, signs that this is finally changing. The cost of launching a kilogram of anything into low-Earth orbit (LEO) has now dropped from some US$ 20,000 to about US$ 3000, with the advent of the reusable Falcon 9 rocket from SpaceX (https://www.spacex.com/media/Capabilities&Services.pdf). There are at least three private companies developing their own space stations to lease out to a rapidly growing number of start-ups that plan to manufacture products in space that require micro-gravity or a virtual absence of gravity for their construction; these products range from greatly improved optical fibers to human organs, even human hearts (Elvis 2021). Other customers for private space stations include research organizations, be they space agencies from smaller countries, rich universities and foundations, or biotech companies. The first paying customers are already known; they are tourists (Kooser 2021). Not all of these ventures will succeed, and there are good reasons to be cautious about participating in first waves of economic activity. Nonetheless, the move toward a large-scale and thriving space economy now looks more promising than ever. At this point, no government or international body could stop it from continuing, even if they wanted to do so as a matter of policy or principle.

Such a thriving in-space economy will need supplies of raw materials. The most obvious of which is water. Tourists paying millions for a space ride are likely to choose the space companies with the best safety record and the space station with the best facilities. That includes showers and (at some point) flushing toilets. Water is also effective as protection against radiation and can become rocket fuel when electrolyzed into hydrogen and oxygen. Conveniently, it turns out that water can be found abundantly in some locations in space, and energetically it is cheaper to supply that water from space than to bring it from Earth; it may even become financially cheaper although it will initially be more expensive. As the in-space economy expands, other raw materials will be needed in quantities that are prohibitive to bring up from the ground. Methane for rocket fuel can be made from the large organic molecules found in space. Iron for large-scale construction can be sourced from asteroids. Even ordinary silicate rock can be a source of oxygen as well as being used for radiation shielding. These are *in situ* resources which are also likely to be used *in situ*. The only space resource worth bringing down to Earth, at least in the near future, is precious metals, such as platinum and palladium. But, eventually, some of the dozens of minerals found only in asteroids (the "meteorite minerals," named for how they were discovered) may turn out to have economic value.

Where will all these resources come from? The Moon and, as indicated, the asteroids. Though Mars is also a prime target for human exploration, at the moment it does not seem to be a likely source of resources that could easily be used elsewhere, and any time soon. Below we look at each source location more closely.

20.2.1 The Moon

The Moon is an appealing source for resources. We have already mapped the whole Moon in quite some detail, down to kilometer resolution in most cases (e.g. the compilation in Elvis, Krolikowski, and Milligan 2020, EKM20). As a result, we already know the best places to go prospecting for the richest deposits. The Moon is also only about 400,000 km from Earth; a three-day trip for astronauts, (NASA 2021) a roughly one second trip for light and radio waves. If there is an emergency involving astronaut miners or prospectors on the Moon, they can be rescued and may get home in a time that could be medically useful. On the other hand, the two second round trip delay in getting radio signals out and back is short enough that humans may not need to be physically present on the Moon in order to operate machinery there. Efficient tele-operation is plausible once the equipment is smart enough to avoid dangers which lie only a few seconds ahead.

An advantage of the Moon over asteroids for mining is that the Moon has stronger gravity, albeit only 1/6 of Earth's gravity. While that does mean expending more rocket fuel to bring supplies from the Moon to where they will be used in space, gravity makes lunar mining operations much more similar to those on Earth than they would be for the asteroids. Gravity is also useful to separate out the valuable materials from the tailings (Lewis 1996), and implicitly to settle the dust and debris from mining, which does not happen on asteroids.

What are the lunar resources that we might use? Water has already been mentioned, and many players have begun to explore options for use (Wall 2021; Foust 2021; EKM20). However, over most of the Moon, water is present only in trace quantities (ppm, Milliken and Li 2017). However, there are a few locations where water seems to be present in large-ish quantities (Li et al. 2018). Places where water and volatiles deposited by comets and asteroid impacts remained in total darkness and did not heat up enough to escape (Berezhnov et al. 2012, Lawrence 2017). These "permanently shadowed regions" (PSRs) are primarily located near the poles. The obliquity of the Moon – the inclination of its axis to the plane of its orbit around the Sun – is just 1.5°, compared with 23.5° for the Earth (Allen 1973). As a result, there are almost no seasons on the Moon and at the poles the Sun traces a course around the horizon, rising and falling by only 1.5° each day. As at sunset or sunrise on Earth, any peak casts long shadows at the lunar poles. The raised walls of craters cast shadows against the opposite side of the crater rim, leaving the floors of the craters in permanent shadow. The crater floors are illuminated only by starlight and by light reflected off the rims. This makes these regions very cold. Below 110 K (−163 °C) water ice is stable even in a vacuum (Paige et al. 2010). Other volatiles are stable at lower temperatures (Paige et al. 2010).

Lunar resources are cataloged by Crawford (2015) and EKM20. Besides water they include: ^{3}He, a fuel for potential future nuclear fusion reactors; uranium and thorium that are fuels for current nuclear fission reactors; iron, probably remnants of asteroid impacts; and a unique location rich in rare earth elements, the KREEP terrane [KREEP stands for K (the atomic symbol for potassium), REE (rare-earth elements) and P (for phosphorus).] This KREEP terrane was likely the last place on the Moon to solidify after its creation from the impact of a Mars-sized planet with the proto-Earth (Ćuk and Stewart 2012).

For the near future lunar resources will be plentiful relative to the size of the nascent in-space economy. Measurements indicate that as much as a billion metric tons of water may be present in the PSRs (Crawford 2015). That is a lot compared with the ~50 mt that would be needed to refuel a rocket (Kutter 2018), or the 1 mt/year used by an astronaut aboard the International Space Station (Jonathan McDowell 2020, private communication), but is still only the volume of a small lake.[1] New York City uses about this much water every day.[2] Lunar resources are plentiful, relative to the anticipated near-term needs of economic expansion, but they are still finite, and far from inexhaustible. They are also highly concentrated into a modest number of locations that are typically only a few kilometers in extent (EKM20). For example, there are just a half dozen PSR areas 50 km across that are colder than 110 K, and a just few dozen larger than about 5 km diameter (Paige et al. 2010). What we draw from this is a simple point: as the in-space economy grows,

1 Lake Mead, behind the Hoover dam, has a capacity about 30 times larger.
2 "nyc's reservoir system". http://www.nyc.gov/html/nycwater/html/drinking/reservoir.shtml (accessed 30 May 2021).

lunar resources will prove to be inadequate. Although asteroid mining presents considerable logistical difficulties, sooner or later, the asteroids will beckon as alternative storehouses of resources.

20.2.2 Asteroids

20.2.2.1 Near-Earth Asteroids

The next accessible source for space resources is the population of near-Earth asteroids (NEAs[3]). There are about 20,000 NEAs larger than 100 m diameter (Mainzer et al. 2011). NEAs have orbits within that of Mars and, as the name implies, come close to the Earth, some of them crossing the Earth's orbit, occasionally coming close enough to impact the Earth. Small ones do not reach the ground intact, such as the 17-m sized asteroid that exploded above the Siberian city of Chelyabinsk in 2013 (Kring and Boslough 2014), but they do lead to showers of small remnant rocks that we call meteorites. From the analysis of the composition of meteorites we know that some asteroids contain large concentrations of valuable materials, including water and the precious platinum group metals (PGMs[4]). There are three main types of meteorite and asteroid. Each has its own value as resources:

1. First, a metallic nickel-iron asteroid 100 m in diameter that is PGM-rich may contain 20 mt of platinum, about 10% of the yearly production on Earth, with a nominal sales value of US$ 1 billion (Elvis 2014). The million tons of iron left behind might be slowly moved to a high orbit around the Earth; there this iron could provide a relatively cheap source of building material for large-scale habitats, or space hotels, or for the long-dreamed of space-based solar power.
2. Second, a carbonaceous water-rich NEA just 10 m across could have a similar mass of water. Its value, once comparable to the US$ 50 M per ton of platinum, goes down as launch costs per kilogram decrease. Presently in LEO it would cost about US$ 20 M to launch from the ground. In a high-Earth orbit, or anywhere else near the top of, or outside, the Earth's gravity well, it could be worth twice as much.
3. The third class of asteroids are called "stony" and are made mostly of silicate rocks. Their value is less obvious. Like lunar regolith, 40% of their mass is made up of oxygen atoms. Solar power could be used to liberate this oxygen for use as life support or the oxidizer of rocket propellant. The rock itself, perhaps turned into concrete and formed into convenient shapes, can be used for radiation shielding.

Unfortunately, only about 1 in a hundred NEAs is rich in either water or PGMs (Elvis 2014). The number that we can reach with a payload large enough to undertake mining is limited by our current rockets to a few percent of that, leaving only a

3 Strictly they are the near-Earth objects (NEOs) because the population includes an ill-determined number of dead comets besides true asteroids.

4 The PGMs are: platinum, palladium, rhodium, osmium, iridium, and ruthenium. The six are called siderophiles, as they dissolve in liquid iron. This property is why they are scarce in the Earth's crust as they were efficiently carried down into the iron core of the Earth when the Earth was still largely molten (REF).

few dozen that might be mineable at present and worth a billion dollars each (Elvis 2014). With more powerful rockets, notably with the SpaceX Starship, this mineable number could become several hundred. However, the crucial point is that NEAs in themselves do not hold greater water resources than the Moon.[5] NEAs could, however, be a profitable training ground for mining the far larger asteroid resources found farther from the Sun. There is much to be learned. Mining without gravity will require new techniques, from attaching to an asteroid and excavating the loosely bound rubble that most asteroids appear to be made of, through to drilling and blasting in order to extract the valuable ore while keeping the dust and debris generated out of the machinery.

20.2.2.2 Main Belt Asteroids

Beyond the orbit of Mars there is a far larger store of resources. There are a million of asteroids larger than about 300 m in size in the "Main Belt" of asteroids that lies between the orbits of Mars and Jupiter (Jedicke et al. 2015). As on the Moon, most of the resources lie in a relatively small number of larger asteroids. The resources of the Main Belt dwarf those available on Earth or the Moon. To give one measure of this abundance, the total amount of iron in the Main Belt asteroids is some 10 million times that of Earth's proven reserves (Elvis and Milligan 2019). Economically, the special challenge of mining Main Belt asteroids is that they are so far from Earth. Transit times on energetically efficient Hohmann transfer orbits are a year or so, and in commerce time is money. It may be that, with the profits from NEA mining, companies will be able to afford to set up a shipping system that is economically viable for this more ambitious project. Without it, any space economy would be severely restricted. With it, the physical resources would still be exhaustible, even within a 500 year period if exponential growth, at even modest rates, is not kept within control (Elvis and Milligan 2019). However, for a time, this would allow for a closer approximation to the kind of expansive commercial presence, superprofits, and rapid economic growth that make space mining such an attractive prospect. It is, however, difficult to imagine large-scale Main Belt mining as a standalone process, rather than part of a system of activities encompassing both lunar presence and the use of Mars or its (gravitationally weaker) moons as a base for operations (Milligan and Elvis 2019). Periodic raids upon Main Belt resources will hardly overcome the combined resource limitations of the Earth, Moon and NEAs. A functioning space economy capable of overcoming these resource limitations is far more likely to be systemic than opportunistic, i.e. something that involves sustained and expansive presence rather than mere periodic raiding to correct resource deficits closer to Earth.

20.2.3 Mars

Mars itself has many resources, including water, and the red planet is the main hope of those who wish to establish self-sustaining settlements beyond Earth. There

5 "Accessible NEAs", NASA, https://cneos.jpl.nasa.gov/nhats/ (accessed 3 June 2021).

is, however, no obvious material resource on Mars that is not found among the asteroids, although exploration may reveal some in the future. The 1/3 of Earth gravity on Mars also imposes an energy penalty on exporting resources from there, compared with the negligible gravity of the asteroids. Nonetheless, it may be that mining the asteroids and the settlement of Mars will be mutually enabling processes which develop as parts of a systemic, rather than opportunistic space economy, for three reasons.

First, among the near-Earth asteroids we already know of there are a dozen or so that we could move into Mars cycler orbits (i.e., a spacecraft trajectory that encounters Earth and Mars regularly) with today's technology (Strange et al. 2015). A spacecraft in a Mars cycler orbit crosses Earth orbit when the Earth is there, and Mars orbit when Mars is there, without need for any expenditure of propellant (Byrnes et al. 1993). A traveler to Mars would be ferried to the Mars cycler as it passed by Earth, then would take another ferry down from the Mars cycler on arriving at Mars. As no fuel is required[6] the Mars cycler spacecraft can be enormously massive, and that mass can be used for radiation protection. That solves the biggest problem for Mars-bound travelers. A large spacecraft also allows for more space per traveler and a richer more entertaining ride; it could be closer to a somewhat constrained cruise ship rather than a lifeboat. A thriving asteroid mining business could well have a Mars cycler building, first as a way to transport large masses to cis-lunar space, and later as a personnel carrying sideline.

Second, following on from the theme of systematicity above, the two moons of Mars, Phobos and Deimos, may be valuable as bases for Main Belt mining (Taylor et al. 2022). They may host repair and supply depots for expensive mining equipment, or they may become sites for refining partially beneficiated ore into purer form. Once there is a profitable human presence on Phobos and Deimos, the cost and difficulty of reaching Mars' surface itself will be much reduced. Explorers and, later, settlers will have an easier journey with much more equipment than higher cost dedicated missions. Miners may want to take rest and relaxation trips to the surface to enjoy real gravity for a while, and to see the sites. They would then provide customers for services provided by Mars settlers.

Third, in order to mine the Main Belt profitably, faster rockets powered by nuclear fusion would be highly desirable because time is money. At least two designs for near-term fusion rockets are now being developed[7] (Cohen et al. 2015; Thomas et al. 2019). They would cut the journey time to Mars to about one month, making radiation much less of an issue. This point is a little more speculative given the slow pace at which we are advancing toward economically viable fusion technologies, but it is included because it helps to draw out an important point about the prospective space economy. However, it is a further way to highlight the systematicity point, given that an economically viable large-scale return of Helium 3 for nuclear fusion power from the Moon or from asteroids to Earth, is difficult to achieve and may be impractical for the foreseeable future. Earlier use of fusion technologies in space is far more likely,

6 No doubt this is an approximation. Perturbations to the orbit will occur and will need to be corrected. This may be done slowly using ion engines to avoid carrying large masses of fuel.
7 NASA MSFC, Pulsed Fission Fusion: https://www.nasa.gov/puff (accessed 3 June 2021).

being considerably more viable from an economic point of view (Crawford 2015). This would again involve a systematic linkage of human activities between Earth, the Moon, Mars, and the Main Belt. Effective use of the available, and limited, resources points toward a broader systematicity than that between the Earth and a single location from which materials would then be extracted. The limitations of accessible resources, and the combination of (i) their dispersal across multiple locations; and (ii) concentration *at* those locations, are drivers for systemic organization.

20.3 Policy Choices and the Emergence of Markets for Space Resources

As this survey of resources on the Moon, near-Earth and Main Belt asteroids, and Mars suggests, they are both abundant and becoming accessible enough to one day sustain an economic system that comprises significant nodes off Earth. While these factors provide the conditions that make such a long-term outcome possible, in the short and medium terms, whether, when, and how markets for space resources in fact develop will depend to a significant degree on policy choices by national governments and international bodies.

Some of the relevant policies are already being developed and adopted by the world's leading spacefaring countries, creating elements of national legislation and policy upon which future generations may build a global regime to govern space resources. Such a regime would consist in a configuration of national and international policy measures governing conduct directed at space resources, including various regulatory and legal instruments. In addition to national measures, international efforts at articulating principles and concepts for governing conduct directed at space resources have also made headway, creating the basis for future global deliberation and negotiation in this issue-area (ISECG International Architecture Working Group 2010; *Security and sustainability in Outer Space* 2015; *Vancouver Recommendations on Space Mining* 2020; Neto *et al.* 2020). Given the founder effect discussed above, or the path-dependent nature of regimes and associated institutions, early policy choices are likely to have outsized effects on long-term outcomes: decisions being made today could shape the prospects for space resource markets for many decades.

In this section, we consider some of the most fundamental and consequential policy choices before governments and international bodies for the development of markets for lunar and near-Earth asteroid resources in the medium term. As we have seen above, within this timeframe, which we define as the next 30 years, the Moon and proximate asteroids are the most likely targets of resource exploitation. Moreover, the policies adopted to govern resource activities at these initial sites are likely to inform, if not provide templates for, those eventually adopted for Mars, the Main Belt, and other destinations beyond the orbit of Mars.

We understand "markets" here to mean figurative spaces where supply and demand for space resources meet, creating exchanges where at least one party to the transactions is a private, or non-governmental, for-profit actor. Within the medium term we have specified, we postulate as unlikely the emergence of so-called normal markets, i.e. those gathering sellers and buyers so numerous that no single actor or subset of actors has a significant influence on price. Instead, within this timeframe, we are more likely to observe the emergence of oligopolies, monopolies, oligopsonies, and monopsonies in contexts where government actors still play significant, if not decisive, roles. For example, we envision a scenario in which the U.S. Artemis program helmed by NASA, a European Moon Village managed by European Space Agency, and a Chinese government-led lunar base program will form the backbone of exploration and settlement activities. These activities would in turn drive demand for lunar resources and associated industries, such as solar power and regolith processing, supplied by perhaps one to at most three firms per product.

An important feature of such strategic markets is that they lack normal competition, the force touted as fostering efficiency and propelling innovation in ideal-typical markets on Earth. Instead of prices set by the interplay of supply and demand as in competitive terrestrial resource markets, in markets for space resources we may, for at least a period, witness administrative pricing by government monopsonies, prices distorted by generous subsidies to national champions in states with active programs, price gouging by monopolies, and/or price setting by cartels. In addition, during at least an initial period of growth, some of these emerging firms will likely operate at the edge of commercial viability, their business hinging on tenuous models that are highly sensitive to new costs or risks. Insofar as policies add or diminish the costs, risks, or other burdens facing firms, even if only slightly, they could prove to be the difference between the emergence or not of these new industries.

Four sets of policy choices will likely influence whether, when, and with what characteristics markets for space resources, however imperfect, emerge in the medium term.

20.3.1 Government Programs that Create Anchor Demand

The first set of policy factors hinges on the extent to which public agencies pursue ambitious exploration and settlement programs that create anchor demand for space resources. Given the resources characterized in the previous section, it is plausible that the first large-scale buyers of lunar resources, such as power harvested from the Peaks of Eternal Light on the Moon or in-space water retrieved from asteroids, will be long-term government-sponsored exploration programs. While private space resource ventures exist, there is scant evidence that they are opening new frontiers. Rather, the limited private efforts attempted so far have responded to government-led initiatives. Indeed, even in more mature areas of commercial space exploration activity, such as the U.S. launch and transportation industries, private firms' projects are closely tied and responsive to government programs. For example, Space X's commercial cargo and crew businesses for now serve mainly NASA's International Space Station program, though the company has plans for space tourism.

20.3.2 Government Support to Emerging Firms

A second set of policy choices revolves around state support to emerging space resource industries (Krolikowski and Elvis 2019, p. 13). Many observers regard space resource firms as "strategic," or capable of contributing to a country's national interests in space beyond generating economic value, so enhanced state support in the future is plausible. State support to budding firms can take multiple forms, both indirect and direct. Indirect measures aim at creating a landscape of supportive policy, legal, and regulatory measures for all firms active within a jurisdiction. An example of indirect support that observers frequently describe as a precondition to the emergence of a viable asteroid mining industry is national legislation to confer upon private firms' property rights over abiotic materials they recover from asteroids, subject to certain limits. Property rights of this sort would entail, at minimum, the freedom of disposition over recovered materials, including the right to buy and sell these at will. In the absence of international agreement, the main instruments through which elements of such property rights have been established to any extent are national laws, enforceable in domestic courts. An example is found in the *U.S. Commercial Space Launch Competitiveness Act* of 2015 (U.S. Congress 2015). A similar form of policy support might be the creation of a licensing body to regulate efforts at space resource extraction, a step that may pave the way for indemnification or insurability, either or both of which might also prove necessary to the emergence of a viable industry.

In contrast to these indirect measures, direct forms of support entail conditional transfers to specific companies. Such transfers could consist of research and development grants to companies creating technologies needed for mining, comparable to the development programs that nurtured the U.S. commercial space launch and transportation industries. These direct forms of support may also entail government agencies' promises to purchase given volumes of space resources at set prices, providing emerging firms with anchor clients for their otherwise too-risky business.

20.3.3 Trade-Offs in the Design of New Regimes

A third set of policy choices concerns the character of the national and international policy regimes that governments may establish to govern activities targeting space resources, another process already underway. Different activities directed at space resources will survive and thrive under distinct policy regimes. We distinguish between three types of activities directed at space resources, set apart by their primary goal: activities aiming at scientific discovery, at establishing a human presence at or settlement of new destinations in the solar system, and at for-profit sales of space resources. For short, we label these "science," "settlement," and "sales" activities (Krolikowski and Elvis 2019, pp. 10–14). While particular programs or missions will often aim at a combination of these objectives, they will nonetheless tend to be guided by a primary goal that dictates their overall orientation. On one level, these different activities might present technological synergies if, for example, they jointly benefit from the development of a common core of new systems. Specific groups of missions might also be complementary in other ways. For instance, scientific discoveries could help locate resources for commercial exploitation, while commercial activities could drive down the cost of science missions. However, in a more general sense, these different types of activity in practice make distinct and often incompatible demands on policymakers, presenting tensions and trade-offs, rather than complementarities.

In this sense, science, settlement, and sales activities are distinct endeavors facilitated by distinct policy regimes (Krolikowski and Elvis 2019, pp. 10–16). Science activities are best served by a policy regime that fosters the responsible sharing of space resources, both protecting them from casual or unnecessary modification to preserve their integrity as research materials and enhancing their accessibility to a global community of researchers. Strict protocols for how to modify or act upon asteroid or lunar material will benefit science activities over the long run. In contrast, settlement activities are best served by a policy regime that subordinates the protection of space resources to the goal of supporting a human presence at new celestial destinations. Space resources have value insofar as they serve such missions – by, say, acting as cycler transports or supplying fuel for human habitats – rather than their intrinsic potential to yield scientific knowledge. Restrictions intended to limit the human impact on space resources to preserve their scientific value might in fact complicate or limit human settlement projects. Finally, a policy regime supportive of sales activities would be one that maximizes net rewards to firms in this emergent industry. One means to this end is granting these firms enhanced proprietary claims over extracted resources. Another is limiting protocols for the protection of space resources as scientific assets, since such requirements are likely to inflict burdens on resource-extraction operations. For example, requirements to adopt modes of conduct toward resources that limit contamination and maximize protection of nearby environments, as well as rules requiring extensive documentation or data sharing, are likely to inflict costs, delays, or risks on operations that imperil new companies' already tenuous business prospects. In short, a policy regime that is most conducive to for-profit resource exploitation will likely have features at odds with a regime most supportive of scientific activity.

As policymakers in various countries adopt policies to govern conduct toward space resources, they may create regimes that are more conducive to either science, settlement, or sales activities. In some respects, policies may serve all three of these activity types, but in other ways they are likely to advantage some forms of activity over others. Governments that prioritize nurturing nascent domestic commercial space-resource industries are more likely to adopt policies that benefit a handful of early-entrant firms, based in rich countries with large space industries, at the expense of a global scientific endeavor to study the Moon or asteroids. On the other hand, governments that prioritize scientific activities will likely support the establishment of global governance structures and mechanisms aimed at preserving the value of space resources to research (Krolikowski and Elvis 2019, pp. 14–15). Acknowledging and negotiating the trade-offs presented by the distinct policy demands of scientific, settlement, and sales activities will be a precondition to effective long-term international policymaking for space resources.

20.3.4 The Collective Management of Crowding and Interference

A fourth set of policy questions confronting decision makers in various countries revolves around how to manage crowding and interference challenges at resource-rich sites in space. As discussed in Section 20.2.1, this challenge likely presents in its most significant and pressing form on the Moon, where attractive resources and topographical features appear concentrated in relatively small areas, but also exist for the asteroids. This concentration has attracted multiple forthcoming lunar missions to a handful of finite areas, creating the risk of crowding and interference between them. Unmanaged crowding and interference could result in outcomes that leave all actors worse off. An example of such an outcome consists of a scramble for a finite, high-value lunar resource between private firms that unnecessarily degrades the surrounding environment, reduces the value of the resource to science, and inflicts losses of opportunity upon emergent commercial firms.

The most promising means to addressing these risks lies in the creation of collective rules for conduct at these resource-rich sites, yielding a form of governance (Elvis, Krolikowski and Milligan 2020, pp. 12–15). Governance mechanisms could include a registry to coordinate a timeline of activities by multiple actors that target the same thorium deposits or cold traps on the Moon. They might also include rules for actors' placement of solar panels to harvest power from the Peaks of Eternal Light, such that these do not interfere with each other or with human habitats installed nearby (Elvis, Milligan and Krolikowski 2016, Ross et al. 2022). If crowding and interference begin to present appreciable risks to new space-resource firms, then concerted action to address them may become a crucial enabling factor for this new industry.

In sum, then, in the medium term, the governments of major spacefaring states, in potential partnership with other stakeholders, are poised to decide four sets of questions that will help shape whether, when, and how markets for space resources emerge. *First*, they will decide to what extent their public agencies will pursue ambitious exploration and settlement programs that create anchor demand for space resources. *Second*, they will consider whether to lend public support to emerging space resource industries and what forms this support will take. *Third*, participants in the creation of emerging regimes for space resources will negotiate the trade-offs between measures support of science, settlement, and sales activities. *Fourth*, these actors will decide whether and how to create governance arrangements to manage crowding and interference challenges at resource-rich sites in space, including on the Moon and at near-Earth asteroids.

20.3.5 Unfolding Policy Developments

At present, these different policy questions range from hypothetical to actual. National laws, such as the 2015 *U.S. Commercial Space Launch Competitiveness Act*, and Luxembourg's 2017 law on the exploration and exploitation of space resources, indicate that governments around the world are already contemplating and even taking steps to foster the emergence of these new industries (U.S. Congress 2015; Le Gouvernement du Grand-Duché de Luxembourg 2017). Moreover, the U.S.-led Artemis Accords illustrate one pathway by which the elements of a global regime for space resources could coalesce. The Accords are a collection of bilateral agreements between the United States and other governments that outline "Principles for Cooperation in the Civil Exploration and Use of the Moon, Mars, Comets, and Asteroids for Peaceful Purposes" (NASA 2020). Although the Artemis Accords have multiple purposes, they commit parties to an understanding of space resource exploitation that supports the emergence of commercial firms engaging in these activities. This position is at odds with a more restrictive understanding of what forms of space resource exploitation are permissible under the Outer Space Treaty, promulgated by other governments. While reaffirming a commitment to the non-appropriation principle in the Outer Space Treaty, the Accords also define a space for national legislation and instruments that would support such private-sector activities. To date, 10 countries with active space programs have become signatories to the Accords, with Brazil and New Zealand reportedly intending to join the group (Park 2021). As the numerous international lunar missions planned for the next decade develop and their implications for space resources grow more apparent, it is likely that more governments will begin an active consideration of these policy questions.

Out of such national and bilateral efforts at governing conduct toward space resources, a global regime may emerge. This regime will not have evolved in the same way as in the past, when the locus of governance development was in United Nations bodies focused on establishing landmark international treaties, such as the Outer Space Treaty. Instead, a coming regime for governing space resources may emerge out of an initial accretion of national and bilateral measures, evolving to include multilateral processes only later, if ever.

In this sense, in responding to the four types of medium-term policy questions discussed above, the governments of key spacefaring states may already be shaping the foundations of an emerging regime. Not only will their choices help shape whether, when, and how markets for space resources evolve, but they will also steer these potential markets toward outcomes that are more or less legitimate and desirable to a global community of stakeholders.

These medium-term choices will influence whether publics, both within these major spacefaring states and in other societies and across multiple generations, come to regard the exploitation of space resources as just and ethically defensible or as an extension of the injustices and harms wrought by market-driven resource exploitation on Earth. It is to these considerations that we turn in Section 20.4.

20.4 Two Problems of Ethics for an Expanding Space Economy

Because of the founder effect, decisions taken at the beginning of a venture can have long-lasting effects, and the processes envisaged are not those guided only by the hidden hand of the market but are influenced by governmental agencies and policy.

We also raise some ethical issues that should be addressed to limit the negative effects of humans expanding into space to live. These do not constitute a critique of space expansion, but an indication that it is a normal process with a downside as well as several upsides. Our presupposition is that it is also a process which can be deliberately and favorably influenced *to some degree*. If such deliberate influence were impossible, and the negative outcomes inevitable, talk about its ethical dimensions would have at best a diminished and indirect standing. However, given the assumptions of a founder effect, together with the picture of state involvement and policy choice, we will take it that ethical questions are salient, practical and can be of direct significance as one consideration among many during discussions about policy. Here, we focus upon two ethical problems in particular. Both concern justice, and both have a clear economic dimension.

The first is the "tenure and entitlement problem." This is a familiar problem, attention to which should help to emphasize that we are not attempting to present a generalized ethical critique of mining, but rather focusing upon the ethical problems which occur in the course of mining. The problem is internal to the practice, and not an outside judgement upon it. Justice in this particular case will concern fairness, and so we are not stretching the concept of justice beyond familiar grounds (Rawls 1985). But neither are we denying that justice has other dimensions.

The second problem is what we term the "near-term justice problem," and is a newer ethical problem, although it draws upon familiar themes of finite resources and distributive justice. Fairness is again an issue.

20.4.1 The "Tenure and Entitlement" Problem

The high costs and high risks of investment in space, and in mining (asteroid mining especially) raise concerns about "piggy backing" upon the hard work and investment of others. It takes a good deal of time to identify a suitable target for mining, to map its trajectory, check its composition, even (given technologies available in the near future) to alter its trajectory. Because no state is allowed to appropriate an asteroid, under space law [Outer Space Treaty (OST), Article II[8]], others may take advantage of all of this expensive hard work. At first, this may appear to be only an economic issue, or an economic issue with a marginal ethical component. A lack of security in investments may make it harder to grow the space economy; however, there is no guarantee that what is economically optimal and what is ethical will coincide. Sometimes they diverge. And even if they did not diverge here on Earth, there is no ultimate guarantee that they will continue to coincide in space. Economic rationality and ethics might coincide *here*, but not *there*. And so, a brief thought experiment to draw out what is ethically wrong about such free-riding may be a useful addition to the point that free-riding is economically problematic.

8 United Nations Treaties and Principles on Outer Space, Treaty on Principles Governing the Activities of States in the Exploration and Use of Outer Space, including the Moon and Other Celestial Bodies ("OST", 1967), Article II: "Outer space, including the Moon and other celestial bodies, is not subject to national appropriation by claim of sovereignty, by means of use or occupation, or by any other means." http://www.unoosa.org/pdf/publications/STSPACE11E.pdf (accessed 28 May 2021).

Imagine a scenario in which space miners face a heartless corporation. Let us call it "the company." The miners take the bulk of the risks, but the company takes the lion's share of the profits. This is, of course, unlikely to occur; we are unlikely to see anything like a space proletariat emerging. But the scenario may still be useful as a thought experiment, and useful because we have a strong sense that *something* is wrong here, but may be unsure about exactly what. Would the wrongness be a matter of "the bosses" exploiting "the workers"? That is possible, but may overstate the continuity of space concerns and twentieth century politics. Moreover, there is another way of seeing the issue: what is unfair in our thought experiment is simply the way in which risks and rewards have become divorced from one another (Athanassoulis and Ross 2010). Under a fairer (more just) arrangement, those who take the risks should at least share significantly in the rewards. And the reason offered is independent of questions about what is economically optimal. The scenario would still describe an ethically problematic state of affairs even if it turned out to be economically optimal for a space economy. But this is also not to suggest that ethics and what is economically advantageous must conflict. Rather, the relation between the two is contingent. Abiding by ethical constraints can be economically advantageous *under some circumstances, but not all.*

Framed in these terms, the point is also no longer just about traditional patterns of labor relations. Rather, it generalizes out and applies as readily to mining interests, to one company in its relations with other companies. It is *unfair*, from an ethical point of view, if company x puts in the investment, does the research and initial work, and risks its investment in some mining venture but company y then takes advantage of this *at the expense of* x. Risk takers and the beneficiaries of risks should not be two entirely different groups of people. Tackling this ethical problem will involve not just claims upon extracted materials, but also claims upon secure *tenure*. With security of tenure understood in the light of the insecurities of space, rather than as something absolute. (Over-riding considerations may still apply, cases in which compensation may be due for lost opportunities if entitlement to mine is withdrawn.) These matters are all the more important given the insecurity of start-ups in the space sector. Justice points toward securing tenure in cases where other normal ethical conditions for mining have been met, and where the preliminary work has been put in by the company in question.

20.4.2 The Near-Term Justice Problem

Even if we are happy about the manageability of ethical concerns about the commercial use of resources in the near term, there remains a strong likelihood of conflict between the interests of near generations and those of future generations who will live at a time closer to resource depletion or exhaustion. Assuming that the latter are likely to live at a point in time that we can reasonably worry about (say within a 500 year period), then there may be a case for saying that we should not always prioritize near-term justice. Or, that our deliberation and decisions ought to be framed by an awareness of possible later harms. Near-term rules we establish to encourage a nascent space resource industry characterized by strictly limited available resources, may be carried forward into the completely transformed context of a mature industry in which the abundance of the Main Belt asteroids is available.

The problem itself arises in the following way. Access to space is currently limited to a small number of major powers. This is problematic in terms of entitlements. The Outer Space Treaty (1967) captures a strong moral intuition when it classifies "the exploration and use of outer space" as the "province of all mankind." (Article I.[9]) Here, we may set aside the question of whether or not outer space can be an actual commons, and focus upon the idea that all are entitled, but few can go. The entitlement remains a dead letter unless there is a "democratization" of space, not simply in terms of the number of economic players, but in terms of access for all nations and peoples. (Even if access for each and every individual remains permanently beyond us.) To help achieve democratization, we might (in theory) radically redistribute resources in order to set access on a fairer footing. We are by no means opposed to this, but it is easier said than done. It would certainly be unfortunate if the democratization of space (however limited) had to wait until we have achieved some broader system of distributive justice on Earth. This idea also seems to be back to front given that it is familiar for newer fields of human activity to offer greater (if still constrained) opportunities for justice. Opportunities which may be more readily denied in established fields (such as those of heavy engineering or political representation) where there is a pre-existing structure geared to resisting certain kinds of change.

Another way to bring about broader access would be to accelerate the process of economic expansion and the process of cost reduction. Doing so would increase the range and extent of opportunities, as well as the range of participants, broadening the space economy out beyond its initial big players. This also looks like the most plausible pathway toward democratization. Near-term justice would be served.

However, this would occur at the expense of a more rapid use of the large, but (as we have seen) both finite and exhaustible resources that the Solar System has to offer. The more rapidly expansion occurs, the more rapidly the resources begin to be used up. And while recycling may alleviate the problem (and would presumably be standard practice in space) it will also become incrementally harder except on a local level, as we expand over larger areas of space. Recycling on Earth is comparatively easy because we are gathered together, in a small area with nowhere beyond a short flight away. The area bounded by Earth, the Moon, Mars, and the Main Belt, is simply not like that. The more rapidly we bring resources into use, the fewer new resources there are for future generations to call upon. In the absence of a stable

9 OST Article I: The exploration and use of outer space, including the Moon and other celestial bodies, shall be carried out for the benefit and in the interests of all countries, irrespective of their degree of economic or scientific development, and shall be the province of all mankind.
Outer space, including the Moon and other celestial bodies, shall be free for exploration and use by all States without discrimination of any kind, on a basis of equality and in accordance with international law, and there shall be free access to all areas of celestial bodies.
There shall be freedom of scientific investigation in outer space, including the Moon and other celestial bodies, and States shall facilitate and encourage international cooperation in such investigation.

state economy, a point of "super-exploitation" will eventually be reached, a point at which new resources are no longer accessible in sufficient quantities to meet the needs which the economic system has helped to generate. Rapid expansion will take us toward this problematic state of affairs more rapidly, but it is a state of affairs which will be far more problematic for those who face it than for those of us who merely worry about it.

20.5 Conclusions

There are, of course, multiple unknowns about space expansion and the development of an extensive space economy. Unpredicted developments will, no doubt, transpire, and we may expect surprises along the way. In spite of this, we have tried to show that the *in situ* space economy has the greatest chance of developing sustainably if it evolves in a systemic way, rather than a merely opportunistic way.

We would do well to anticipate systematic interconnection of ongoing activities across several in-space environments, rather than periodic raids upon NEAs or the Main Belt. The large scale of infrastructure already points toward processes and expenditure which are likely to remain well beyond the means of any group of private players, and will involve strong state involvement, at least during an initial phase. (Whatever occurs later on.)

These conditions point toward the salience of policy making. The magnitude and distribution of space resources raises both policy and ethical issues, to which the priorities of different stakeholders dictate different approaches. Recognizing these tensions and trade-offs is a precondition to making effective policy for space resource exploitation. In the medium term, whether, when, and with what characteristics markets for space resources emerge will to a significant degree depend on key policy choices, some already under way in major spacefaring countries. More fundamentally, ethical choices embedded within these policy measures will influence whether eventual space resource activities are regarded as just and defensible or, instead, as the next iterations of a historical pattern of harmful exploitation established over centuries on Earth.

The salience of policy is strengthened if we assume a "founder effect" such that those of us at the start of the process are likely to have a disproportionate influence upon the later, overall, pattern of development. The prospect of such influence, though far short of making actual decisions *for* future generations, will strongly influence them, pointing to the need to take our ethical obligations toward those future generations seriously, and not simply to dismiss them as beyond our control. This awareness raises unsettling questions that demand our attention today. With regard to the pace of expansion of the space economy, and with regard to the use of the finite resources of the Solar System, we face an ongoing trade-off. A trade-off between the interests of near-term justice (which may be best served by rapid expansion and resource utilization), and justice toward those generations of humans who will come afterwards, and who will live closer to the edge of expansion's physical limits.

References

Allen, C.W. (1973). General astronomical constants. In: *Astrophysical Quantities*, 26–26. The Athlone Press.

Athanassoulis, N. and Ross, A. (2010). A virtue ethical account of making decisions about risk. *Journal of Risk Research* 13 (2): 217–230.

Berezhnoy, A.A., Kozlova, E.A., Sinitsyn, M.P. et al. (2012). Origin and stability of lunar polar volatiles. *Advances in Space Research* 50: 1638–1646.

Byrnes, D.V., Longuski, J.M., and Aldrin, B. (1993). Cycler orbit between Earth and Mars. *Journal of Spacecraft and Rockets* 30 (3): 334.

Cohen, S.A. et al. (2015). Reducing neutron emission from small fusion rocket engines. In: *66th International Astronautical Congress 2015, IAC 2015, Space - The Gateway for Mankind's Future*, 7749–7759. International Astronautical Federation, IAF.

Crawford, I.A. (2015). Lunar resources: a review. *Progress in Physical Geography* 39: 137–167.

Ćuk, M. and Stewart, S.T. (2012). Making the Moon from a fast-spinning Earth: a giant impact followed by resonant despinning. *Science* 338: 1047–1052.

Elvis, M. (2014). How many assay probes to find one ore-bearing asteroid? *Planetary and Space Science* 91: 20.

Elvis, M. (2021). *Asteroids: How Love, Fear, and Greed will Determine our Future in Space. Yale University Press*, ISBN 978-0-300-23192-2.

Elvis, M. and Milligan, T. (2019). How much of the Solar System should we leave as wilderness? *Acta Astronautica* 162: 574–580.

Elvis, M. and Milligan, T. (2019). How much of the Solar System should we Leave as Wilderness? *Acta Astronautica* 162: 574.

Elvis, M., Milligan, T., and Krolikowski, A. (2016). The peaks of eternal light: a near-term property issue on the moon. *Space Policy* 38 (November): 30–38. http://www.sciencedirect.com/science/article/pii/S0265964616300194 (aAccessed 18 April 2017).

Elvis, M., Krolikowski, A., and Milligan, T. (2020). Concentrated lunar resources: imminent implications for governance and justice. *Philosophical Transactions of the Royal Society A: Mathematical, Physical and Engineering Sciences* 379 (2188): 1–21. https://doi.org/10.1098/rsta.2019.0563.

Foust, J. (2021). "Companies and government agencies announce plans for lunar rover projects", https://spacenews.com/companies-and-government-agencies-announce-plans-for-lunar-rover-projects/ (accessed 30 May 2021).

ISECG International Architecture Working Group (2010). *The ISECG Reference Architecture for Human Lunar Exploration*. International Space Exploration Coordination Group Available at: https://https://www.lpi.usra.edu/lunar/strategies/ISECGLunarRefArchitectureJuly2010.pdf.globalspaceexploration.org/wordpress/wp-content/uploads/IAC61/IAC-10.A5.1.7-Lunar-ISRU.pdf.

Jedicke, R., Granvik, M., Micheli, M., Ryan, E., Spahr, T., and Yeomans, D.K., 2015 Surveys, astrometric follow-up, and population statistics. In *Asteroids IV* (P. Michel et al., eds.), pp. 795–813. Univ. of Arizona, Tucson, DOI: https://doi.org/10.2458/azu_uapress_9780816532131-ch040.

Kooser, A. (2021). SpaceX signs 'blockbuster deal' to send space tourists to the ISS, c|net, https://www.cnet.com/news/spacex-signs-blockbuster-deal-to-send-space-tourists-to-the-iss/#ftag=CADf328eec (accessed 3 June 2021).

Kring, D.A., and Boslough, M., Chelyabinsk: portrait of an asteroid airburst, *Physics Today* 67(9), 32 (2014); doi: https://doi.org/10.1063/PT.3.2515

Krolikowski, A. and Elvis, M. (2019) 'Making policy for new asteroid activities: in pursuit of science, settlement, security, or sales?', *Space Policy*, 47(February), pp. 7–17. doi: https://doi.org/10.1016/j.spacepol.2018.04.005.

Kutter, B. (2018). [Slide resource]. Transportation and Propellant Resources in the Cislunar Economy, https://www.docdroid.net/zRCttF4/2-1-transportation-propellant-resources-in-the-cislunar-economy-kutter-pdf#page=4.

Lawrence D.J., 2017, A tale of two poles: toward understanding the presence, distribution, and origin of volatiles at the polar regions of the Moon and Mercury. *Journal of Geophysical Research, Planets* 122, 21–52. (doi:https://doi.org/10.1002/2016JE005167)

Le Gouvernement du Grand-Duché de Luxembourg (2017). *Loi du 20 juillet 2017 sur l'exploration et l'utilisation des ressources de l'espace*. Available at: http://legilux.public.lu/eli/etat/leg/loi/2017/07/20/a674/jo/en (accessed 29 May 2021).

Lewis, J.S. (1996). *Mining the Sky: Untold Riches form the Asteroids, Comets, and Planets*. Addison-Wesley. ISBN: 0-201-47959-1.

Li S, Lucey PG, Milliken RE, Hayne PO, Fisher E, Williams JP, Hurley DM, Elphic RC. 2018 "Direct evidence of surface exposed water ice in the lunar polar regions." *Proceedings of the National Academy of Sciences of the United States of America* 115, 8907–8912. (doi:https://doi.org/10.1073/pnas.1802345115)

Mainzer, A. et al. (2011). NEOWISE observations of near-Earth objects: preliminary results. *The Astrophysical Journal* 743: 156.

Milliken, R.E. and Li, S. (2017). Remote detection of widespread indigenous water in lunar pyroclastic deposits. *Nature Geoscience* 10 (8): 561–565.

NASA (1976). *Space Shuttle*, prepared by Lyndon B. Johnson Space Center, https://history.nasa.gov/SP-407/foreword.htm

NASA (2020). *The Artemis Accords: Principles for Cooperation in the Civil Exploration and Use of the Moon, Mars, Comets, and Asteroids for Peaceful Purposes*. U.S. National Aeronautics and Space Administration. Available at: https://www.nasa.gov/specials/artemis-accords/img/Artemis-Accords-signed-13Oct2020.pdf (Accessed: 29 May 2021).

NASA (2021). Apollo 11 Mission Overview: https://www.nasa.gov/mission_pages/apollo/missions/apollo11.html (accessed 1 June 2021).

Neto, de O.B. *et al.* (2020) *Building Blocks for the Development of an International Framework for the Governance of Space Resource Activities: A Commentary*. The Hague, The Netherlands: Eleven International Publishing. Available at: https://boeken.rechtsgebieden.boomportaal.nl/publicaties/9789462361218#152.

Paige, D.A. et al. (2010). Diviner lunar radiometer observations of cold traps in the Moon's South Polar Region. *Science* 330: 479–482.

Park, S.-S. (2021). *South Korea signs Artemis Accords; Brazil, New Zealand likely next, SpaceNews*. Available at: https://spacenews.com/south-korea-signs-artemis-accords/ (accessed 29 May 2021).

Rawls, J. (1985). Justice as fairness: political not metaphysical. *Philosophy and Public Affairs* 14: 223–251.

Ross, A., Ruppert, S., Gläser, P., and Elvis, M. (2022). Towers on the peaks of eternal light: quantifying the available solar power. *Acta Astronautica*, in preparation.

Security and sustainability in Outer Space (2015). *European External Action Service, European Commission*. Available at: https://eeas.europa.eu/headquarters/headquarters-homepage/8466/security-and-sustainability-outer-space_en (accessed 24 August 2020).

Sonter, M.J. (1997). The technical and economic feasibility of mining the near-Earth asteroids. *Acta Astronautica* 41: 637–647.

Strange, N., Landau, D., and Longuski, J. (2015). Redirection of asteroids onto Earth-Mars cyclers, advances in the astronautical sciences. *AAS Spaceflight Mechanics* 155: 15–462. available at http://www.univelt.com/book=5109.

Taylor, A., McDowell, J.C., and Elvis, M. (2022). Phobos and Mars orbit as a base for asteroid mining, planetary and space science. *Acta Astronautica* 214: 105450.

Thomas, S.J. et al. (2019). Fusion propulsion and power for extrasolar exploration. In: *IAF Space Propulsion Symposium 2019, 70th International Astronautical Congress (IAC)*, 679–696. ISBN: 978-1-7138-1491-7.

U.S. Congress (2015). *U.S. Commercial Space Launch Competitiveness Act*. Available at: https://www.congress.gov/114/plaws/publ90/PLAW-114publ90.pdf (accessed 29 May 2021).

United Nations General Assembly (1967).Outer Space Treaty (Treaty on Principles Governing the Activities of States in the Exploration and Use of Outer Space, including the Moon and Other Celestial Bodies). [cited 2020 Aug 24].

Vancouver Recommendations on Space Mining (2020). Vancouver, Canada: Outer Space Institute. Available at: http://www.outerspaceinstitute.ca/docs/Vancouver_Recommendations_on_Space_Mining.pdf (accessed 26 August 2020).

Wall, M. (2021). "Canada to launch moon rover by 2026", Space.com. https://www.space.com/canada-launch-moon-rover-2026 (accessed 30 May 2021).

Zelinsky, W. (1972). *The Cultural Geography of The United States*. Prentice Hall: Hoboken.

The Future of Space Exploration: A Young Perspective

Space is the future of our civilization. We as a society are increasingly valuing sustainability and innovative uses of resources to ensure the preservation of humanity. Many of our most well-known innovators, like Elon Musk and Jeff Bezos, have set their sights on interplanetary exploration and ultimately habitation for the preservation of our global society. It's an exciting prospect and hence will become a major sector for both government and privately owned space organizations. Yes, the space race will be on again but with technology more advanced, more powerful, and more efficient than ever before. As a young female interested in these areas, I am so excited about the prospects of space exploration and interplanetary settlement in our near future. I am passionate about making this future a reality.

In situ resource utilization on other planets, I believe, is the way of the future. Some may say that employing these resources for the expansion of our human civilization is not ethically viable considering we have over-exploited the resources on our own planet, Earth. However, it needs to be considered how different the environments are. The Moon does not have an atmosphere, which is one of the major reasons why our Earth is undergoing an accelerated greenhouse effect, and as a result, accelerated climate change. We cannot combust fuels on the Moon without a chamber as there is no oxygen, we cannot pollute the oceans as there are no seas. The Moon is a harsh environment that is different to Earth but can be utilized in so many ways.

Going to the Moon will allow us to take the first step in the exploration of deep space. It is a proof of concept that we can live somewhere other than Earth. It is theorized we will be able to find every element we have found on Earth in volatiles on the Moon. However, it is not economically viable to extract these materials and bring them back down to Earth; instead, we can utilize them *in situ* to thrust us forward and accelerate exploration opportunities. In deep space, the Moon is the closest celestial body to Earth (384 400 km away). This could allow the Moon to become a training base of sorts for astronauts who are planning on living off-Earth

In-Space Manufacturing and Resources: Earth and Planetary Exploration Applications, First Edition.
Edited by Volker Hessel, Jana Stoudemire, Hideaki Miyamoto, and Ian D. Fisk.

for a prolonged period of time. They can adapt to the challenges of the Moon's harsh environment and understand what it is like to live without the Earth's "goldilocks environment." Furthermore, while on the Moon, inter-planetary inheritance training will also allow for trainees to help increase the infrastructure and the extraction processes of the key resources on the Moon itself. This will not only allow preparation of technologies for these space expeditions to be done more efficiently, but it also allows for the astronauts themselves to learn how to do these processes on other planets where they are less likely to be able to carry resources and must instead extract them from the planet itself.

Travelling to celestial bodies beyond the Moon will take many months or even years. Hence, when crewed missions are ready to be launched, there will need to be a high amount of cargo for life support throughout the trip. This is also why utilizing the Moon as a "Petrol Station" for these expeditions could be vital. Although generating the infrastructure needed to produce these materials would also be an expensive endeavor, these facilities could be utilized by different missions and decrease their individual costs overtime significantly. Hence, implementing this practice will be both economically and environmentally sustainable.

With all these exciting opportunities potentially accessible in the near future, there will be a subsequent increase in the number of jobs required in the space sector. It is not just rocket science or aeronautics that is going to be important here, all aspects of STEM will be crucial in allowing for these expeditions to be effective and efficient. This is why innovations like the Andy Thomas Centre and Lot 14 in South Australia are so important in getting younger generations excited about the prospects of exploring space. Areas like this have allowed me as a 17-year-old to be inspired about the limitless opportunities for me in this area of STEM. The things we are planning on doing are exciting, it is now our job to share this with our community and get them excited too.

If we can get these areas up and running and start getting our community excited about the prospects of our space sector, we will be able to do things that have never been seen in the history of the world. I believe I will see a human set foot on Mars in my lifetime and show that what we once thought was impossible, is something we can, and will, achieve. The collaboration of our global space organizations can create a regime for equitable sharing of in situ resources that will allow us to achieve interplanetary settlement most effectively. Instead of this being a competition, we should recognize our common goal and work together. I am excited to see how far we can go, what limits we can push, and what opportunities will arise from us just taking the first step in our space exploration journey.

Zoe Nunn
Student, Scotch College Adelaide, Australia

Index

In-Space Manufacturing and Resources: Earth and Planetary Exploration Applications, First Edition.
Edited by Volker Hessel, Jana Stoudemire, Hideaki Miyamoto, and Ian D. Fisk.

d

e

f